D0163537

DIGITAL SYSTEMS FUNDAMENTALS

McGRAW-HILL SERIES IN ELECTRONIC SYSTEMS
John G. Truxal and Ronald A. Rohrer, Consulting Editors

Chua INTRODUCTION TO NONLINEAR NETWORK THEORY
Director and Rohrer INTRODUCTION TO SYSTEM THEORY
Huelsman THEORY AND DESIGN OF ACTIVE *RC* CIRCUITS
Meditch STOCHASTIC OPTIMAL LINEAR ESTIMATION AND CONTROL
Melsa and Schultz LINEAR CONTROL SYSTEMS
Motil DIGITAL SYSTEMS FUNDAMENTALS
Peatman THE DESIGN OF DIGITAL SYSTEMS
Ramey and White MATRICES AND COMPUTERS IN ELECTRONIC CIRCUIT ANALYSIS
Rohrer CIRCUIT THEORY: AN INTRODUCTION TO THE STATE VARIABLE APPROACH
Schultz and Melsa STATE FUNCTIONS AND LINEAR CONTROL SYSTEMS
Stagg and El-Abiad COMPUTER METHODS IN POWER SYSTEM ANALYSIS
Timothy and Bona STATE SPACE ANALYSIS: AN INTRODUCTION
Truxal INTRODUCTORY SYSTEM ENGINEERING

Digital Systems Fundamentals

John M. Motil
School of Engineering
San Fernando Valley State College

McGraw-Hill Book Company
NEW YORK ST. LOUIS SAN FRANCISCO
DÜSSELDORF JOHANNESBURG KUALA LUMPUR
LONDON MEXICO MONTREAL
NEW DELHI PANAMA RIO DE JANEIRO
SINGAPORE SYDNEY TORONTO

DIGITAL SYSTEMS FUNDAMENTALS

Library of Congress Catolog Card Number 75-172029

07-043515-4

1 2 3 4 5 6 7 8 9 0 K P K P 7 9 8 7 6 5 4 3 2

This book was set in Times Roman by Textbook Services, Inc., and printed and bound by Kingsport Press, Inc. The designer was Merrill Haber; the drawings were done by John Cordes, J & R Technical Services, Inc. The editors were Charles R. Wade and Barry Benjamin. Matt Martino supervised production.

TO
my parents,
and my family
Raymonde
Jan
Jacqueline
Monique

Contents

Preface xv
Summary of Notation xx

Chapter 0 An Introduction and Overview 1

Survey, Philosophy, Limitations

GOALS AND RELEVANT ATTRIBUTES 3
ORIENTATION OF A SYSTEM 3
ASSIGNMENT OF DIGITAL VALUES TO ATTRIBUTES 4
INTERACTION BETWEEN ATTRIBUTES 5
BEHAVIORAL AND STRUCTURAL VIEW OF SYSTEMS 6
HIERARCHY OF SYSTEMS 6
PROPERTIES OF SYSTEMS 8
SYSTEM EQUIVALENCE 9
GENERAL SYSTEMS PROCESSES 9
APPLICATION TO COMPUTATION, COMMUNICATION, AND CONTROL 10
Summary

Chapter 1 Deterministic Digital Systems 12

Instantaneous Switching Systems

MODELING 13
- Modeling a Relay
- Modeling an Interconnection of Relays

REPRESENTATION 14
Representation as an Attached Diagram
Representation as a Detached Diagram
Representation as an Algebra
From Networks to Algebra
Precedence of Operators
From Algebra to a Network
Representation as a Table-of-Combinations
ANALYSIS 21
Complete Induction
Switching (Boolean) Algebra
Properties of the Algebra
SYNTHESIS 28
OPTIMIZATION 30
DESIGN 34
Majority Indicator
Design from Incomplete Specifications
Multi-output Networks
IMPLEMENTATION 40
Flow Networks
Level Networks
Electric Level Networks
Mechanical Level Networks
IDENTIFICATION 49
APPLICATION TO COMPUTATION 52
Numbers, Positional Notation
Addition of Digits
Addition of Numbers
Design of an Adder
EXTENDING THE SYSTEM 58
Other Binary Systems
The EXCLUSIVE-OR operator
The NOR operator
The NAND and INHIBIT-AND
*Non-binary Systems
Summary
General System Processes
Problems

Chapter 2 Sets and *n*-tuples 74

Classifying, Enumerating, Counting

SETS: REPRESENTATION AND ANALYSIS 74
Representation of Sets
Representation by Extension

Representation by Intension
Representation by Set Diagram
Set Diagrams in Karnaugh Map Form
Fundamental Concepts of Sets
Operations on Sets
Operations on Set Diagrams
Algebra of Sets
APPLICATIONS OF SETS 84
Set Diagram for Optimizing Switching Networks
Set Diagrams and Tables-of-Combinations
*Optimizing Incompletely Specified Systems
*Extending Set (Karnaugh) Diagrams
Applications of Sets to Classifying, Counting, Checking
Classifying Blood
Size of Sets
Conclusions from Sizes
n-TUPLES (ORDERED ARRANGEMENTS) 95
Equality of *n*-tuples
Sets of *n*-tuples and *n*-tuples of sets
Set Product
Applications of Set Product
Extending the Set Product
Representation of Set Product
COUNTING AND COMMUNICATING 99
Counting by Enumeration
OPTIMIZING COMMUNICATION 104
Summary
Problems

Chapter 3 Interaction 112

Relations, Functions, Algorithms

RELATIONS 112
Representation of Relations
Representation of a Relation as a Sagittal Diagram
Representation of a Relation as a Set
Representation of a Relation as an Operation
Representation of a Relation as a Matrix (Array)
Representation of a Relation as a Flow Graph
Summary of Representations of Relations
Other Concepts: Domain, Range, Inverse
Composition of Relations
*Relations on One Set and Possible Properties
*Equivalence Relations
Extending and Restricting Relations

FUNCTIONS 123
SYSTEMS 124
FUNCTIONS-ON-RELATIONS 128
Operations on Arrays (Matrices)
Product of Matrices (Arrays)
ALGORITHMS 133
Constructing Complex Algorithms—Sub-algorithms
Summary
Problems

Chapter 4 Probabilistic Systems 143

Instantaneous Nondeterministic Systems

MODELING AND REPRESENTING PROBABILISTIC SYSTEMS 145
Modeling a Crummy Relay
Orientation
Observation
Basic Events
Relative Frequency
Composite Events
Modeling a Two-dice System
General Model of a Probabilistic System
Alternative Notation
Representation of a Probabilistic System
Other Examples of Probabilistic Systems
A Survey
Probabilities in Noncrummy Relays
ANALYSIS OF PROBABILISTIC SYSTEMS 160
Conditional Probability
Conclusions from Conditional Probabilities
Properties of Probabilistic Systems
Restricted Results
A Systems View of Probabilistic Systems
Forward Conditional Probabilities
Sagittal Representation
Reverse Probabilities
A Communication System
SYNTHESIS AND OPTIMIZATION OF PROBABILISTIC SYSTEMS 176
Optimizing Crummy Classification
Interconnection and Equivalence of Probabilistic Systems
Extended Cascading
Symmetric Systems
Multi-input systems

INDEPENDENT SYSTEMS—ANALYSIS AND SYNTHESIS 188
Probabilistic Independence
Independent Relay Networks
Complex Independent Networks
More Reliable Systems from Less Reliable Flow Components
More Reliable Systems from Less Reliable Level Components
*More Optimization
Repeated Independent Trials
Repeated Coin Toss
Independent Arrows
Dependent Missiles
RANDOM VARIABLES, DENSITY FUNCTIONS, EXPECTED VALUE 205
Random Variables
Probability Density Functions
Expected Value
Some Particular Probability Density Functions
Binomial Probability Density Function
*Poisson Probability Density Function
Geometric Probability Density Function
Expected Value of a Function of Random Variables
Sums of Random Variables
Properties of Expected Value
VARIANCE, COVARIANCE, DEPENDENCE 223
Variance
Standard Deviation
Properties of Variance and Standard Deviation
*Covariance and Correlation
Degree of Dependence (Mutual Information)
DECISION MAKING 231
Decision Making Using Information Instruments
Extreme Information Instruments
Oracles
Demons
Decision Making Using Combinations of Instruments
Summary
Problems

Chapter 5 Sequential Systems 247

Dynamic, Deterministic Digital Systems

MODELING AND REPRESENTATION 248
Modeling a Relay
Behavior
Remodeling the Relay
Ternary Counter

A General Model of a Sequential System
Sequential Behavior

APPLICATION TO COMPUTATION, COMMUNICATION, CONTROL 256
Synchronous and Asynchronous Systems
Application to Computation: The Binary Sequential Adder
*Application to Communication: Comma Coding
Application to Control: Coin-operated Dispenser

ANALYSIS AND SYNTHESIS OF SEQUENTIAL SYSTEMS 259
Ideal Delay
Networks Having Ideal Delays
Multiple Delay Elements
Synthesis: From State Diagrams to Networks
State Assignment
General Behavioral Properties
Causality
The Response Separation Property
The State Separation Property
*Particular Properties of Sequential Systems (a Digression)
Finite Memory
Linearity

OPTIMIZATION OF SEQUENTIAL SYSTEMS 274
State Equivalence
Optimizing States by Method of Merging
Optimizing States by Method of Partitioning

IMPLEMENTATION OF SEQUENTIAL SYSTEMS 281
Some Practical Components
Implementation of Sequential Systems
Other Possible Components
Timing Considerations

LARGER INTERCONNECTIONS OF SEQUENTIAL SYSTEMS 296
Interconnections without Feedback
Shift Registers
A Binary Coded Counter
Interconnections with Feedback

Summary

Problems

Chapter 6 Stochastic Systems 306

Dynamic Nondeterministic Systems

MODELING, IDENTIFICATION, AND REPRESENTATION 306
Modeling a Relay with Error Bursts
Identification

Representation of Stochastic Systems
Examples of Stochastic Systems
Transportation System
Replacement System
Queues
Modeling of Stochastic Systems
Another Example: Time-dependent Activity
ANALYSIS OF STOCHASTIC SYSTEMS 312
Transition Equations
Analysis by Iteration
*Matrix Interpretation
Analysis by Inversion
Comparison of Iteration and Inversion Methods
Analysis by Simulation (Monte Carlo Method)
Random Numbers
Ergodicity
Nonergodicity
Maintenance System
Separate Ergodic Systems
Systems with Output
Stationarity
*Extending Memory Length
OPTIMIZATION OF STOCHASTIC SYSTEMS 330
Stochastic Systems with Inputs
Decision Making in Stochastic Systems
Control of Multi-stage Stochastic Systems
Dynamic Programming
CONCLUSION 337
General Model of Stochastic Systems
The Relay Revisited
Summary
Problems

Chapter 7 Large Digital Systems 344

Digital Computation, Communication, and Control

REGISTERS 345
Manipulation of Registers
Readout
Transfer
Transforming
DIGITAL COMPUTERS 348
The Operation Unit
The Memory Unit
Single-address Organization
The Control Unit

PROGRAMMING THE COMPUTER 353
BEHAVIOR OF THE COMPUTER 355
The Timing Network
Sequencing the Register Transfers
STRUCTURE OF THE COMPUTER 360
EXTENDING THE COMPUTER 364
Summary
Problems

Appendix: Electronic Digital Systems 367

BINARY DIGITAL LEVELS 367
LOGIC CONVENTION 367
BASIC ELECTRONIC COMPONENTS 368
BASIC DIGITAL COMPONENTS 369
CONFIGURATIONS (OR FAMILIES) 370
INTEGRATED CIRCUITS 372
LOADING: FAN-IN AND FAN-OUT 373
INTERCONNECTING DIGITAL COMPONENTS 373
FLIP-FLOPS 374
SHIFT REGISTERS 377
Summary

References 379
Index 381

Preface

This book is an introduction to the fundamental concepts of systems which are used for computation, communication, and control. The systems are considered from a discrete, finite, or digital point of view.

WHY? (The Goals)

My goal, in writing this book, is to provide a unified introduction to systems which may be deterministic, probabilistic, sequential, or stochastic. I have tried to write this at the lowest level at which it is possible to synthesize, optimize, and design systems.

Most of the systems studied here are man-made and motivated by man's desire to communicate and control. Although man uses natural systems involving materials and energy, his problems are often in the man-made interaction and complexity he forces on them. As a result, the study of the natural systems, in terms of physics, chemistry, and biology, is only of secondary concern here.

Of greater importance is the concept of structure, organization, form, or pattern, which is often called information. Our intuition about information is usually less developed than our intuition about matter or energy; therefore it may appear more abstract. Here we will try to develop an intuition and insight into information systems in general and digital systems in particular.

Binary digital systems are ideal for such a study because their basic elements are extremely simple; they have only two values of yes, no or on, off or high, low. To create anything useful we must consider complex interconnections of such simple elements. The study of such interactions can then be extended to the more complex systems which we encounter in real life.

Digital systems are not only conceptually simple, but they are also computationally convenient and practical. Perhaps most importantly, the concepts of digital systems are portable; they can be readily extended to nondigital systems.

Despite its simplicity the digital view is quite general. The concepts are not limited by mathematical assumptions (such as linearity) or philosophical assumptions (such as two-valued logic) or engineering restrictions of any particular device (such as a relay).

Ultimately our goal is to act: compute, communicate, control, or make decisions. This action can be taken despite information which is nonquantitative, incomplete, uncertain, or changing. Nonquantitative information, such as male–female, may be as precise as any quantitative information. Incomplete information is encountered in many problem specifications. Uncertain information comes from noise, loss, or error. Changing information results from aging, evaluating, and modifying. To take action amid such complexity would be much more difficult without the simplicity of the digital approach.

WHEN? (The Level of Maturity)

This book is intended for a first course in systems at an intermediate undergraduate level. It could be used at a lower level if the emphasis is on description alone; but it is meant to be used at a level requiring some creativity in synthesis and design.

WHO? (The reader)

Although this book was written mainly for students in all fields of engineering and computer science, it could be read profitably by others in mathematics, business, biology, ecology, and psychology.

WHERE? (The Place in the Curriculum)

This book could be used as an introduction (prerequisite) to many areas, such as switching theory, automata theory, information theory, computational linguistics, data structures, computer organization, operations research, and statistics. It could also be used in a course which complements the more continuous oriented courses in circuits, controls, and communications. It could even serve as a survey of man-made systems. It is also suitable for self-study.

WHAT? (The Outline of the Content)

The book consists of eight chapters, a digital half dozen. They are labeled from 0 to 7, again a digital way of counting.

Chapter 0, the introduction, provides an overview of four main types or models of digital systems: deterministic, probabilistic, sequential, and stochastic. It illustrates how an electromagnetic relay may be viewed in any of these four ways depending on the application or goal. The relay reappears throughout the book to provide unity and continuity, but the concepts are applied to other devices regardless of their physical form. It is important to realize that the relay is used simply as a "vehicle" to convey significant systems concepts.

Chapter 1 is concerned with instantaneous deterministic systems and their application as switching networks. It introduces the concept of Boolean algebra (and logic), which is also the algebra of probabilistic events. More importantly, it introduces the behavioral and structural views, as well as the processes of modeling, representing, analyzing, synthesizing, optimizing, designing, implementing, and evaluating systems.

Chapter 2 (on sets and n-tuples) and Chapter 3 (on interaction) are brief but basic presentations of sets, functions, relations, systems, and algorithms; they make the book self-contained. These concepts are motivated by the first chapter, and they are used in all the following chapters. These two chapters may be reviewed quickly if the reader has a background in modern or abstract mathematics.

Chapter 4, on instantaneous probabilistic systems, involves digital or discrete systems with nondeterminism (randomness, uncertainty, noise, error, or unreliability). The emphasis is not on the mathematically convenient systems which are symmetric, uniform, independent, or mutually exlcusive) because practical systems may not have such convenient properties. The emphasis is instead on evaluating, optimizing, synthesizing, and decision making in spite of the nondeterminism.

Chapter 5, on sequential systems, involves the concept of time in a digital or discrete way, leading to dynamic but deterministic systems having time delay, memory, or feedback. It introduces the general concepts of state, state equivalence, and causality, as well as the more restrictive concepts of linearity, superposition, and finite memory. All these concepts are applied to the design of systems such as adders, subtractors, comparators, and counters.

Chapter 6, on stochastic systems, involves both uncertainty (probability) and change (time). These dynamic probabilistic systems are studied from both a microscopic and macroscopic viewpoint and are optimized by dynamic programming.

Chapter 7, on large digital systems, indicates how the smaller systems of the previous chapters can be interconnected to make much larger and more practical systems for computation, communication, and control. As an example, a simple digital computer is developed.

The Appendix describes some common electronic devices which are used in digital systems. However, such a field develops so rapidly that this survey can quickly become obsolete. The concepts in the rest of the book are more basic, fundamental, and general. They apply equally to relays, integrated circuits, hydraulic networks, mechanical components, and fluidics, and therefore are less likely to become obsolete.

HOW? (The Method)

This book may be used in many different ways. It is not intended to be covered entirely in any one course. Sections, usually marked with a star, could be left for the student or omitted entirely. Similarly entire chapters could and should be left for self-study. For example, the chapter on stochastic systems could be read by the student while paralleling the development of sequential systems in the class.

There is considerable flexibility possible in selecting topics from this book. It evolved and was used most often as a general introduction to systems, emphasizing the deterministic and probabilistic aspects equally. This was done by selecting Chapters 1, 2, 3, 4, and 6.

Perhaps a more unified approach would be to consider all four models of Chapters 1, 4, 5, and 6. This would require a prerequisite of modern mathematics (from high school possibly) and some computing experience, along with a quick review of the fundamental mathematics of Chapters 2 and 3.

The book could also be used in a more traditional way as an introduction to deterministic switching systems by selecting Chapters 1 (with Appendix), 2, 3, 5, 7. Of course there are many other meaningful paths through this book.

NOTATION

The concepts of information systems originate in many fields such as operations research, statistics, linguistics, computability, engineering, circuits, psychology, sociology, and biology. The mathematics is that of set theory, symbolic logic, Boolean algebra, abstract algebra, combinatorics, topology, and probability (as opposed to calculus, differential equations, analysis, complex variables and transform methods which are used for studying natural systems). This diversity of origin leads to an even greater diversity of notation. For example, we can find all the following occurences or interpretations of the symbol +:

$$1 + 1 = 2 \qquad 1 + 1 = 1 \qquad 1 + 1 = 0 \qquad 1 + 1 = 10$$

Since we will use all of these interpretations (sometimes two or three in one equation), there could be considerable confusion. As a result it was necessary to introduce some new symbols such as

$$1 + 1 = 2 \qquad 1 \vee 1 = 1 \qquad 1 \underline{\vee} 1 = 0 \qquad (1 + 1)_2 = (10)_2$$

It is reassuring to know that 1 plus 1 still equals 2. But then one haystack added to another haystack yields one haystack!

To avoid excessive notation, when an operation such as multiplication (denoted $A * B$) is understood, it is written more briefly as AB. For consistency and unity, especially in the later models, inputs are denoted by X_i and outputs by Y_j. A summary of the notation follows this preface.

The unified notation of this book has evolved as a natural outgrowth of the preference of the majority of students. Some students and instructors may prefer other notations, especially if they do not consider probabilistic systems.

Much of the representation of systems has been diagramatic (involving Karnaugh maps, Sagittal diagrams, state diagrams, flowgraphs, etc.), for this was found to be a great aid in gaining an insight into systems.

EXERCISES, PROBLEMS, PROJECTS

The main goal of this book is to develop in a reader the insight and ability to creatively face any situation or problem. This ability can be developed through the active effort of overcoming barriers, of solving problems; it does not come from the passive agreement with other people's solutions.

For this reason many problems are provided, but certainly not all need be done. There is a set of exercises after each section within a chapter and a larger set of problems at the end of each chapter. A few problems are of the drill type, but most require some degree of creativity. The problems marked with a dagger are either complex or require mathematics not contained in this book. Many of the problems are provided with answers. Most problems are sufficiently short and simplified to illustrate the basic ideas behind real-world problems, but they can readily be extended to the more complex problems.

Some of the longer problems labeled as projects may be done as term papers. It is hoped that a student can extend and apply some of the systems concepts to things that interest him personally, since the concepts are portable. The projects provide such an opportunity.

Some problems involving algorithms are computer-oriented. The construction or synthesis of algorithms, especially in some programming language, is not emphasized although it could be in a course having such a prerequisite. But the representation and analysis of flow diagrams of algorithms is important. The algorithms are more important than the numbers which they produce!

ACKNOWLEDGMENTS

Many people made this book possible. I am greatly indebted to my teachers, W. Ross Ashby, Richard Bellman, Jack Carlyle, Seymour Ginsburg, Franz Hohn, Jacob Marshak, Atwel Turquette, Thomas Saaty, Sundaram Seshu, and Lofti Zadeh. They shared with me their ideas of systems, but they are not responsible for how I may have altered their general ideas to fit my particular digital unity. Of course I am indebted to Aristotle, Claude Shannon, Percy Bridgman, and many others whose ideas I may be using without being aware of it.

I wish to thank my friends and colleagues, especially Edward Hriber, Raymond Davidson, and George Harness, for their suggestions and encouragement, but particularly for the freedom they gave me to experiment. Thanks also to the many students for their criticisms, discussions, and recommendations.

The detailed publisher's reviews from John Truxal and especially Ronald Rohrer are greatly appreciated.

I am also grateful to Sergene Zimmerman, Rita Hary, and especially Phyllis Osborn for their patient typing of the manuscript.

But I am most grateful to my wife and family for their patience, understanding, humor, and love.

JOHN M. MOTIL

Summary of Notation

Name	Notation	Read
Notation involving the ordinary algebra, arithmetic:		
Addition	$A + B$	"A plus B"
Subtraction	$A - B$	"A minus B"
Multiplication	$A * B$	"A times B"
Division	A / B	"A divided by B"
Exponentiation	A^n	"A to the nth power"
Notation involving switches, events, and propositions A, B		
OR operation	$A \vee B$	"A or B" or "A and/or B"
AND operation	$A \cdot B$	"A and B"
NOT operation	$\bar{A}$	"not A" or "A not"
NOR operation	$A \downarrow B$	"neither A nor B"
NAND operation	$A \uparrow B$	"A nand B"
EXCLUSIVE-OR	$A \underline{\vee} B$	"A or else B" or "A exclusive-or B"
INHIBIT-AND	$A \nearrow B$	"A but not B" or "A inhibited by B"
Biconditional	$A \leftrightarrow B$	"A if and only if B"
Implication	$A \rightarrow B$	"if A then B" or "A implies B"
Notation involving sets S, T and relations Q, R		
Union of sets	$S \cup T$	"S union T"
Intersection	$S \cap T$	"S intersect T"
Complement	$\tilde{S}$	"S complement"
Product	$S \times T$	"product of S and T" or "S cross T"
Composition	$Q \circ R$	"Q composed with R"
Inverse	R^{-1}	"R inverse"
Other notation involving relations and events E, F		
Subset	$E \subseteq F$	"E is a subset (or subevent) of F"
Conditional	$E \mid F$	"E given F"
Mutual exclusiveness	$E \text{ M } F$	"E is mutually exclusive of F"
Independence	$E \perp F$	"E is independent of F"
Notation involving systems		
Inputs	X_i	(for symbols) and
	T	(for sequences or tapes)
Outputs	Y_j	(for symbols) and
	R	(for response sequences)
States	S	
Next states	S'	
Output functions	f	(relates symbols),
	F	(relates sequences)
Transition functions	g	(relates symbols),
	G	(relates sequences)

Chapter 0: An Introduction and Overview

Survey, Philosophy, Limitations

This book is concerned with systems, and in particular the unity of systems. Philosophical, mathematical, and engineering aspects of systems all contribute to this unity. However, these aspects and their contribution to the unity may be lost in the details of any one chapter. For that reason this chapter outlines the basic idea of systems and indicates how the various chapters add to the unity. The first half of this chapter (up to Section 6) is introductory and should be read carefully; the rest may be read quickly at first and reread after any chapter as a review.

A *system* is any set of objects related by some interaction. The two concepts of set and interaction (or relation) are so important that a chapter has been devoted to each of them. However, these concepts are also so natural that we may conceive many examples of systems, such as:

Electrical devices interconnected in a network
Components of a digital computer interacting to compute
A computer controlling some process
Objects manufactured on a production line

Items classified by an inspector
Words in a language united by some grammar
Symbols transmitted on a channel interacting with noise
Industries interrelated economically
People in an organization, working and communicating
A society interacting with an environment
Variables related mathematically by a function

It appears that everything is a system, because systems theory is essentially a way of thinking—a philosophy. The unity behind such diverse systems comes from a way of looking at the sets and relations involved. To illustrate this "systems view" we will use the electromagnetic relay, although we could just as well use a rock, a resistor, a railroad, or a robot.

An electromagnetic relay is a device used to control mechanically a switch in one circuit with a current or voltage in another circuit. It could, for example, be used to enable a small current at one location to control a much larger current at some other remote location. Relays may have many shapes, forms, and sizes, as shown in Figure 1.

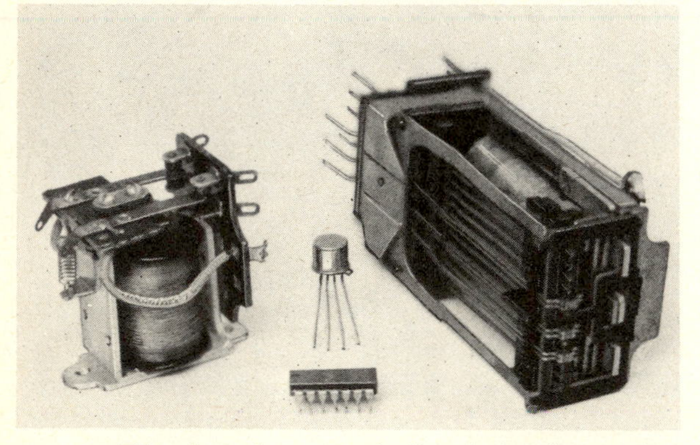

Figure 1

All relays have a similar behavior. The typical relay of Figure 2 consists essentially of a coil of wire and some flat metal springs. When a voltage or current is applied to the coil, it attracts one of the metallic flat springs, causing this spring to break its electric contact with one spring and to make contact with another spring. Removing the voltage from the coil releases the movable spring, which returns to its original position.

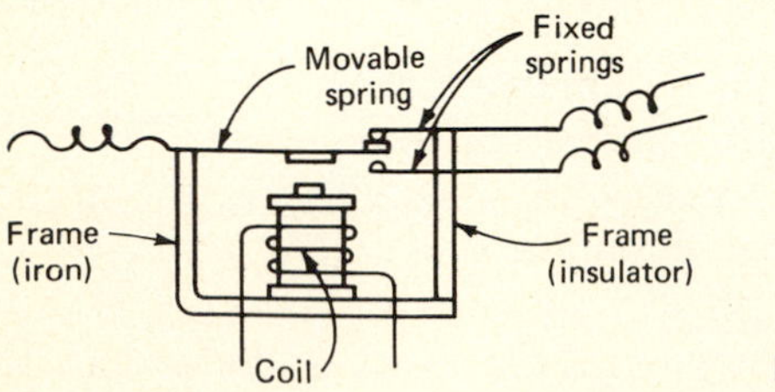

Figure 2

Any such system has associated with it many properties, variables, parameters, or values, which will be called *attributes*. A relay has the attributes of weight, size, cost, temperature, color, beauty, smell, resistance, voltage to operate, time to release, average lifetime, time to manufacture, place of manufacture, noise of operation, number of turns, maximum current-carrying capacity of the contact, metallurgical structure of the springs, chemical composition of the springs, the brand name, and many more.

This list of attributes illustrates that even a small system consisting of one relay is complex. To reduce this complexity so that it is manageable, we must simplify the system. The resulting simplified system is called a *model* of the real system.

1. GOALS AND RELEVANT ATTRIBUTES

The modeling of a system begins by making assumptions. Some of the attributes can be assumed irrelevant and thus ignored. However, relevance depends on the goal or purpose or use of the system.

For example, if the relay is to be used in a low-speed switching system such as a telephone network, the relevant attributes are the operating current, number of contact springs, and cost. If it is to be used in a higher-speed and more accurate system, such as a computer, the relevant attributes are mass, time to operate, inductance, probability of failure, and cost. However, if the relay is to be used as a paperweight, the relevant attributes are weight, size, beauty, color, and cost.

The goals are often expressed explicitly in terms of maximizing or minimizing such factors as cost, weight, stability, reliability, sensitivity, time, safety, or convenience. Practical goals are usually combinations of these individual goals. The specification of the goals is not always simple; factors may be nonmeasurable (beauty, convenience), and individual goals may conflict (increasing one decreases another). Even when the goals are established, we need not blindly select the optimum value but may use it to see how some nonoptimum value compares with the optimum.

2. ORIENTATION OF A SYSTEM

A system can be simplified by ignoring many attributes which are assumed irrelevant to the goal. Some attributes can be expressed in terms of others and thus can be suppressed. The remaining relevant attributes can be "oriented" into inputs and outputs. Those which can be controlled or manipulated (such as the voltage) are called *inputs*; those which can be

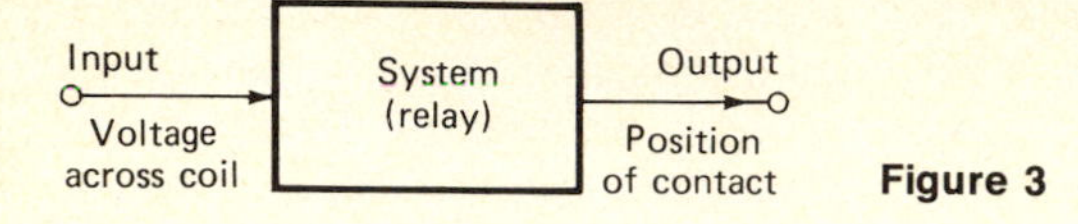

Figure 3

observed (such as the contact position) are called *outputs*. The relay thus could be represented by the "black box" or "systems" form of Figure 3.

3. ASSIGNMENT OF DIGITAL VALUES TO ATTRIBUTES

Now that the relevant attributes have been selected, they can be assigned values so that later these values may relate the outputs to the inputs in some (usually mathematical) form. The attribute of voltage in the relay example could have any value between, say, 0 and 3 volts. On such a continuous scale there are an infinite number of values of voltage. However, it is physically impossible to distinguish between all such values. An alternative therefore is to quantize, partition, or split up the values into a finite number of intervals that can be distinguished usefully from one another.

In the case of the relay, there are two natural values of voltage: (1) a voltage which causes the movable spring to move (and make contact with the other spring) and (2) a voltage which does not cause this spring to move. These two values are usually represented by the two digits 0 and 1, although other symbols such as a square and a triangle could be used but with greater inconvenience. Similarly a contact could be assumed as open or closed and represented as the digits 0 and 1 respectively.

Such a system with two values is called a *binary system*. Other binary systems involve logic (with values true, false), sets (an element is in a set or not in a set), as well as sex (male, female). Our greatest emphasis will be on such binary systems.

Many physical systems appear more naturally to be many-valued. For example;

Polarities are positive, negative, or zero.
Rotation may be clockwise, counterclockwise, or absent.
Variables may increase, decrease, or remain constant.
Qualities may be graded as A, B, C, or D.
Dice have up to six dots on a face.
Decimal digits range from 0 to 9.
The English alphabet has 26 letters.

The values in such systems could be represented by digits 0, 1, 2, . . ., *n* and also studied by the methods of this book. However, we will find ways of coding such values as combinations of the binary values. For example, the integer 5 could be coded as 11111 or possibly as the combination 101. At present the binary systems appear to be most common, most reliable, and most economical. But nonbinary digital systems are possible.

4. INTERACTION BETWEEN ATTRIBUTES

After the attributes are oriented and assigned values, they can be related in some way. This interaction or interdependence between the inputs and outputs takes the mathematical form of a function or a relation. Again the form of this relation will depend on the goal or purpose of the system. Some possible models for a relay are suggested below.

When sufficient current is applied to the coil, the following four types of behavior (models) are possible:

1. *Deterministic model*: the contacts always close, doing so instantly.
2. *Probabilistic model*: the contacts usually close, doing so instantly.
3. *Sequential model*: the contacts always close, but after some delay.
4. *Stochastic model*: the contacts usually close, but after some delay.

Each of these models corresponds to a chapter. The models are usually represented in some graphical manner as in Figure 4. Notice that these

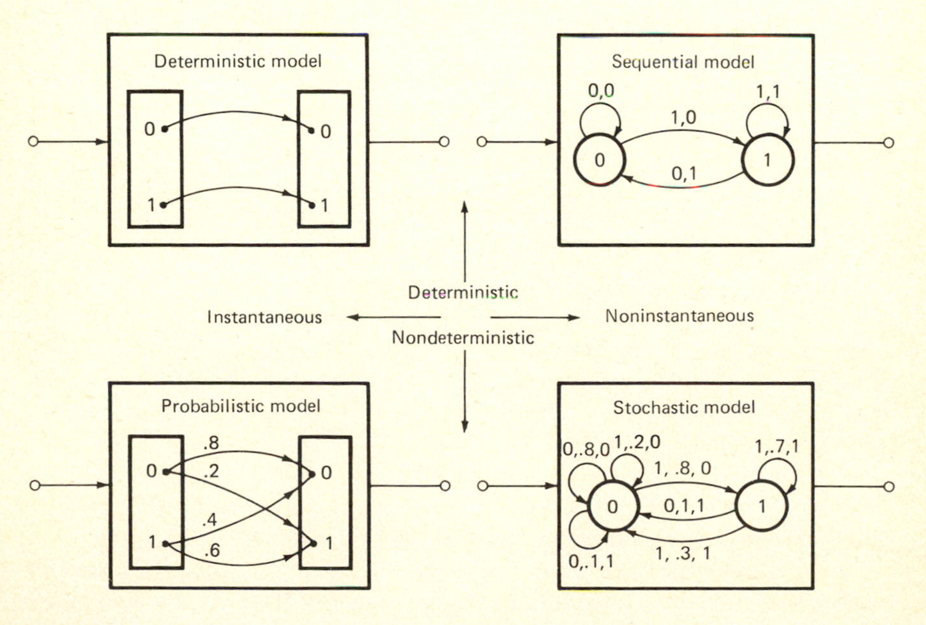

Figure 4

four models represent the four possible ways of looking at time (instantaneous or noninstantaneous) and correspondence (deterministic or nondeterministic). The meaning of each diagram will be clarified in the corresponding chapter.

5. BEHAVIORAL AND STRUCTURAL VIEW OF SYSTEMS

The preceding view of systems is called the *behavioral view*. It emphasizes the black box and the relation between its input and output. But black boxes can be interconnected in many ways to form larger systems. This leads to the *structural view*.

The structural view of systems emphasizes the internal organization or interconnection of systems. This view gives us insight into complex systems. The remainder of this introduction is concerned with the structural view. It may be read quickly now and reread after any chapter as a review.

The behavioral and structural views of systems are complementary, for the behavioral view stresses the external aspects, whereas the structural view stresses the internal aspects. The behavioral view would be most useful for using, operating, or controlling a system (computer, car, or process); whereas the structural view would be most useful in designing, constructing, optimizing, or modifying a system. The difference between the behavioral and structural views is analogous to the difference between psychology and anatomy. Both views are important and necessary.

6. HIERARCHY OF SYSTEMS

A significant way of understanding and constructing complex systems is to view any system as being composed of smaller systems. The nature of the larger system can be determined by understanding the nature of each smaller system and the interaction or interconnection between these smaller systems. Then this large system may be connected to a number of other such systems, thus forming a larger system. Again this larger system can be a component of an even larger system, and this process repeated, so forming a hierarchy of systems. At each level of the hierarchy a large system can be readily studied or synthesized from interconnections of smaller systems.

This book proceeds according to such a hierarchy, starting from relays (or transistors or mechanical components), building NORs, adders, memory devices, shift registers, and finally ending with computers.

In Chapter 1, on instantaneous deterministic systems, we consider relays and their interconnections to form digital components known as

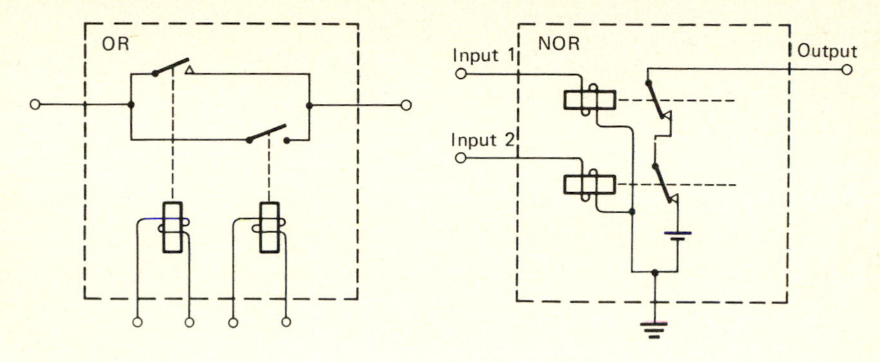

Figure 5

ANDs, ORs, NOTs, and NORs, as in Figure 5. We will also consider briefly how such digital components can be made from hydraulic and mechanical devices. (In the Appendix these components are constructed from electronic devices such as transistors.) Then these digital components are used to construct larger functional units such as adders and majority indicators as shown in Figure 6.

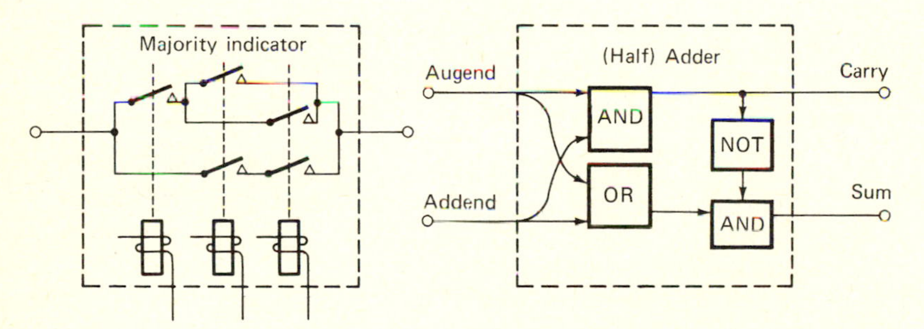

Figure 6

In Chapter 4, on instantaneous probabilistic systems, we will consider these components when their probability of failure is important, and we will find ways of interconnecting them, as in Figure 7, to obtain greater reliability. We will also see, for example, that connecting relay contacts simply in parallel does not in general increase reliability, contrary to our intuition. This probabilistic view is particularly important if the best available components are insufficient to attain the goal. But after they are interconnected properly to attain the required reliability, the resulting system is often considered to be deterministic.

In Chapter 5, on sequential systems, we will consider interconnections of systems involving time delays and feedback, as shown in Figure 8. We will see how we can use the delays and feedback to construct memory elements, called flip-flops, as well as other larger sequential systems.

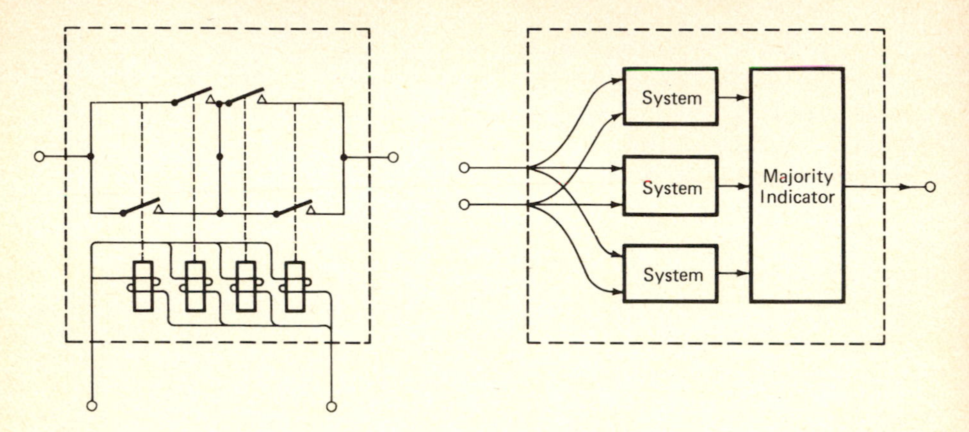

Figure 7

Some of these components can be interconnected as shift registers, which are used to store and manipulate sequences of digits.

In Chapter 6, on stochastic systems, we will consider systems involving both probabilities and time, which describe phenomena such as error bursts or decaying or changing probabilities.

Finally in Chapter 7 we use the components of all the chapters to construct a simple digital computer. These concepts can also be applied to communicate and control, but more detailed applications are left for other books.

It is important to realize that a system at a high level of the hierarchy (such as a computer) is best studied in terms of an interconnection (of registers) at a slightly lower level of the hierarchy, rather than as a very complex interconnection (of ANDs, etc.) at a very low level.

7. PROPERTIES OF SYSTEMS

When small systems are interconnected to form larger systems, the larger systems may or may not have the same properties as the component systems. For example, it may be possible that two systems both having a

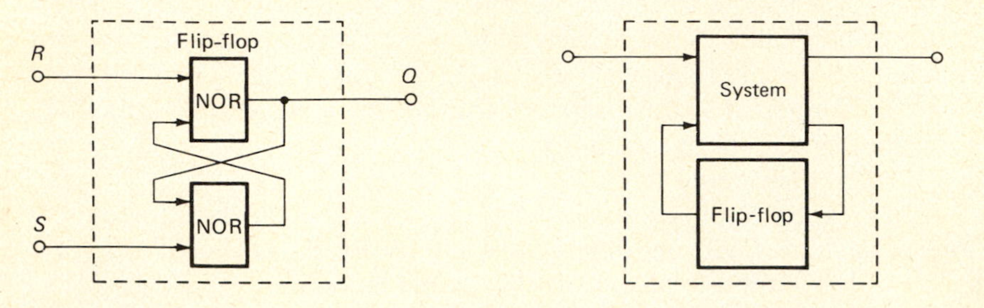

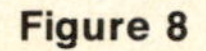

Figure 8

property such as reliability (or ergodicity, whatever that is) may be interconnected and that resulting system does not have this property. On the other hand, two systems neither of which has a property such as reliability (or functional completeness) may be interconnected in such a way as to have this property.

Thus the whole system may have all the properties of its parts, or it may have none of these properties, or more surprisingly, it may have more properties than the parts. In essence, the properties of a large system depend not only on the properties of the smaller components but also on the interconnection.

8. SYSTEM EQUIVALENCE

One of the interesting and useful consequences of the systems approach is that two systems consisting of different numbers and types of components with differing interconnections may have the same properties and identical input-output behavior. This means that if each such system were literally enclosed within a black box, it would be impossible to distinguish between the two systems by applying inputs and observing outputs of each black box. In such a case the systems are equivalent in a behavioral sense but not in a structural sense. In any applications the simpler, usually more economical system could replace its equivalent system.

In other words, there may be many different ways of doing the same thing; we can choose the way that best satisfies our goal.

9. GENERAL SYSTEMS PROCESSES

Certain general processes may be applied to any system, namely, modeling, identification, representation, analysis, synthesis, implementation, optimization, and evaluation. We will apply each of these processes to all four models.

Modeling is the process of abstracting relevant attributes from a physical system. It yields a simplified version of reality which is similar to the original physical system in important respects. The model (instantaneous, deterministic, sequential, probabilistic, stochastic, or other) depends on the goal and application of the system and is subject to later evaluation.

Identification is the process of determining the characteristics, parameters, or structure of a system from its input-output behavior. The structure is usually mathematical, with the form and accuracy dependent on the purpose or goal.

Representation is the process of describing a system in a convenient, useful, powerful, and insightful way. This may take various diagrammatic, tabular, algebraic, or symbolic forms. One representation may be more convenient than another, again depending on the goal.

Analysis is the process of studying the properties and the input-output relations of a system from a knowledge of its component systems, their interconnection, and properties.

Synthesis is the process of creating or indicating a structure (of interconnected systems) to exhibit a specified input-output behavior.

Implementation is the process of converting a synthesized structure (model) into a physical device (reality). Implementation, or realization, is dependent on economic factors as well as other goals.

Optimization of either structure or behavior is a process of improving or simplifying which involves taking some action to attain a goal.

Evaluation is the process of appraising a system to determine how well the goal has been satisfied. If the goal is not satisfied, some or all of the above processes must be reconsidered.

10. APPLICATION TO COMPUTATION, COMMUNICATION, AND CONTROL

The three important applications of systems to computing, communicating, and controlling are briefly described as follows.

A computing system can be abstractly considered as a single black box of Figure 9, acting upon a coded input sequence of symbols, processing these sequences in some specific way to yield an output sequence.

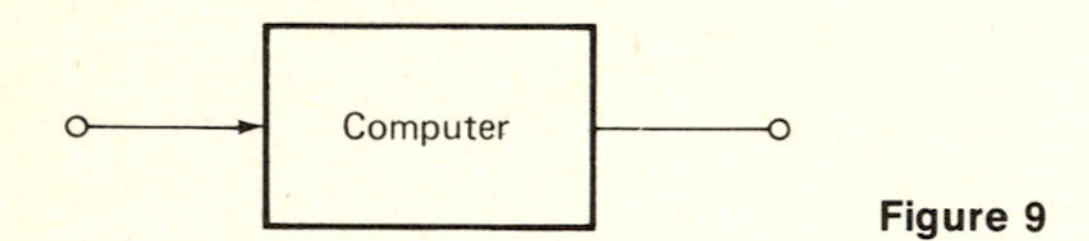

Figure 9

A communication system consists of a channel which is to transmit information from some source to a destination. The channel is usually limited in some way (reliability, speed, etc.). This limitation can be overcome or decreased by properly encoding the signal at the input of the channel and then decoding it at the output. This more complex system, shown in Figure 10, then becomes superior in some sense to the original channel.

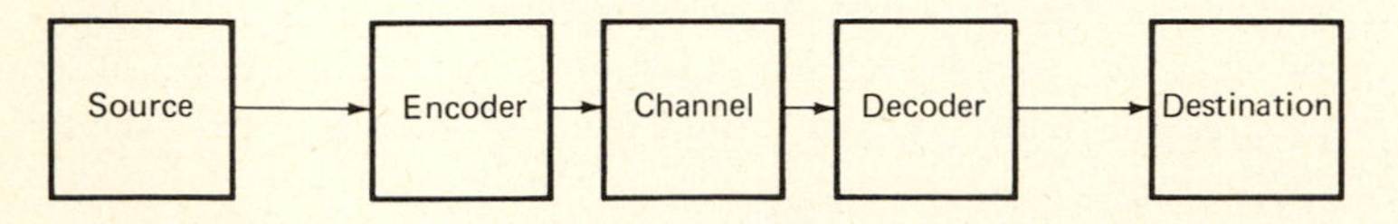

Figure 10

A control system in essence consists of a controlled system (or plant) and a controlling system (or controller). The controlled system has various properties and limitations (constraints). The controller on observing the system acts upon the controlled system in such a way as to attain a specified goal. An abstract diagram of this interaction is shown in Figure 11.

Of course each of the three above types of systems may be studied at various levels of complexity. Each back box may contain multitudes of smaller systems; each connecting line may represent a bundle of many lines. A decoder, for example, may consist of a computation system; a control system may involve within it a number of communication systems.

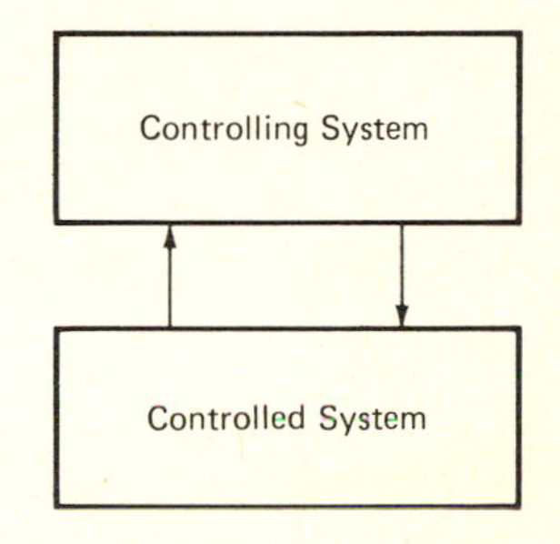

Figure 11

SUMMARY

Real systems are extremely complex. To achieve understanding or use of them requires modeling, which ignores all but the most relevant attributes. But relevance depends on the goals, purposes, or applications. Since there may be many goals, there may be many models of any one system.

In this book we study digital models which may be deterministic, probabilistic, sequential, or stochastic. Other models are possible (continuous, nonlinear, fuzzy, etc.) but are outside the realm of this book. However, the general concepts and processes (such as representation, analysis, synthesis, and optimization) are portable, and can be applied to other models.

There is a conceptual significance to representing a model as a black box. In every case the same complex reality is in the black box, but the interaction between the inputs and outputs differs depending on the goal. It is very important to realize that because of complexity a model can never describe reality exactly. However, this need not prevent us from attaining our goals! Hopefully, we have established goals that are worth attaining.

Chapter 1
DETERMINISTIC DIGITAL SYSTEMS

Instantaneous Switching Systems

In this chapter we will study one type of system, a binary deterministic switching system. Such switching systems are found in telephone networks, computers, spacecraft, factories, traffic controllers, and dispensing machines.

The components of such systems may be switches, relays, diodes, transistors, hydraulic valves, or levers. We will consider relays first, since they are simple to understand and once they are understood, the principles are easily extended to other devices.

Any relay consists essentially of a coil and an arrangement of flat metal springs, some fixed and some movable, as in Figure 1. When the coil has no current flowing in it, the movable spring rests in contact with the upper fixed spring, called the *break spring*. The electric circuit through the break spring is said to be *closed*. This condition is known as the *unenergized condition* or sometimes the normal position.

When current flows in the coil, it acts as an electromagnet attracting the movable spring, which breaks contact with the upper, or break, spring and makes contact with the lower, or *make* spring. The circuit through

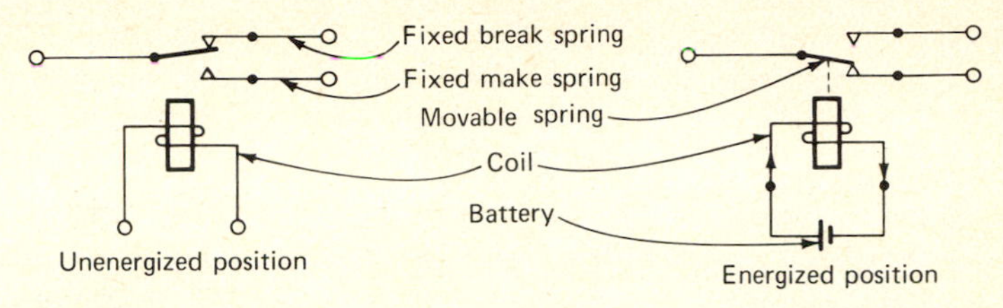

Figure 1

the break spring is now *open*, and the circuit through the make spring is closed. This condition is known as the *energized condition* or operated position. When the current in the coil ceases, the movable spring again returns to the upper spring.

A combination of springs is often called a *contact*. The above arrangement of three springs is known as a *transfer contact*. Some relays are constructed having many contacts, with all the movable springs mechanically connected together. This enables all the contacts to open and close together. An ordinary telephone relay may have as many as twelve such gangs of contacts. The relay in Figure 2 has three gangs. In addition to a transfer contact (three springs), it has some springs arranged to form a make contact (two springs) and others arranged to form a break contact (two springs). A make contact is often called a *normally open* (no) contact and a break contact is called a *normally closed* (nc) contact. The complexity of a relay, which is a measure of its price and installation cost, is often given by the total number of springs, which in this case is seven $(2 + 2 + 3)$.

1. MODELING

Modeling a Relay

On observing the operation of a relay, the following assumptions seem natural. (1) Although we know that the movable spring can occupy an infinite number of positions between the make and break positions, such intermediate positions are difficult to maintain, and so we will assume that

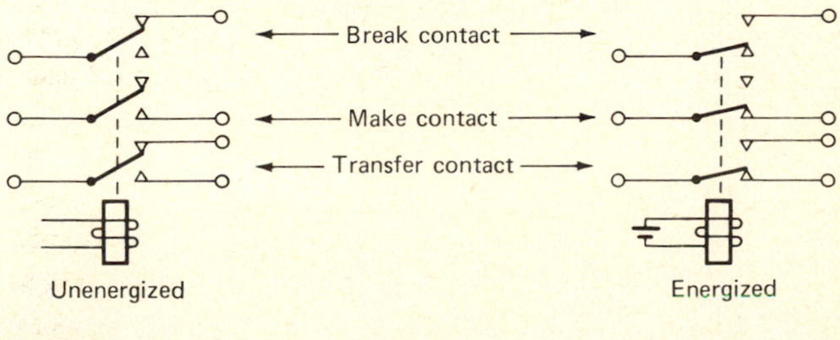

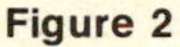

Figure 2

there are only two positions (*binary*). (2) Although we know that there is a delay between the time of energizing a coil and the operating of a contact, this delay time is short and we will assume no delay, i.e., *instantaneous action.* (3) Although we know that an actual relay may malfunction, the malfunction is occasional and will be ignored. This exact relation between the coil condition and the contact position is called the *deterministic* assumption; no uncertainties or probabilities are involved.

The resulting simplified view of a real relay is called a *binary, instantaneous, deterministic model.* Because of the assumptions, it appears to be very limited. However, it is extremely useful provided that the assumptions are respected. If any of these three assumptions does not hold in a practical situation, the model may be extended to other models such as many-valued, probabilistic, sequential, or stochastic. These models will be discussed in later chapters, but we should keep in mind that *all* models have their limitations. Reality is too complex to know in all details.

Modeling an Interconnection of Relays

So far we have considered only a single relay. However, when relays are interconnected, they interact to yield interesting systems which may be used for computation, communication, and control.

Consider the system of Figure 3. It will be used as a running example to illustrate ways of representing systems, analyzing their behavior, and optimizing their structure. Later, in Section 4, we will also see its application to selecting people who meet certain conditions.

2. REPRESENTATION

There are four common representations of relay networks: an attached diagram, a detached diagram, an algebra, and a table-of-combinations.

Representation as an Attached Diagram

The network of Figure 3 is given in an inconvenient though common

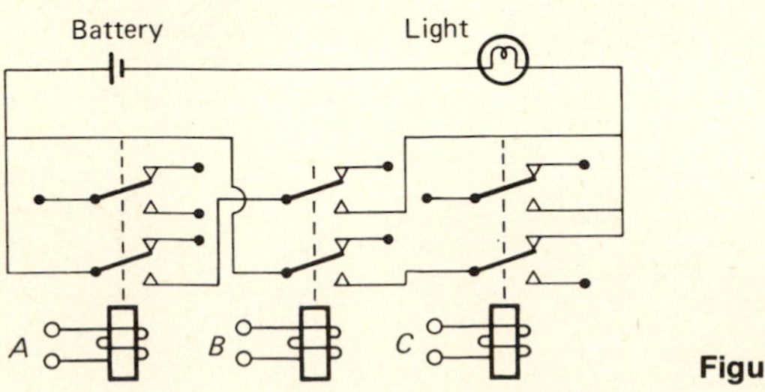

Figure 3

form. The contacts are conventionally drawn for relays in the unenergized position. Some contacts of the relays are not used; some circuits meander everywhere. This corresponds to the actual physical configuration of the components and hides the basically simple operation of the network. This representation is known as the *attached representation* since the contacts are shown near to (or attached to) their coils.

Representation as a Detached Diagram

As an aid in describing the operation of a network, it may be convenient to rearrange the diagram. The contacts can be drawn detached from their coils, with the physical connection being understood by labeling all contacts of a relay by a symbol associated with that relay. For example, all the make contacts of relay A could be labeled A and all the break contacts labeled $\overline{A}$. Thus when relay A operates, all its make contacts close and its break contacts open, regardless of where on the diagram they may be located. Of course physically the contacts are still attached to the coils; it is only diagrammatically that they are detached.

The rearrangement of the previous network into a detached-contact form results in the simple series-parallel network of Figure 4. From this detached diagram the behavior can be indicated immediately by inspection. By following all (two) paths through the network we see that the network is closed (and the light goes on) when A and B are operated or when B is operated and C is not operated.

This detached diagrammatic representation can be simplified further by not drawing the contact but replacing it with its algebraic symbol. If the symbol has a bar over it, it represents a break contact; otherwise it is a make contact. Such a more abstract detached-contact diagram is shown in Figure 5. Note also that the coils are not shown.

Representation as an Algebra

Relay networks can easily be described symbolically by algebraic expressions. Three basic networks are given in Figure 6. A network N consist-

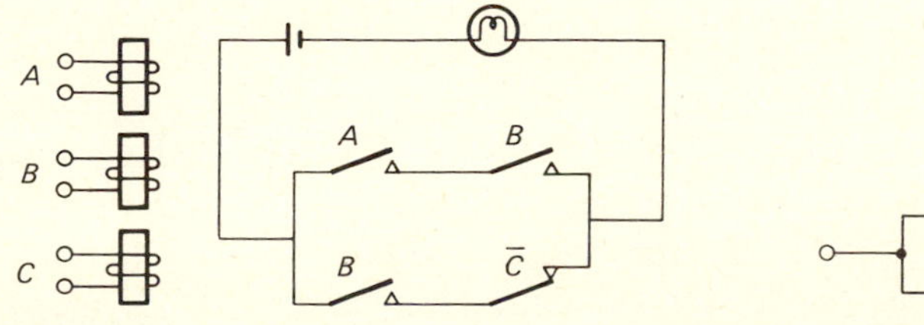

Figure 4

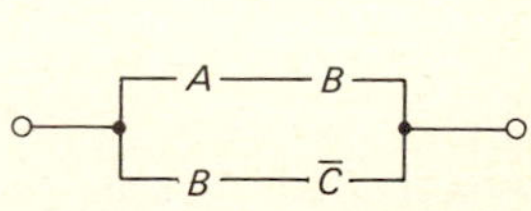

Figure 5

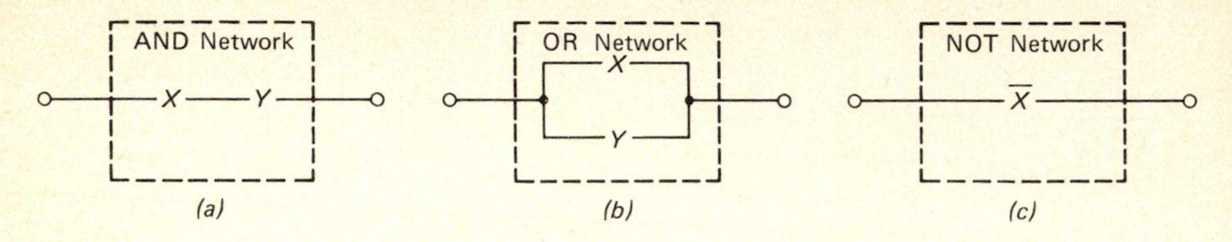

Figure 6

ing of a series connection of contacts X, Y as in Figure *6a* is called an AND *network* since it is closed whenever both relays X and Y are operated. This network will by symbolized by

$$N = X \cdot Y$$

and read as "X and Y." When no confusion can result, the dot is omitted, to yield $N = XY$.

A network N consisting of the parallel connection of contacts X, Y as in Figure 6*b* is called an OR *network* since it is closed whenever relay X or Y is operated. This network is symbolized algebraically as

$$N = X \vee Y$$

and is read as "X or Y." This symbol, taken from the first letter of the Latin *vel*, means and/or and is often called the INCLUSIVE-OR. The choice of symbols is arbitrary; what is important is that some algebraic expression can correspond to a network. Other alternative notations that have been used for the OR network are

$$X \cup Y \qquad X + Y \qquad XY \qquad XAY \qquad X \oplus Y \qquad X \mathbf{+} Y$$

A network N consisting of a single break contact of a relay X as in Figure 6*c* is known as a NOT network since the network is closed when the relay is not operated. The NOT network is denoted by the symbol

$$N = \bar{X}$$

read as "not X" or "X not."

The three symbols for AND, OR, and NOT have described interconnections of contacts (such as X, Y), but they can also be extended to interconnections of networks of contacts (such as N) since networks are also either open or closed and may be connected in series or parallel. This enables the algebra to represent complex networks.

From Networks to Algebra. Consider again the original switching network, which is redrawn in Figure 7 as network N_3. It is also shown

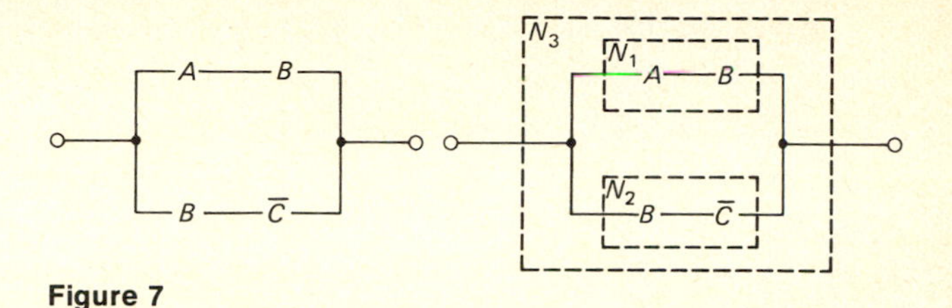

Figure 7

decomposed into networks N_1 and N_2. The network N_3 consists of the parallel connection of networks N_1 and N_2, and so we may write

$$N_3 = N_1 \vee N_2$$

The network N_1 is a series connection, which is written

$$N_1 = A \cdot B$$

Similarly network N_2 can be written

$$N_2 = B \cdot \bar{C}$$

Finally the network N_3 can be written directly in terms of the contacts within it as

$$N_3 = N_1 \vee N_2 = (A \cdot B) \vee (B \cdot \bar{C})$$

This algebraic expression $AB \vee B\bar{C}$ should be read

"A and B or B and not C"

which in more detail describes the behavior of the network: "the network is closed when relays A and B are operated or when relay B is operated and relay C is not operated."

Precedence of Operators Notice that the expression $X \vee Y \cdot Z$ could be ambiguous, for it may be $X \vee (Y \cdot Z)$ or it may be $(X \vee Y) \cdot Z$. In such cases we will use the convention that the NOT operation precedes the AND, which precedes the OR. This is similar to the ordinary algebraic operations where exponentiation precedes multiplication which precedes addition. Thus the expression $X \vee Y \cdot Z$ will mean $X \vee (Y \cdot Z)$ by analogy to the fact that $X + Y * Z$ means $X + (Y * Z)$ in the ordinary algebra.

From Algebra to a Network. The algebraic expression

$$N = (X \vee \bar{Y})(\bar{X} \vee Y(X \vee Z))$$

can be rewritten in terms of other larger networks as

$$N = P(Q \vee R)$$

where $P = X \vee \bar{Y}$

$Q = \bar{X}$

$R = Y(X \vee Z)$

Similarly network R can be decomposed as

$$R = Y\,S$$

where

$$S = X \vee Z$$

Finally, the various networks may be drawn from the expressions in a step-by-step manner, as in Figure 8. First, the simpler connection of the larger nets P, Q, and R is drawn (as dotted lines); then the smaller ones are substituted for P, Q, and R. Usually this entire process of decomposition of the algebra and construction of the larger networks can be done conceptually as one large step. However, for large complex networks the above method is convenient.

Example: Intermediate Relays Write an algebraic expression corresponding to the network of Figure 9. First,

$$N = \bar{X} \vee \bar{Z}$$

But

$$Z = X \vee Y \qquad \text{and} \qquad \bar{Z} = \overline{X \vee Y}$$

And so

$$N = \bar{X} \vee (\overline{X \vee Y}) \qquad \text{by substitution}$$

It should be noticed that the NOT of an expression is achieved by the use of a break contact of an intermediate relay such as Z.

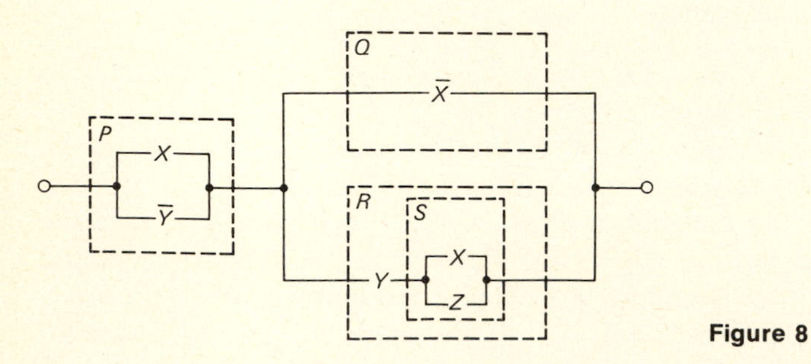

Figure 8

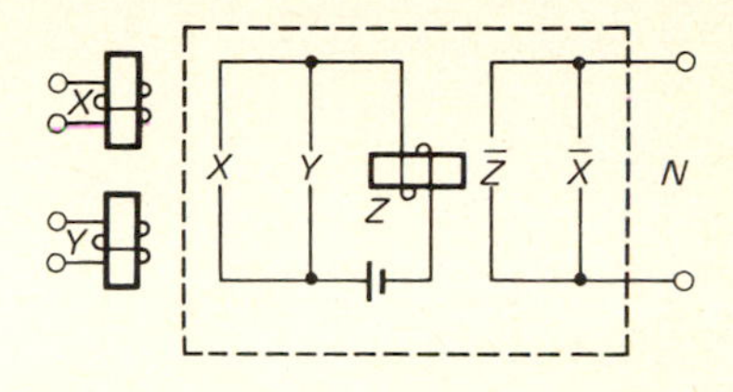

Figure 9

Representation as a Table-of-Combinations

In the algebraic representation all the variables (such as A, B, Y, Z) can have only two values corresponding to an open circuit or a closed circuit. For convenience the two values may be represented by the symbols 0 and 1.

An open circuit is represented by value 0.
A closed circuit is represented by value 1.

These values are not numeral quantities in the ordinary sense since they are not operated on by such operators (operations) as addition or division but by the switching operators of AND, OR, and NOT. These switching operators are described by the tables of Figure 10. The table for the OR operator, for example, is interpreted as follows:

$0 \vee 0 = 0$ indicates that an open circuit in parallel with an open circuit is still an open circuit.
$0 \vee 1 = 1$ indicates that an open circuit in parallel with a closed circuit is a closed circuit,
$1 \vee 0 = 1$ indicates that a closed circuit in parallel with an open circuit is a closed circuit,
$1 \vee 1 = 1$ indicates that a closed circuit in parallel with a closed circuit is a closed circuit.

Two other ways for describing an OR operator are given in Figure 11. The correspondence between the values of the variables and the value of the operator should be evident.

X	Y	$X \cdot Y$
0	0	0
0	1	0
1	0	0
1	1	1

(a)

X	Y	$X \vee Y$
0	0	0
0	1	1
1	0	1
1	1	1

(b)

X	$\bar{X}$
0	1
1	0

(c)

Figure 10

$X \backslash Y$ ($X \vee Y$)	0	1
0	0	1
1	1	1

X	$\vee$	Y
0	0	0
0	1	1
1	1	0
1	1	1

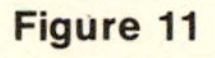

Figure 11

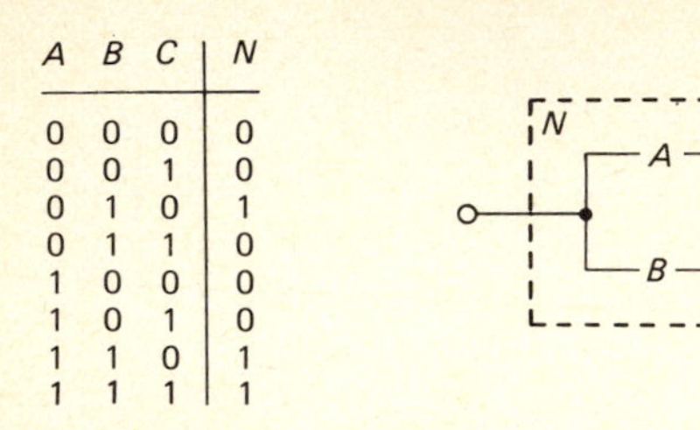

A	B	C	N
0	0	0	0
0	0	1	0
0	1	0	1
0	1	1	0
1	0	0	0
1	0	1	0
1	1	0	1
1	1	1	1

Figure 12

A table-of-combinations for an algebraic expression is a list of all possible combinations of values for the variables along with the corresponding value of the expression. It is commonly called a truth-table. Physically it shows, for all possible relay positions, whether a network is closed or not. The table-of-combinations for the previous network is shown in Figure 12. On the left is a list of all eight possible combinations of values of the three variables A, B, C. On the right is the value of the network for each combination. This value can be determined by inspection. For example, in the third combination only relay B is operated, but there is a path through B and the break contact of relay C so that the network is closed ($N = 1$) in this case.

The tables-of-combinations for other expressions having two and four variables are given in Figure 13.

These examples indicate that for an expression having n switching variables, there are 2^n possible combinations of values. The order of listing these combinations is not significant; however, it is convenient to list

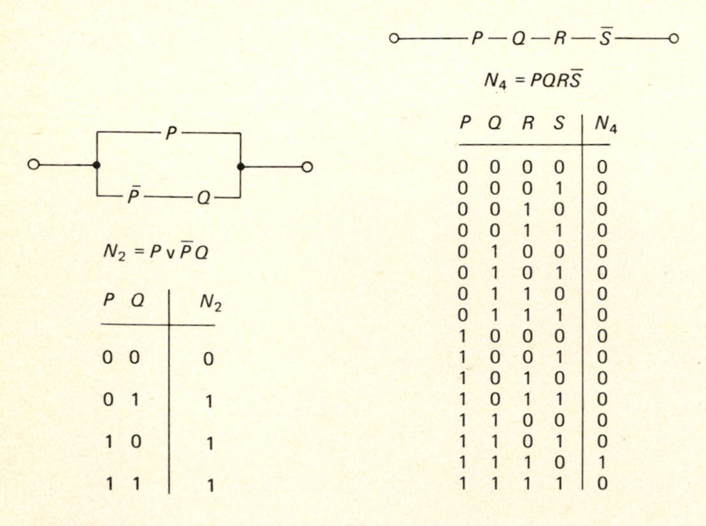

$N_2 = P \vee \bar{P}Q$

P	Q	N_2
0	0	0
0	1	1
1	0	1
1	1	1

$N_4 = PQR\bar{S}$

P	Q	R	S	N_4
0	0	0	0	0
0	0	0	1	0
0	0	1	0	0
0	0	1	1	0
0	1	0	0	0
0	1	0	1	0
0	1	1	0	0
0	1	1	1	0
1	0	0	0	0
1	0	0	1	0
1	0	1	0	0
1	0	1	1	0
1	1	0	0	0
1	1	0	1	0
1	1	1	0	1
1	1	1	1	0

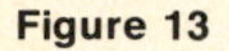

Figure 13

them in the given consistent (or standard) form. In the first column, the first half of the values are 0s, and the remaining half are 1s. In the second column, the first one-quarter of the values are 0s, the next one-quarter of the values are 1s. This halving process continues until at the nth column the 0s and 1s alternate. Later we will see that this order corresponds to the binary numbers.

EXERCISE SET 1

1. Represent the attached diagram of Figure 14 by a detached diagram, write its switching algebraic expression, and draw the table-of-combinations. *Ans.*: $(A \vee B)\bar{B}(A \vee C)$
2. Draw a detached diagram corresponding to the switching algebraic expressions:
   (*a*) $(A \vee B)(C \vee D)(E \vee F)$ (*b*) $(A \vee B)(C \vee D)(E \vee F))$
   (*c*) $A \vee B(C \vee D)(E \vee F)$ (*d*) $(A \vee B(C \vee D)\ (E \vee F)$
   (*e*) $A \vee B(C \vee D(E \vee F))$ (*f*) $A \vee (BC \vee D)(E \vee F)$
   (*g*) $A \vee BC \vee DE \vee F$
3. Write switching algebraic expressions for the networks of Figure 15.

3. ANALYSIS

So far we have considered many representations of a relay switching network: an attached diagram, a detached diagram, an algebra, and a table-of-combinations. These representations will be useful in analysis, the study of properties of interconnected systems. The detached diagram is simply a convenience for reading and drawing networks. However, the algebraic representation is considerably more useful because it not only represents entire networks conveniently but suggests properties and allows operations on the networks. The table-of-combinations can be used to prove or verify the properties.

Complete Induction

Consider again the previous running example, described algebraically as

$$AB \vee B\bar{C}$$

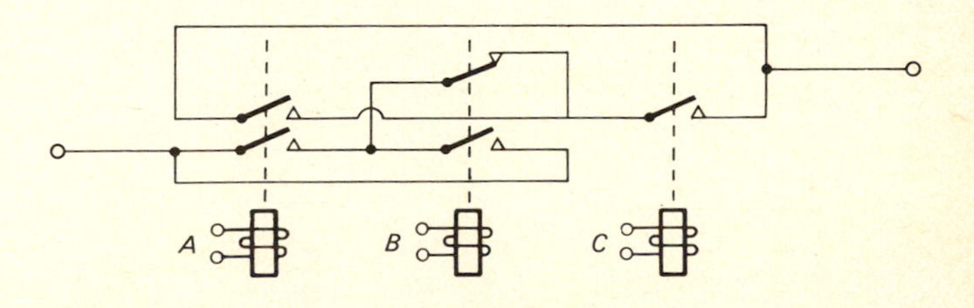

Figure 14

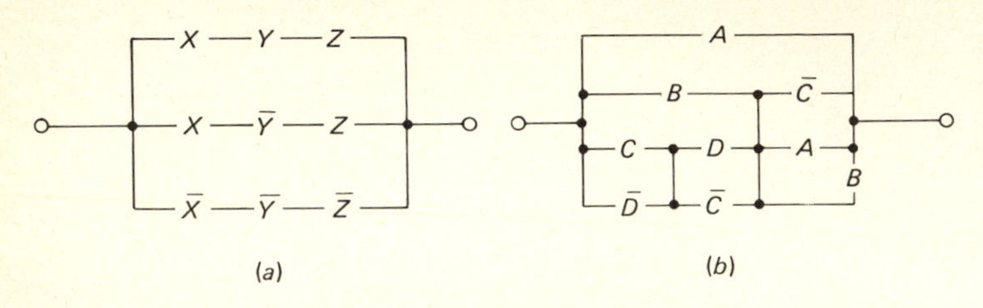

Figure 15

Analogy to the ordinary algebra suggests the factoring (or distribution) of the common variable B, resulting in the expression

$$B(A \vee \bar{C})$$

The equivalence can be verified in this case by comparing the corresponding networks of Figure 16.

Two algebraic expressions are defined as *equivalent* if they assume identical values for all possible values of their variables. But in the switching algebra, all possible combinations of values are listed in the table-of-combinations representation; therefore, we need simply compare the table-of-combinations to determine equivalence. This is done in Figure 17 for the networks at the left N_L and at the right N_R of Figure 16. From the table-of-combinations we see that both networks open and close identically for all possible combinations. The networks are equivalent and may be denoted as

$$AB \vee B\bar{C} = B(A \vee \bar{C})$$

This method of "trying all possibilities" in a table-of-combinations form is known as the method of *complete induction*. It is an exhaustive method which is sometimes exhausting also. However, since it is a powerful method which can be applied to many types of digital systems, we will study more convenient methods of complete induction after this next generalization.

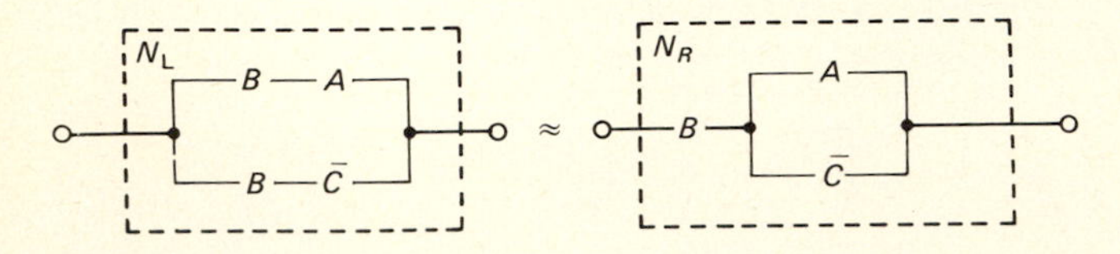

Figure 16

X	Y	Z	N_L	N_R
0	0	0	0	0
0	0	1	0	0
0	1	0	1	1
0	1	1	0	0
1	0	0	0	0
1	0	1	0	0
1	1	0	1	1
1	1	1	1	1

Figure 17

$0 \cdot 0 = 0$ $\quad 0 \vee 0 = 0$ $\quad \bar{0} = 1$
$0 \cdot 1 = 0$ $\quad 0 \vee 1 = 1$
$1 \cdot 0 = 0$ $\quad 1 \vee 0 = 1$ $\quad \bar{1} = 0$
$1 \cdot 1 = 1$ $\quad 1 \vee 1 = 1$

Figure 18

Switching (Boolean) Algebra

We have started from a physical system and have arrived at a mathematical model of this system in the form of an algebra, often called a *Boolean algebra*. The algebra has variables such as A, B, C, X, Y, Z, but they have only two values, 0 and 1. There are also three operators defined in Figure 18. We have started to investigate some properties or laws such as the factoring (or distributive) property. However, before going further with the relay motivation, it would be well to realize that systems other than relays may be described by the same algebra. Therefore, rather than doing the analysis in terms of relay networks, we will do it in general algebraic terms; then later we can relate the general results to relays and also to other devices. This is a minor change in emphasis, with a resulting major change in the applicability of the algebra.

Properties of the Algebra

As a first step in our more general analysis we must be able to construct a table-of-combinations from the algebra alone. This is easily done by using the operations given in Figure 18.

Consider, for example, the expression $N = P \vee \bar{P} \cdot Q$. For the combination $P = 0$ and $Q = 1$ this expression becomes

$$N = P \vee \bar{P} \cdot Q$$

$$= 0 \vee \bar{0} \cdot 1 \qquad \text{by substituting } P = 0 \text{ and } Q = 1$$

$$= 0 \vee 1 \cdot 1 \qquad \text{by definition of NOT, i.e., } \bar{0} = 1 \text{ above}$$

$$= 0 \vee 1 \qquad \text{by definition of AND, i.e., } 1 \cdot 1 = 1$$

$$= 1 \qquad \text{by definition of OR, i.e., } 1 \vee 1 = 1$$

Thus, the expression has value 1 for this combination. Similarly this method can be applied to the remaining three combinations as follows (in less detail):

For $P = 0$ and $Q = 0$: $\quad N = 0 \vee \bar{0} \cdot 0 = 0 \vee 1 \cdot 0 = 0 \vee 0 = 0$

P	Q	N
0	0	0
0	1	1
1	0	1
1	1	1

Figure 19

	P	Q	$\bar{P}$	$\bar{P}$	$\cdot$	Q	P	$\vee$	$\bar{P} \cdot Q$	N
	0	0	1	1	0	0	0	0	0	0
	0	1	1	1	1	1	0	1	1	1
	1	0	0	0	0	0	1	1	0	1
	1	1	0	0	0	1	1	1	0	1
Stage	1		2		3			4		5

Figure 20

For $P = 1$ and $Q = 0$: $N = 1 \vee \bar{1} \cdot 0 = 1 \vee 0 \cdot 0 = 1 \vee 0 = 1$
For $P = 1$ and $Q = 1$: $N = 1 \vee \bar{1} \cdot 1 = 1 \vee 0 \cdot 1 = 1 \vee 0 = 1$

All these results can now be summarized on a table-of-combinations as in Figure 19.

This table-of-combinations was completed by substituting one row at a time. The algebraic substitution can also be done one column at a time. This is illustrated in Figure 20, which is constructed in five stages:

Stage 1: *All* 2^n combinations of the n variables are written out.
Stage 2: A column corresponding to $\bar{P}$ is constructed and filled in with the opposite values to those in column P.
Stage 3: The two columns $\bar{P}$ and Q are rewritten and compared and a value 1 placed under the AND symbol when both values are 1.
Stage 4: The AND column from stage 3 is compared to the P column with the value 1 placed under the OR symbol if either value in the row is a 1.
Stage 5: The value of the network is recopied from stage 4.

This method can be shortened by not rewriting any columns, as shown in Figure 21. Otherwise, the stage-by-stage procedure is identical to that outlined above.

The procedure can be further extended to prove the property

$$P \vee \bar{P}Q = P \vee Q$$

by constructing another table-of-combinations for $P \vee Q$ to the right of

	P	$\vee$	$\bar{P}$	$\cdot$	Q
	0	0	1	0	0
	0	1	1	1	1
	1	1	0	0	0
	1	1	0	0	1
Stage	1	4	2	3	1

Figure 21

P	Q	P	$\vee$	$\bar{P}$	$\cdot$	Q	$=$	P	$\vee$	Q
0	0	0	0	1	0	0	T	0	0	0
0	1	0	1	1	1	1	T	0	1	1
1	0	1	1	0	0	0	T	1	1	0
1	1	1	1	0	0	1	T	1	1	1

Figure 22

the original table-of-combinations. Then, the two tables-of-combinations are compared as in Figure 22 with a T (true) marked under the equality sign if it is true that both sides are equal; otherwise, an F (false) is marked.

Since the equality in this case is true for all possible combinations of values, this property has been proved. It has no formal name, but it is still useful.

Expressions involving only one variable can be proved similarly, but they require only two rows in the table-of-combinations. For example, a property called *idempotence*

$$X \cdot X = X$$

is proved in Figure 23*a*. Idempotence, meaning equal power, describes the fact that all exponents in this algebra are of the same power, the first power. Another significant property proved in Figure 23*b* is

$$X \vee \bar{X} = 1$$

which is called the property of *complementarity*.

X	X	$\cdot$	X	$=$	X
0	0	0	0	*T*	0
1	1	1	1	*T*	1

(*a*)

X	X	$\vee$	$\bar{X}$	$=$	1
0	0	1	1	*T*	1
1	1	1	0	*T*	1

(*b*)

Figure 23

Properties involving three variables require 2^3 or 8 rows to prove. For example, an "unusual" distributive property

$$(X \vee Y)(X \vee Z) = X \vee YZ$$

is proved in Figure 24. The numbers below the columns indicate the order of evaluating the columns. This distributive property was indicated

X	Y	Z	$(X$	$\vee$	$Y)$	$\cdot$	$(X$	$\vee$	$Z)$	$=$	X	$\vee$	Y	$\cdot$	Z
0	0	0	0	0	0	0	0	0	0	T	0	0	0	0	0
0	0	1	0	0	0	0	0	1	1	T	0	0	0	0	1
0	1	0	0	1	1	0	0	0	0	T	0	0	1	0	0
0	1	1	0	1	1	1	0	1	1	T	0	1	1	1	1
1	0	0	1	1	0	1	1	1	0	T	1	1	0	0	0
1	0	1	1	1	0	1	1	1	1	T	1	1	0	0	1
1	1	0	1	1	1	1	1	1	0	T	1	1	1	0	0
1	1	1	1	1	1	1	1	1	1	T	1	1	1	1	1
1	2	3	4	6	5	10	7	9	8	16	14	15	11	13	12

Figure 24

X	Y	$\overline{X}$	v	$\overline{Y}$	¬	≠	$\overline{X}$	v	$\overline{Y}$
0	0	0	0	0	1	T	1	1	1
0	1	0	1	1	0	F	1	1	0
1	0	1	1	0	0	F	0	1	1
1	1	1	1	1	0	T	0	0	0
1	2		3		4	8	5	7	6

Figure 25

as unusual because the corresponding property in the ordinary algebra does not hold true; i.e.,

$$(X + Y)(X + Z) \neq X + YZ$$

Notice the following tempting expression, which does not hold in this algebra:

$$\overline{X \vee Y} \overset{?}{=} \bar{X} \vee \bar{Y}$$

This can be shown in Figure 25. Actually the entire table-of-combinations is not necessary to disprove such an expression. Any one combination leading to a false result disproves this. Instead, there is a property of

$X \cdot X = X$	1. *Idempotence*	$X \vee X = X$
$1 \cdot X = X$	2. *Operation of unit (1)*	$1 \vee X = 1$
$0 \cdot X = 0$	3. *Operation of zero (0)*	$0 \vee X = X$
$X \cdot \overline{X} = 0$	4. *Complementarity*	$X \vee \overline{X} = 1$
$X \cdot Y = Y \cdot X$	5. *Commutativity*	$X \vee Y = Y \vee X$
$(\overline{X \cdot Y}) = \overline{X} \vee \overline{Y}$	6. *Dualization*	$(\overline{X \vee Y}) = \overline{X} \cdot \overline{Y}$
$X(X \vee Y) = X$	7. *Absorption*	$X \vee XY = X$
$X(YZ) = (XY)Z$	8. *Associativity*	$X \vee (Y \vee Z) = (X \vee Y) \vee Z$
$X(Y \vee Z) = XY \vee XZ$	9. *Distributivity*	$X \vee YZ = (X \vee Y)(X \vee Z)$

Figure 26

dualization, sometimes known as *De Morgan's law*, which is

$$\overline{X \vee Y} = \bar{X} \cdot \bar{Y}$$

The properties of this Boolean algebra are summarized in Figure 26 for convenience of reference. Notice that each property holds for both the AND and OR operator, and so the properties are listed in pairs known as *duals*. The property listed at the left holds for the AND operator; that at the right holds for the OR operator.

Any of these properties can be interpreted as describing equivalent systems. For example, the unusual distributive property

$$(X \vee Y)(X \vee Z) = X \vee YZ$$

indicates the equivalence of the networks of Figure 27.

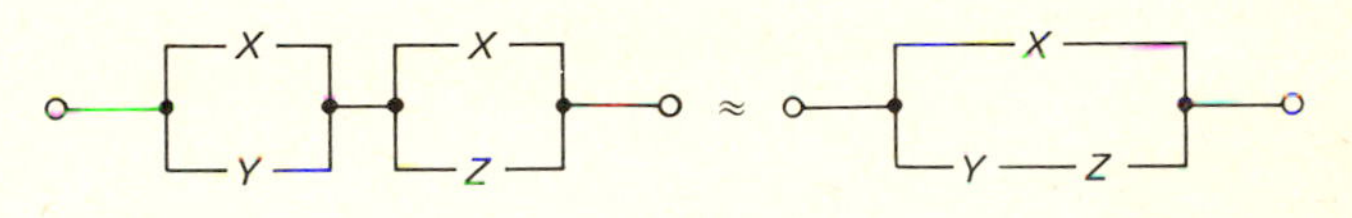

Figure 27

Similarly the dualization property

$$\overline{X \vee Y} = \bar{X} \cdot \bar{Y}$$

indicates the equivalence of the networks of Figure 28. This diagram shows that an intermediate relay such as Z is not necessary for constructing the NOT of an expression.

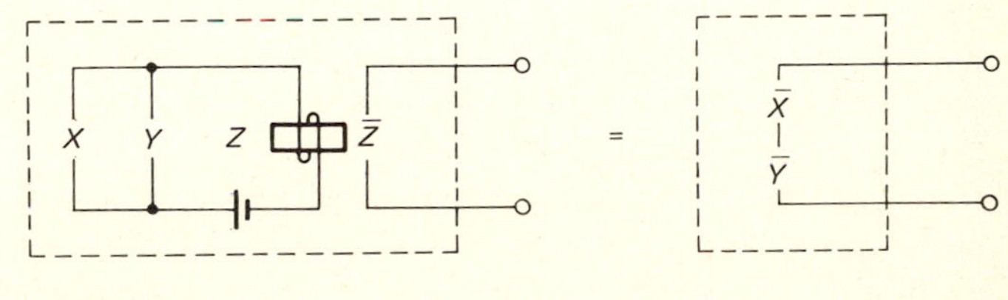

Figure 28

EXERCISE SET 2

1. Check the following equalities by complete induction, and draw the corresponding networks:

(*a*) $I \vee X = 1$ a property of operation of 1
(*b*) $X \cdot \bar{X} = 0$ a property of complementarity
(*c*) $\overline{X \cdot Y} = \bar{X} \vee \bar{Y}$ another dualization property
(*d*) $X(\bar{X} \vee Y) = XY$ a convenient but unnamed property

2. Prove by induction the following extended properties:

(*a*) $1 \vee P\bar{Q} = 1$ an extended property of operation of 1
(*b*) $\overline{ABC} = \bar{A} \vee \bar{B} \vee \bar{C}$ an extended dualization property
(*c*) $XY \vee \overline{XY} = 1$ an extended complementarity property

3. Prove by induction the following properties, which will be useful later:

(*a*) $XY \vee \bar{X}\bar{Y} = \overline{X\bar{Y} \vee \bar{X}Y}$ notice that $XY \vee \bar{X}\bar{Y} \neq 1$
(*b*) $X\bar{Y} \vee \bar{X}Z \vee \bar{Y}Z = X\bar{Y} \vee \bar{X}Z$ notice the common $X\bar{Y} \vee \bar{X}Z$

4. SYNTHESIS

With such a brief introduction to switching networks it is already possible to synthesize or design networks which satisfy given specifications. This constructive activity of synthesis is sometimes called *logical design*. The following example illustrates not only a synthesis procedure but also that although the design may be "logical," the specifications may not be so!

Design a network to select male applicants that satisfy the following specifications:

> "The applicant must be able and bearded, or if not bearded, then certainly still able, but not born under the astrological sign of Cancer."

There are three relays necessary for this network:

A relay *A* which operates if the applicant is able (proficient)
A relay *B* which operates if the applicant is bearded
A relay *C* which operates if the applicant was born under Cancer

Before synthesizing let us digress to see how the relays could be operated. In many cases they are operated from data coded on punched cards. If an applicant has a certain property, a hole is punched in a particular position on this applicant's card. For example, the card in Figure 29 indicates that the applicant is not able, not bearded, and was born under the sign of Cancer, a very bad combination indeed.

Such punched cards when "read" by a machine as in Figure 29 allow springs to make contact through the holes, thus operating the appropriate relays. In other cases beams of light are used to detect the holes by

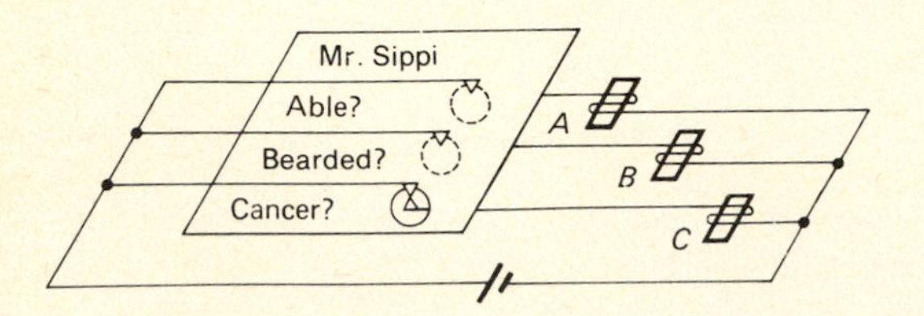

Figure 29

activating light-sensitive electronic devices, which in turn may operate the relays.

Let us now return to the synthesis problem. The contacts of the relays A, B, and C are now used to construct the required network. The verbal specification:

> "Able and bearded, or not bearded but able and not born under the sign of Cancer"

can be translated directly to the algebraic expression

$$A \cdot B \vee \bar{B} \cdot A \cdot \bar{C}$$

or more briefly

$$AB \vee \bar{B}A\bar{C}$$

The required network follows in Figure 30a, having a complexity of 10 springs. This network could operate a magnetic device which separates cards according to whether the specifications are satisfied or not.

This network may not be the simplest, the most economical, or the most reliable network, but it does behave in the required way. The network has been synthesized but not simplified. It can be simplified by the methods of the previous section as follows:

$$\begin{aligned} N &= AB \vee \bar{B}A\bar{C} \\ &= A(B \vee \bar{B}\bar{C}) && \text{by distributive property and commutativity} \\ &= A(B \vee \bar{C}) && \text{by the unnamed property } P \vee \bar{P}Q = P \vee Q \end{aligned}$$

The resulting simplified network is shown in Figure 30b. It has a complexity of six springs over the previous 10 springs.

This direct algebraic translation is only one method of synthesis. Later we will consider other methods, but first let us consider more optimizing or simplifying.

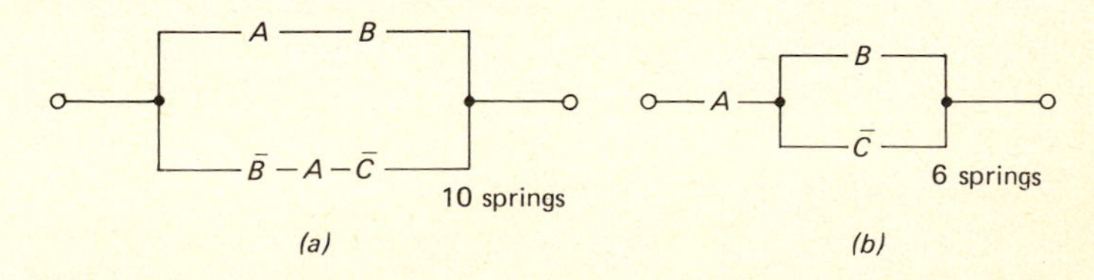

Figure 30

EXERCISE SET 3

Synthesize algebraic expressions corresponding to the following specifications of an applicant. Note that more than one equivalent expression may correspond to any verbal specification.

1. Applicant is able or beautiful but not both. *Ans.*: $(A \vee B)(\overline{A \cdot B})$ or also $A\bar{B} \vee \bar{A}B$
2. Applicant is neither beautiful nor born under Cancer. *Ans.*: $\bar{B} \cdot \bar{C}$ or $\overline{B \vee C}$
3. Applicant is not only beautiful but able.
4. Applicant is ugly as well as unable.
5. Applicant has either all or neither of the properties of beauty and ability.
6. Applicant is beautiful or able, and born under Cancer.
7. Applicant is beautiful, or able and born under Cancer.
8. Applicant is beautiful or able and born under Cancer.

5. OPTIMIZATION

So far we have studied a large number of equalities (summarized in Figure 26) which were often motivated by the ordinary algebra and proved by the table-of-combinations method. In this section we will use these equalities to optimize or simplify networks. As an example, consider the following expression:

$$AB \vee AB\bar{C} \vee A\bar{B}\bar{C} = A(B \vee \bar{C})$$

Of course this can be proved by a table-of-combinations (try it), but in general the right-hand side is not given, and so we may apply equalities as follows:

$$N = AB \vee AB\bar{C} \vee A\bar{B}\bar{C}$$

$$= A(B \vee B\bar{C} \vee \bar{B}\bar{C}) \quad \text{by distributive property (factoring)}$$

$$= A(B \vee (B \vee \bar{B})\bar{C}) \quad \text{by distributive property again}$$

$$= A(B \vee 1 \cdot \bar{C}) \quad \text{by complementarity } (X \vee \bar{X} = 1)$$

$$= A(B \vee \bar{C}) \quad \text{by operation of 1 } (1 \cdot X = X)$$

The proof of this equality corresponds to showing that the two networks of Figure 31 are equivalent. Notice that we have shown that the networks are equal in a behavioral sense only; they are open and closed for the same combinations. However, the networks are not equal in all ways. In fact they are not equal in one most important respect, namely, complexity. The network at the left is considerably more complex than the one at the right since it involves more contacts, more labor to construct it, more possibilities of failure, and more difficulty in locating and repairing any

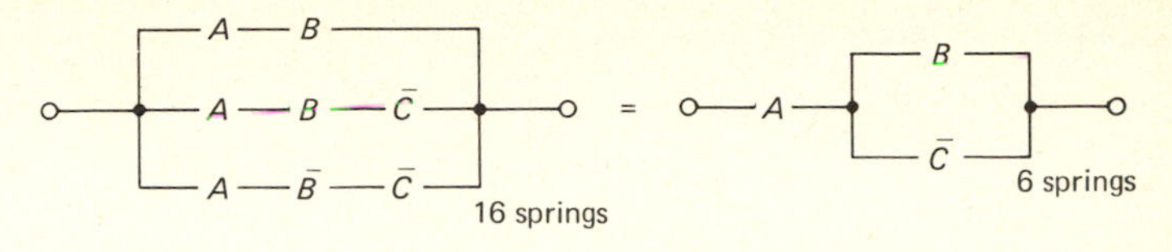

Figure 31

failure. In short, the more complex network is the more costly one.

One conventional measure of the complexity of a network is simply the number of springs in the network. For example, the original network has a complexity of 16 springs, whereas the simplified equivalent network has a complexity of only six springs. The simplification process in this case has resulted in a saving of 10 springs, a saving which can be translated into economic terms also.

Algebraic simplification involves knowing the various switching equalities and recognizing where they can apply. Usually only the original network is given, and there is no indication how simple the final expression can be. Neither is there any way of knowing in general whether your result is the simplest possible. The following examples illustrate the process of simplification or optimization.

Example Optimize $N = (A \vee C)(AB \vee C)$.

$N = (A \vee C)(AB \vee C)$	
$= (A \vee C)(AB) \vee (A \vee C)(C)$	by distributivity
$= AAB \vee CAB \vee AC \vee CC$	by distributivity again
$= AB \vee ABC \vee AC \vee C$	by idempotence (and commutativity)
$= AB\ (1 \vee C) \vee (A \vee 1)C$	by distributivity (or factoring)
$= AB\ 1 \vee 1\ C$	by operation of unit
$= AB \vee C$	by operation of unit

The complexity of the original expression is ten; that of the final expression is six. Note that the final simple expression is actually a part of the original expression, illustrating that some part of a network can be equivalent to the whole network.

This example can illustrate another significant point: most expressions can be simplified in more than one way so that it is often worthwhile to examine the original expression in the hope of finding an easier method of simplifying. For instance, in the above example, recognizing the unusual distributive property leads directly to the solution in many fewer steps.

$N = (A \vee C)(AB \vee C)$	
$= AAB \vee C$	by distributivity (of common C)
$= AB \vee C$	by idempotence

Example Optimize the network of Figure 9.

$$
\begin{aligned}
N &= \bar{X} \vee \overline{X \vee Y} \\
&= \bar{X} \vee \bar{X}\bar{Y} && \text{by dualization} \\
&= \bar{X}(1 \vee \bar{Y}) && \text{by distributivity} \\
&= \bar{X} && \text{by operation of unit}
\end{aligned}
$$

This example illustrates not only that the NOT of an expression need not require an intermediate relay (such as Z) but that sometimes even the original relays (such as Y) are unnecessary or redundant. This situation can be more extreme, as shown in the following example.

Example Optimize $N = \bar{A} \vee \bar{B} \vee (AB \vee C)(B \vee D)$.

$$
\begin{aligned}
N &= \bar{A} \vee \bar{B} \vee (AB \vee C)(B \vee D) \\
&= \bar{A} \vee \bar{B} \vee ABB \vee ABD \vee CB \vee CD && \text{by distributivity} \\
&= \bar{A} \vee \bar{B} \vee AB \vee ABD \vee CB \vee CD && \text{by idempotence} \\
&= \overline{AB} \vee AB \vee ABD \vee CB \vee CD && \text{by dualization} \\
&= 1 \vee ABD \vee \cdots \vee CD && \text{by complementarity} \\
&= 1 && \text{by operation of 1}
\end{aligned}
$$

This network, shown in Figure 32, is always closed. It is a very costly short circuit.

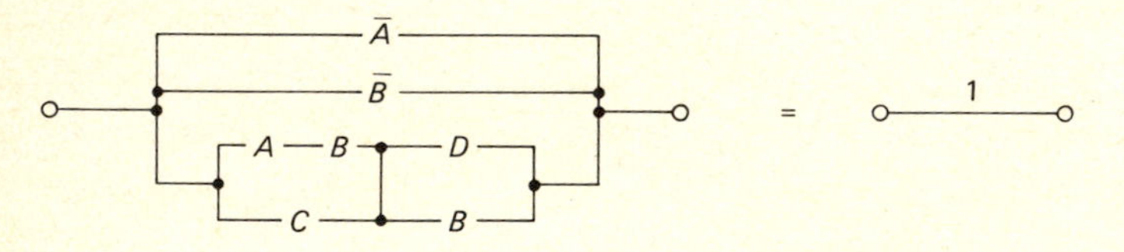

Figure 32

There are some other unusual aspects of optimizing. For example, let us optimize the network.

$$N = AB\bar{C} \vee ABC \vee A\bar{B}C$$

We could start by combining the first two terms to get

$$N = AB(\bar{C} \vee C) \vee A\bar{B}C = AB \vee A\bar{B}C = A(B \vee \bar{B}C)$$

However, we could also combine the last two terms to get

$$N = AB\bar{C} \vee A(B \vee \bar{B})C = AB\bar{C} \vee AC = A(C \vee B\bar{C})$$

What may not occur to us is that we may start with *both* of the above combinations since $ABC = ABC \vee ABC$ (by idempotence). Starting with both these combinations leads to

$$
\begin{aligned}
N &= AB\bar{C} \vee ABC \vee ABC \vee A\bar{B}C && \text{by idempotence} \\
&= AB(\bar{C} \vee C) \vee A(B \vee \bar{B})C && \text{by distributivity} \\
&= AB \vee AC && \text{by complementarity} \\
&= A(B \vee C) && \text{by distributivity}
\end{aligned}
$$

As a final example, let us attempt to optimize the network

$$N = A\bar{C} \vee BC$$

This network appears to be in its simplest form, having a complexity of eight springs. However, by recognizing that the C and $\bar{C}$ can be combined on a transfer contact, the complexity is reduced to seven springs. On a detached diagram a transfer contact is indicated by a small arc, as shown in Figure 33.

This has been a brief introduction to optimization. Later in this chapter and in the next, we will encounter other methods of optimizing.

EXERCISE SET 4

In the following exercises, justify every step with equalities from Figure 26.

1. Prove the following switching equalities algebraically:
   (*a*) $X \vee \bar{X}Y = X \vee Y$
   (*b*) $\overline{X\bar{Y} \vee \bar{X}Y} = XY \vee \bar{X}\bar{Y}$
   *(*c*) $X\bar{Y} \vee \bar{X}Z \vee \bar{Y}Z = X\bar{Y} \vee \bar{X}Z$ *Hint*: $A = (X \vee \bar{X})A$.
2. Optimize the following networks:
   (*a*) $A\bar{B}C \vee A\bar{B}\bar{C} \vee BC$ try for $C = 7$ springs
   (*b*) $ABC\bar{C} \vee AC \vee AB \vee A\bar{C}$
   (*c*) $\overline{(A\bar{B} \vee C)}(A \vee B) \vee C$ try for $C = 4$ springs
   (*d*) $P(\overline{\overline{(Q \vee R)} \vee \bar{P}R})$
   *(*e*) $P\bar{Q} \vee R(S \vee \bar{P} \vee Q)$ try for $C = 6$ springs
3. Optimize the following two similar expressions. One can be simplified to 2 springs and the other to 15 springs.
   (*a*) $\bar{A}\bar{B}C \vee \bar{A}B\bar{C} \vee A\bar{B}\bar{C} \vee ABC$
   (*b*) $\bar{A}\bar{B}C \vee ABC \vee \bar{A}BC \vee A\bar{B}C$
4. Optimize the networks of Figures 8, 14, and 15.
5. Optimize the given network of Figure 34.

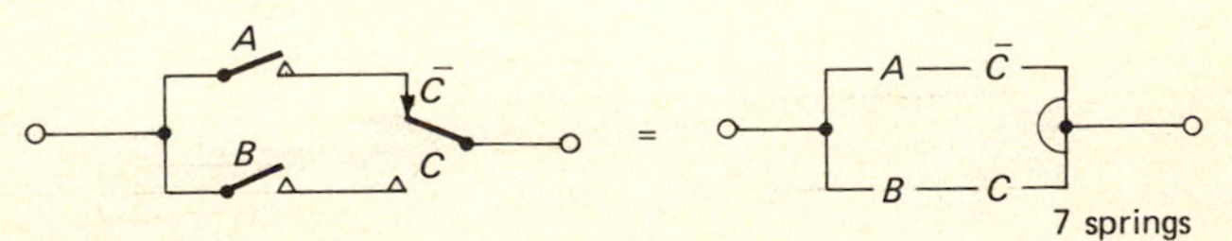

Figure 33

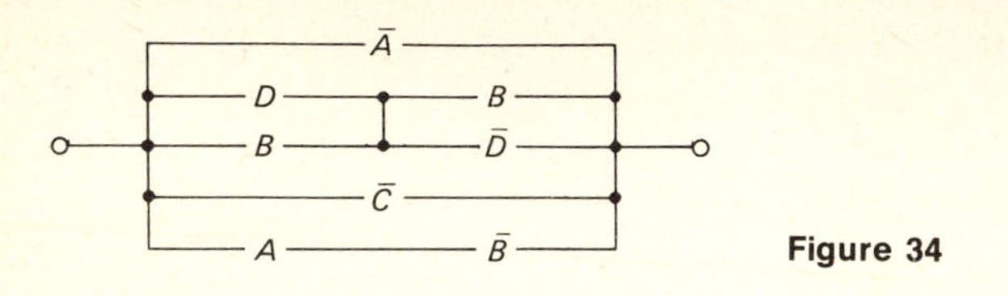

Figure 34

6. DESIGN

For some synthesis problems there may be detailed specifications which can be handled most readily by means of a table-of-combinations. The table lists all possible input combinations, and the designer indicates the corresponding output he desires for each input. In this way the designer can be certain of having accounted for every possible input eventuality. This table can then be transformed into a network by a straightforward procedure. The rather useful MAJORITY network will be used to illustrate this method of synthesis.

MAJORITY Indicator

Let us design a MAJORITY network of three relays X_1, X_2, X_3 which is to indicate (by being closed) when the majority of the three relays are operated. Such a network could be used in a voting system.

First, a table of $2^3 = 8$ combinations of input values is drawn as in Figure 35; then each input combination is checked to see which output value corresponds to it. In this case the output Y has the value 1 whenever two or three of the inputs have the value 1.

The corresponding network is constructed in the following way. For each output that has a value of 1, the make and break contacts of all relays corresponding to this combination are so arranged as to give this required output of 1. For example, input combination 011 (which means $X_1 = 0$ and $X_2 = 1$ and $X_3 = 1$ as appears on the fourth row) will yield this output of 1 if the break contact of X_1 is in series with a make contact of X_2 and X_3. In other words, $\bar{X}_1 X_2 X_3$ has value 1 only if $X_1 = 0$ and $X_2 = 1$

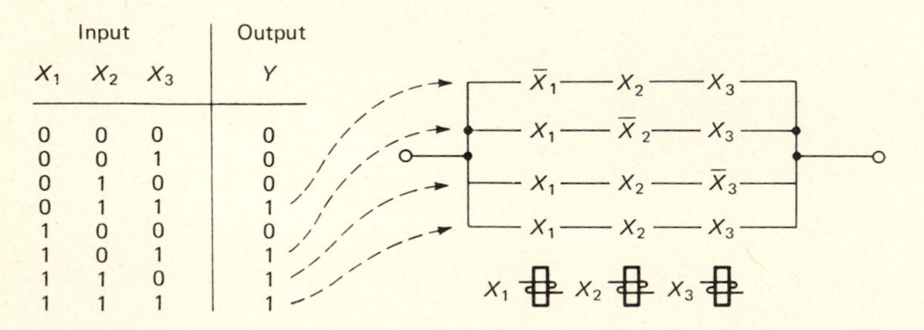

Input			Output
X_1	X_2	X_3	Y
0	0	0	0
0	0	1	0
0	1	0	0
0	1	1	1
1	0	0	0
1	0	1	1
1	1	0	1
1	1	1	1

Figure 35

and $X_3 = 1$, since $\overline{0} \cdot 1 \cdot 1 = 1 \cdot 1 \cdot 1 = 1$. Similarly the other necessary combinations are

$$X_1\bar{X}_2X_3 \qquad X_1X_2\bar{X}_3 \qquad X_1X_2X_3$$

Since the output is 1 when any one "or" the other of these four combinations has the value 1, they are simply connected in parallel as in Figure 35.

The algebraic expression of this network is

$$Y = \bar{X}_1X_2X_3 \vee X_1\bar{X}_2X_3 \vee X_1X_2\bar{X}_3 \vee X_1X_2X_3$$

Notice the form of this expression. It consists of the OR of a number of terms each containing all the n variables, some of which are NOTted. Such a *standard form* is often called minterm form, disjunctive normal form, or canonical product form.

An alternative shorthand form for this expression is

$$Y = V(3, 5, 6, 7)$$

indicating the OR of the third, fifth, sixth, and seventh row of the table-of-combinations. Notice that the counting starts at 0 and goes to 7. This numbering scheme corresponds to binary numbers, which will be considered in Section 9. The system can be optimized as follows:

$$\begin{aligned}
Y &= \bar{X}_1X_2X_3 \vee X_1\bar{X}_2X_3 \vee X_1X_2\bar{X}_3 \vee X_1X_2X_3 && \text{original}\\
&= \bar{X}_1X_2X_3 \vee X_1\bar{X}_2X_3 \vee X_1X_2\bar{X}_3 \vee X_1X_2X_3 && \text{idempotence}\\
&\quad \vee X_1X_2X_3 \vee X_1X_2X_3 &&\\
&= \bar{X}_1X_2X_3 \vee X_1X_2X_3 \vee X_1\bar{X}_2X_3 \vee X_1X_2X_3 && \text{commutativity}\\
&\quad \vee X_1X_2\bar{X}_3 \vee X_1X_2X_3 &&\\
&= X_2X_3(\bar{X}_1 \vee X_1) \vee X_1X_3(\bar{X}_2 \vee X_2) && \text{distributivity}\\
&\quad \vee X_1X_2(\bar{X}_3 \vee X_3) &&\\
&= X_2X_3 \vee X_1X_3 \vee X_1X_2 && \text{complementarity}\\
&= X_2X_3 \vee X_1(X_2 \vee X_3) && \text{distributivity}
\end{aligned}$$

The resulting optimized network of Figure 36 has only 10 springs, compared to the original 24 springs.

The next example is more involved and illustrates some useful synthesis methods.

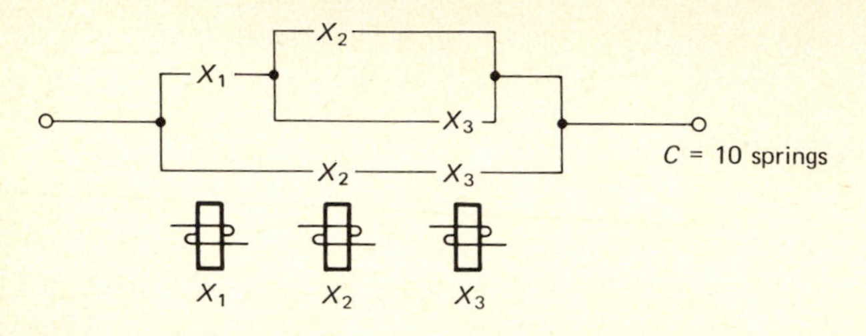

Figure 36

Design a network N consisting of relays A, B, C, and D to be closed under any of the four following conditions:

Condition C_1: all four relays are operated.
Condition C_2: relay A and D have opposite values.
Condition C_3: relay A is operated and either B or C (but not both) is operated.
Condition C_4: exactly two relays, neither of which is B, are operated.

The table-of-combinations for these four variables involves $2^4 = 16$ rows, as in Figure 37. The values for each of the four conditions could be written in separate columns and then all combined by the OR operation in the last column to satisfy the specification

$$N = C_1 \vee C_2 \vee C_3 \vee C_4.$$

Note that there are at least three deliberate errors in this table. Find them to convince yourself that you know this synthesis method. Why do none of these errors change the final result? (*Hint*: The errors are in rows 3, 11, and 14.)

Row	A	B	C	D	C_1	C_2	C_3	C_4	N
0	0	0	0	0					0
1	0	0	0	1		1			1
2	0	0	1	0					0
3	0	0	1	1		1			1
4	0	1	0	0					0
5	0	1	0	1		1			1
6	0	1	1	0					0
7	0	1	1	1		1			1
8	1	0	0	0		1			1
9	1	0	0	1				1	1
10	1	0	1	0		1	1	1	1
11	1	0	1	1			1	1	1
12	1	1	0	0		1	1		1
13	1	1	0	1			1		1
14	1	1	1	0		1	1		1
15	1	1	1	1	1				1

Figure 37

This network could be specified by the shorthand standard form as

$$N = V(1, 3, 5, 7, 8, 9, 10, 11, 12, 13, 14, 15)$$

A simpler and shorter method in this case would be to list the values for which the network is open. This is simply the NOT of the network N.

$$\bar{N} = V(0, 2, 4, 6)$$

This expression $\bar{N}$ may now be written out in detail and simplified.

$$\begin{aligned}\bar{N} &= \bar{A}\bar{B}\bar{C}\bar{D} \vee \bar{A}\bar{B}C\bar{D} \vee \bar{A}B\bar{C}\bar{D} \vee \bar{A}BC\bar{D} \\ &= \bar{A}\bar{B}\bar{D}(\bar{C} \vee C) \vee \bar{A}B\bar{D}(\bar{C} \vee C) \\ &= \bar{A}\bar{D}(\bar{B} \vee B) \\ \bar{N} &= \bar{A}\bar{D} \\ N &= \overline{\bar{A}\bar{D}} \\ N &= A \vee D\end{aligned}$$

The original specification with four conditions is seen to be considerably redundant since it can be replaced by the following simple specification: network N is closed when either relay A or D (or both) operates. It is often the case that such overly specified conditions can be optimized to one rather simple condition.

Design from Incomplete Specifications

A typical engineering design problem can take a form such as

> "Design a network to open and close an elevator door at the proper time."

This specification is not very exact. It requires further investigation, which may indicate that this is a rather simple elevator which is already equipped with four relays:

1. A relay A which is operated when the elevator is between floors
2. A relay B which is operated when either the "up" or the "down" button in the elevator is pushed
3. A relay C which is operated when the "open door" button inside the elevator is pushed
4. A relay D which when energized closes the door and when unenergized opens the door

	A	B	C	D
C_0	0	0	0	*d*
C_1	0	0	1	0
C_2	0	1	0	1
C_3	0	1	1	*d*
C_4	1	0	0	1
C_5	1	0	1	1
C_6	1	1	0	1
C_7	1	1	1	1

A = 1 if between floors

B = 1 if "up" or "down" button is pushed

C = 1 if "open door" button is pushed

D = 1 to close the door

Figure 38

The table-of-combinations method of synthesis is most useful here, since we can check all possible combinations, specifying the behavior of the door for each combination and the reason for this choice. This is done in Figure 38.

Note that for the first combination (labeled c_0), when the elevator is idle on some floor, the door could be open to allow ventilation or it could be closed for more speedy movement to another floor. If the choice between ventilation and speed is quite arbitrary, the choice could be designated as *d*, don't care, and left to be filled in later when cost of the network is considered. For the second combination (labeled 1, or C_1) the elevator is not between floors and so the open button should open the door ($D = 0$). For the combination C_2 pushing the "up" or "down" button should close the door before moving. Combination C_3 is also assigned a don't-care value, since it is unlikely to have the "open" and "up-down" button pushed simultaneously. In the last four cases the elevator is between floors, and so for safety reasons the door is closed.

Now the networks corresponding to all four combinations of don't-cares could be synthesized and optimized ultimately yielding the network

$$N = A \vee B\bar{C} \qquad \text{for } C_0 = 0 \text{ and } C_3 = 0$$

or network

$$N = A \vee B \qquad \text{for } C_0 = 0 \text{ and } C_3 = 1$$

or network

$$N = A \vee \bar{C} \qquad \text{for } C_0 = 0 \text{ and } C_3 = 0$$

or network

$$N = A \vee B \vee \bar{C} \qquad \text{for } C_0 = 1 \text{ and } C_3 = 1$$

The most economical networks are $A \vee B$ and $A \vee \bar{C}$.

Notice that our design of this network incorporated the goals of safety, speed, convenience (ventilation), and ultimately, cost. When such goals conflict, we can leave the combination unspecified until later, when cost can be used to resolve the conflict. In this case two networks, $A \vee B$

and $A \vee \bar{C}$, are equally economical; we can therefore choose between these two by considering some other goals, such as reliability, noise, or further convenience. Actually, after some evaluation (from a larger system viewpoint) we would select network $A \vee B$. This is because we unfortunately had a small-systems viewpoint and considered only what would happen if we were in the elevator pushing buttons, moving, etc. The alternative network, $A \vee \bar{C}$ we would find, gives us no way of getting into this elevator to push the buttons!

Multi-output Networks

In the systems studied so far we have encountered many inputs but only one output. Systems having many outputs are also possible. Consider, for example, the fluid-storage reservoir of Figure 39*a*. This reservoir has two sensors, X_1 and X_2, which operate corresponding relays when the fluid reaches or exceeds their level. The goal of the system is to monitor the fluid and indicate any extreme levels. When the level is above X_2, the output Y_1 operates a major alarm signaling someone to open a relief valve to lower the level. When the fluid level is below X_1 the output Y_2 operates another (minor) alarm. The table-of-combinations for this two-input-two-output system is given in Figure 39*b*. Since it is not possible to have water operating X_2 without also operating X_1, we do not care what value the network has in this case. By inspection we choose the don't care values d to be 1, leading to the simplest networks:

$$Y_1 = X_2$$
$$Y_2 = \bar{X}_1$$

In this simple case the solution may have been intuitively evident without introducing the table-of-combinations and don't-care conditions. However, in more complex cases intuition may not be as helpful. Even in this case intuition may indicate that the major-alarm output Y_1 should directly operate the relief valve, although this could cause the relief valve to oscillate on and off indefinitely for any slight level changes or ripples.

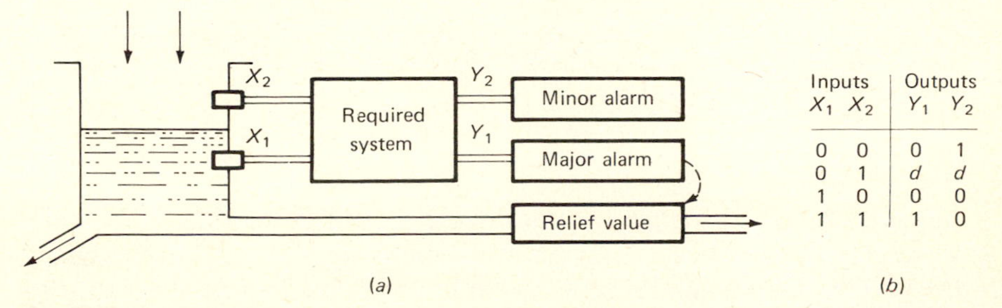

Inputs X_1	Inputs X_2	Outputs Y_1	Outputs Y_2
0	0	0	1
0	1	d	d
1	0	0	0
1	1	1	0

Figure 39

EXERCISE SET 5

1. A system consists of three relays A, B, C. Write an algebraic expression (and the shorthand description) for the system if it is to close when:

(*a*) Exactly two of the relays are closed.
(*b*) At least two of the relays are closed.
(*c*) At most two of the relays are closed.
(*d*) At most two of the relays are open.
(*e*) All the relays are closed.
(*f*) No relays are closed.
(*g*) Some of the relays are closed.
(*h*) An odd number of relays is closed.
(*i*) An even number of relays is open.

Ans. g: $V(1, 2, 3, 4, 5, 6, 7) = \overline{\bar{A}\bar{B}\bar{C}} = A \vee B \vee C$

2. Repeat the above problem if the system consists of four relays A, B, C, and D rather than the three.

3. Synthesize and optimize a network which is closed for either of the following two conditions and open for all others:

Condition 1: Relays A and C are either both operated or both released.
Condition 2: Exactly two of the three relays A, B, C are operated.

Try for a complexity of eight springs.

4. Design a simplified voting mechanism involving four voters A, B, C, D. Persons A and C have four votes each, B has three votes, and D has only one vote. The network is to be closed whenever there is larger than a two-thirds majority.

5. In a certain process the combination of high pressure and high temperature could destroy the batch. Also, if the temperature is low but the rate of flow is high, the result is a defective (raw) product. When the pressure is high with accompanying low rate and high viscosity a defective (burned) product again results. Write a simplified algebraic expression for a system to sound an alarm whenever at least two of these three situations occur. Let the variables T, P, R, and V operate corresponding relays (value 1) when they are excessively high. (*Hint*: There is an easy and a hard way to do this.)

6. In an airplane the pilot can be ejected by pushing one button B. This first ejects the canopy top T and when this is detected by switch C closing, the seat and pilot are ejected. Under no circumstances can the seat be ejected before the canopy is ejected!

(*a*) Draw a table of combinations for this two-input two-output system, indicating which choices are arbitrary (don't-cares). By inspection select the simplest relay network.

(*b*) To prevent accidental ejection when the single button is accidentally pushed, the pilot is now required to push two widely separated buttons, A and B, simultaneously. Construct this newer, safer network.

7. Find another alternative way of obtaining a switching network directly from a table-of-combinations. (*Hint*: For each row having value 0 construct a parallel network of all the contacts; then connect these in series.)

7. IMPLEMENTATION

Implementation, or realization, is the process of converting an abstract structure into a physical form. In this section we will see that the binary algebraic structure applies to nonelectrical systems having hydraulic and mechanical components. We will also consider an alternative view of electric networks.

Flow Networks

The previous switching model was discussed in terms of relays and corresponding flow of current. However, the model can be applied to any networks where any type of flow is controlled by binary or on-off devices. It could apply to the flow of liquids, gas, heat, air, or traffic. Let us consider, for example, the flow of water controlled by two-position valves. Figure 40 illustrates some fluid-flow components and their analogy to electric components.

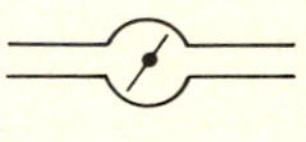
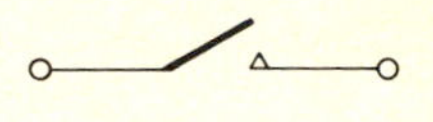

A butterfly valve in this position corresponds to a relay contact in this position

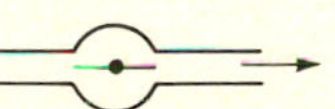
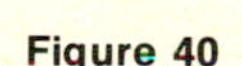

Figure 40

There is a possibly confusing fluid- and electric-flow convention because a closed switch allows flow of current whereas a closed valve prevents flow of fluid. To avoid confusion we will not speak of closed or open but of flow or no flow. The symbol 1 denotes flow, as always.

Hydraulic networks have a form very similar to relay networks. For example, the previously designed majority network is constructed of hydraulic components in Figure 41. It could be used to provide irrigation only when the majority of three people concurrently request it by pushing down on their corresponding lever (which in turns operates the valves).

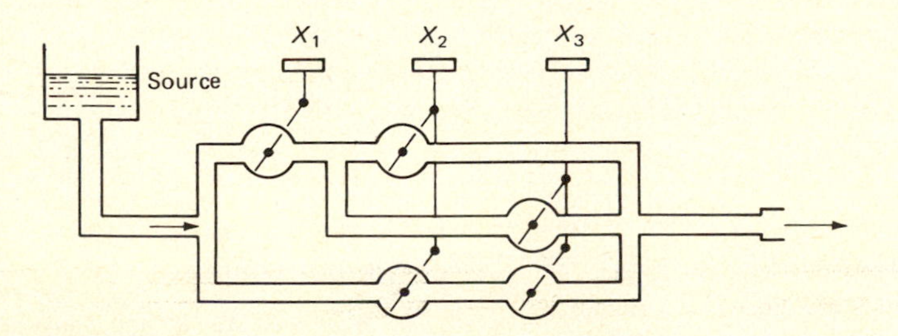

Figure 41

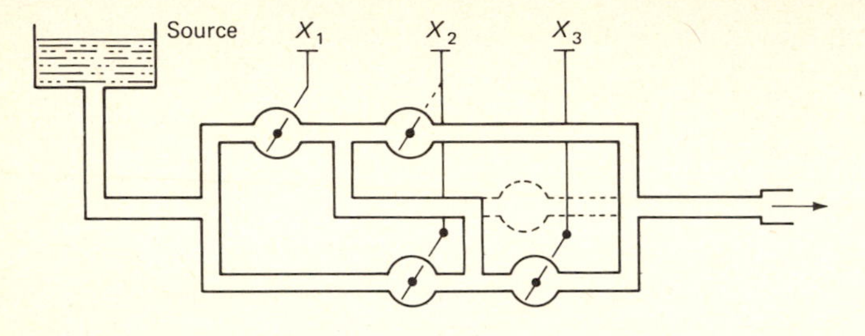

Figure 42

Intuitively, it may appear that this network could be simplified by combining the two valves marked X_3 into one, as in Figure 42. These two networks may appear to be equivalent, but they are not. Why?

The network is no longer a "democratic" majority device since X_2 alone may cause flow. This is because there is now a "sneak path" from the source, through the bottom valve of X_2, back up to the top valve of X_2, and then out.

This problem illustrates some of the pitfalls of unaided intuition. Notice that the algebraic model would not allow such an error, but then the algebraic model has other pitfalls. The main pitfall is that we tend to forget the model is based on the three assumptions of determinism, two-valuedness, and instantaneity.

Level Networks

Until now we have considered only *flows* of quantities such as electric currents, fluids, and vehicles. There is, however, an alternative view of switching systems in terms of *levels* of quantities such as voltage potential, fluid pressure, and traffic density. These two types of networks are similar in many ways since they can be two-valued and they are described by the same algebra. However, they are also different in important ways involving their network structure and simplification. The similarities and differences must be respected, for they represent two alternative interpretations of the same algebra. No other physical network interpretations are known other than the flow and level ones. Flow networks are sometimes called *branch-type circuits*, whereas level networks are called *gate-type circuits*. The following sections describe some electric and mechanical level networks.

Electric Level Networks Usually we associate electric flow with current and electric level with voltage. In this section our emphasis will be on voltage of two levels: (1) a low level denoted by symbol 0 and (2) a high level denoted by symbol 1. We can now view OR, AND, and NOT components from the viewpoint of voltages of these two levels. The diagram of Figure 43 shows a level OR component constructed from

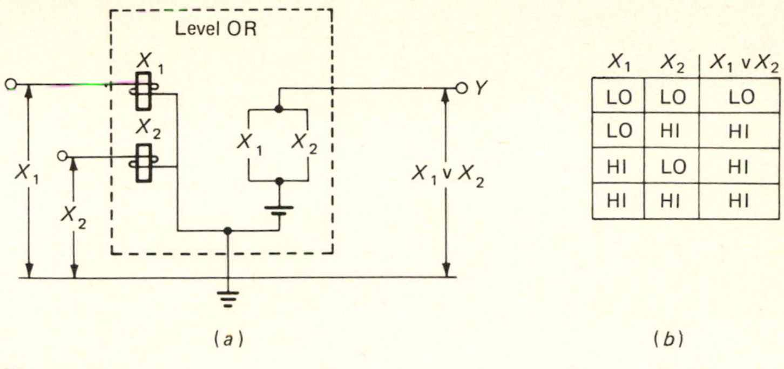

X_1	X_2	$X_1 \vee X_2$
LO	LO	LO
LO	HI	HI
HI	LO	HI
HI	HI	HI

Figure 43

relays. Level components constructed from other electric devices will be introduced later.

In this OR network, the output voltage Y is at a high level when either input voltage level X_1 or X_2 is at a high level (sufficiently high to operate the relays). Such a level OR component is usually symbolized more formally by the block symbol and corresponding table-of-combinations of Figure 44.

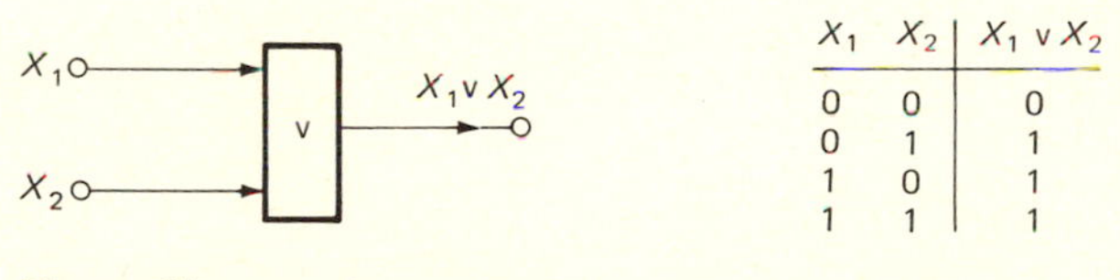

X_1	X_2	$X_1 \vee X_2$
0	0	0
0	1	1
1	0	1
1	1	1

Figure 44

Notice that the lower terminal of the network has been omitted since all blocks will have the same common, or ground, voltage.

The voltage (or level) behavior can be illustrated by time sequences of the inputs and outputs as in Figure 45. These waveforms could be directly observed on an oscilloscope.

Level AND and NOT elements can be constructed, symbolized, and defined as in Figure 46.

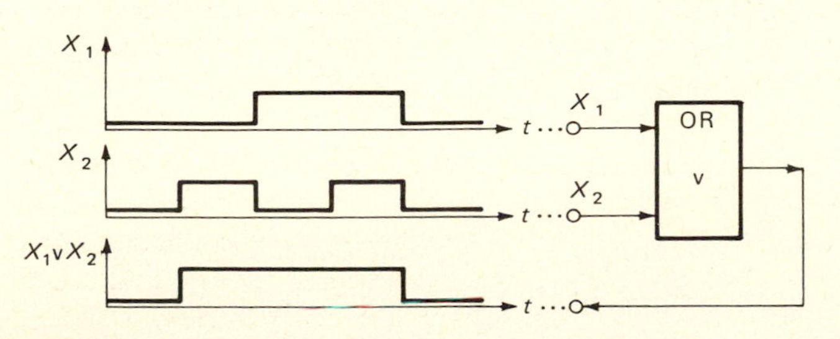

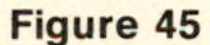
Figure 45

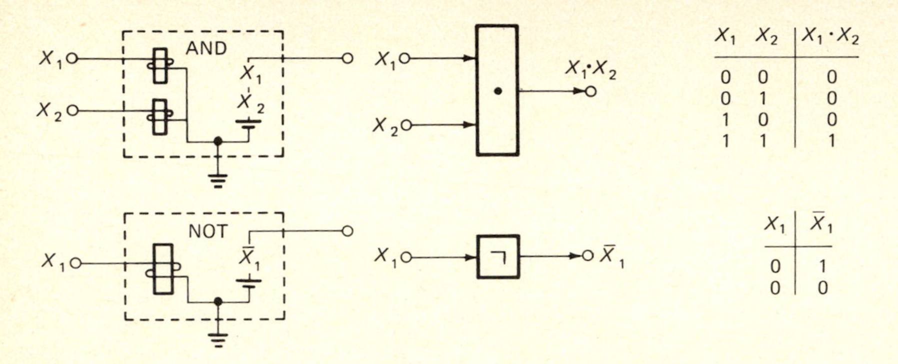

Figure 46

These level elements can be interconnected to form larger networks, as in Figure 47. Since the defining tables for the AND, OR, and NOT elements are the same for both the flow and level components, the switching algebra applies to both.

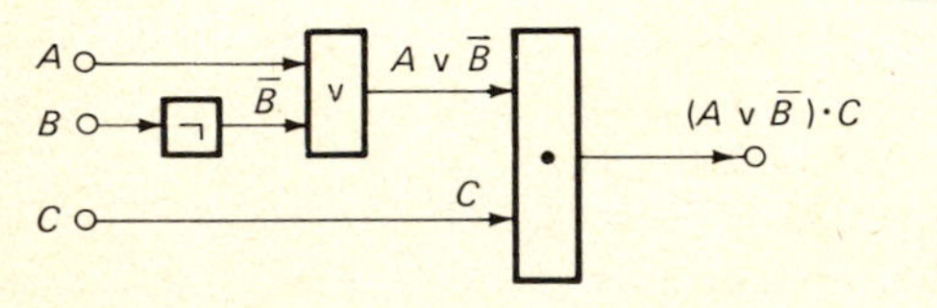

Figure 47

It is possible to construct any switching algebraic expressions as either a flow system or a level system. For example, the previous MAJORITY voting network could be constructed using level elements as in Figure 48. Note that all the input variables marked with the same labels X_i are to be connected together, but showing this connection on the diagram would only complicate it.

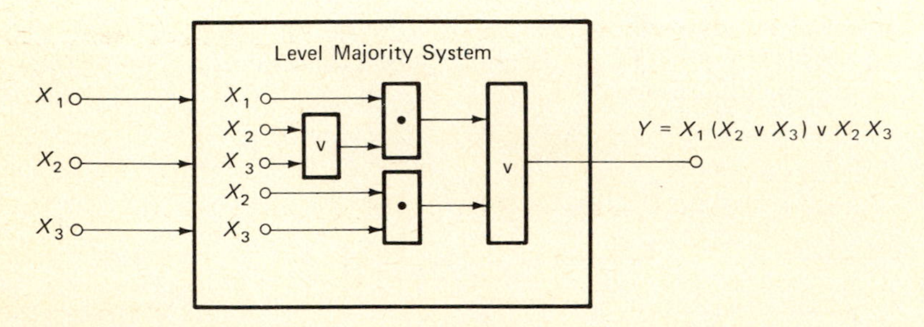

Figure 48

Any electric level component such as the OR can be constructed in many different ways as shown in Figure 49. All these behave the same at

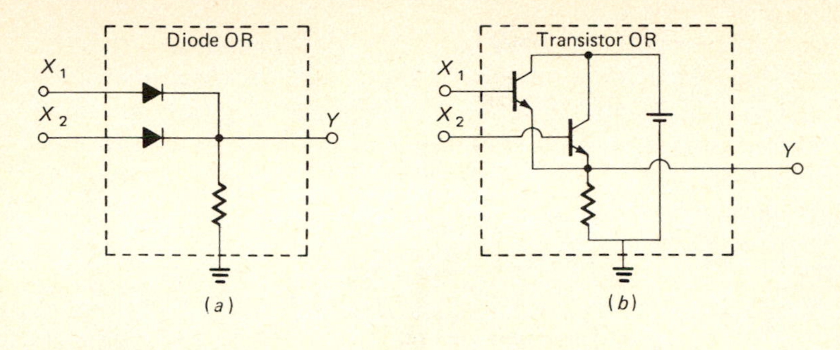

Figure 49

the terminals. This illustrates the power of the abstract, systems, black-box approach; what is within the black box labeled "OR" is behaviorally irrelevant. The physical considerations have been separated from the logical and mathematical considerations. This makes it possible to understand many aspects of switching systems without knowing electronics. A more detailed treatment of such digital electronic systems may be found in the Appendix.

The choice between various behaviorally equivalent black boxes depends on the goal or application. Electronic devices such as transistors and diodes are preferred for their high speed, small size, and low energy consumption, whereas electromagnetic devices such as relays are preferred for their low cost, simplicity, isolation ability, and high current-carrying capacity. The cost or complexity of electric level networks is conventionally taken as the total number of input arrows since this usually corresponds to the number of costly elements. For example, the level majority network of Figure 48 has a complexity of eight arrows; the network of Figure 47 has a complexity of five arrows.

Although all the algebraic switching properties are identical for both level and flow networks, since they are derived from identical tables-of-combinations, their physical interpretation differs. For example, the distributive property

$$(X \vee Y)(X \vee Z) = X \vee YZ$$

indicates that the networks of Figure 50 are equivalent.

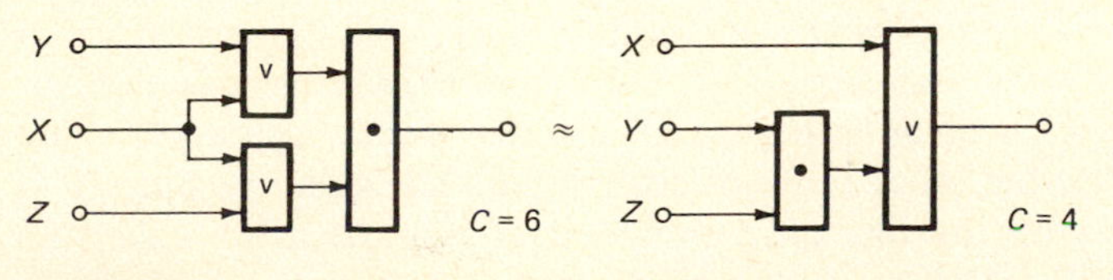

Figure 50

Similarly the property of dualization, $\overline{X \vee Y} = \bar{X}\bar{Y}$ indicates that the networks of Figure 51 are equivalent. Note that for electronic level networks, $\overline{X \vee Y}$ is simpler than $\bar{X} \cdot \bar{Y}$ (since complexities are three arrows and four arrows respectively); whereas for relay flow networks, $\bar{X}\bar{Y}$ is simpler than $\overline{X \vee Y}$ (since complexities are four springs versus six springs and an intermediate relay).

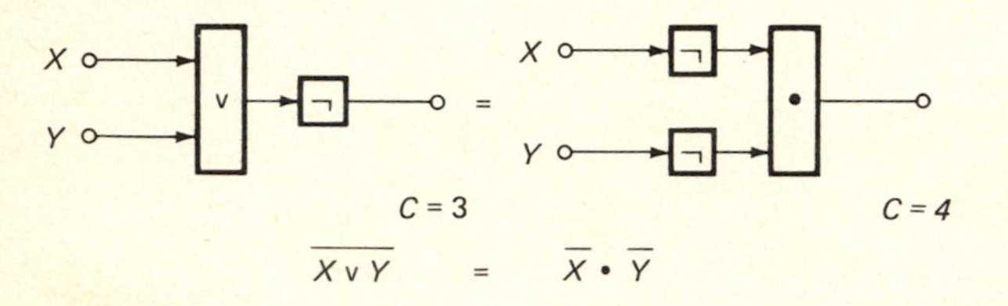

Figure 51

In the level implementation it is possible to have operators with more than two inputs. For example, the two-input diode OR network of Figure 49*a* can be extended by simply adding a single diode to become the three-input OR operator of Figure 52. The output has value 1 if any input *A* or *B* or *C* has value 1. An OR with four or more inputs can be constructed similarly. An AND can also be extended to have more than two inputs. Such operators will be useful later for synthesis and optimization.

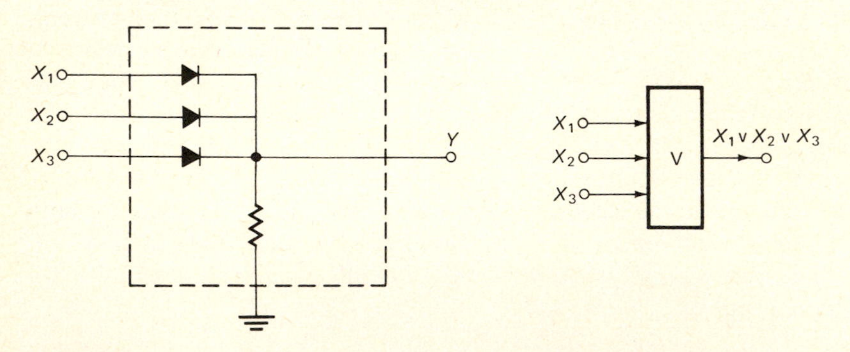

Figure 52

Synthesis of level systems is similar to that of flow systems. As an example, let us synthesize a *disagreement detector* which has output value 1 when its two inputs differ. It could be used as a comparator which sounds an alarm when two signals do not agree. The table-of-combinations for this device is given in Figure 53. Notice that it has output value 1 when one or else the other (but not both) inputs has value 1. Since it is the regular OR with the last case excluded, it is most often called an EXCLUSIVE-OR.

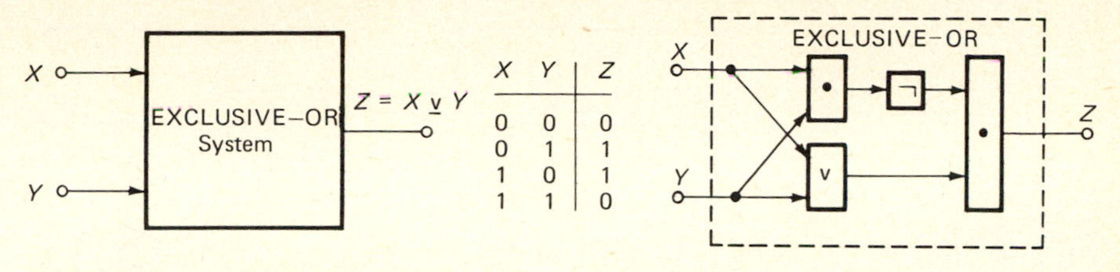

X	Y	Z
0	0	0
0	1	1
1	0	1
1	1	0

Figure 53

From the table-of-combinations of Figure 53 we can write the expression

$$Z = X\bar{Y} \vee \bar{X}Y$$

which can be optimized as

$$\begin{aligned} Z &= (X\bar{Y} \vee \bar{X})(X\bar{Y} \vee Y) && \text{by distributivity} \\ &= (\bar{Y} \vee \bar{X})(X \vee Y) && \text{by } P \vee \bar{P}Q = P \vee Q \\ &= (\overline{XY})(X \vee Y) && \text{by dualization} \end{aligned}$$

The EXCLUSIVE-OR is very useful in computers and communication systems. It is often symbolized as the usual OR symbol with a bar below it and defined as

$$X \veebar Y = X\bar{Y} \vee \bar{X}Y = (X \vee Y)(\overline{XY})$$

A more complex system to synthesize is the multi-output system of Figure 54*a*. First, the network D could be synthesized and optimized

$$\begin{aligned} D &= AB\bar{C} \vee ABC \\ &= AB(C \vee \bar{C}) \\ &= AB \end{aligned}$$

Similarly for network E,

$$\begin{aligned} E &= \bar{A}\bar{B}C \vee \bar{A}BC \vee A\bar{B}C \\ &= C(\bar{A}(B \vee \bar{B}) \vee A\bar{B}) \\ &= C(\bar{A} \vee A\bar{B}) \\ &= C(\overline{AB}) \end{aligned}$$

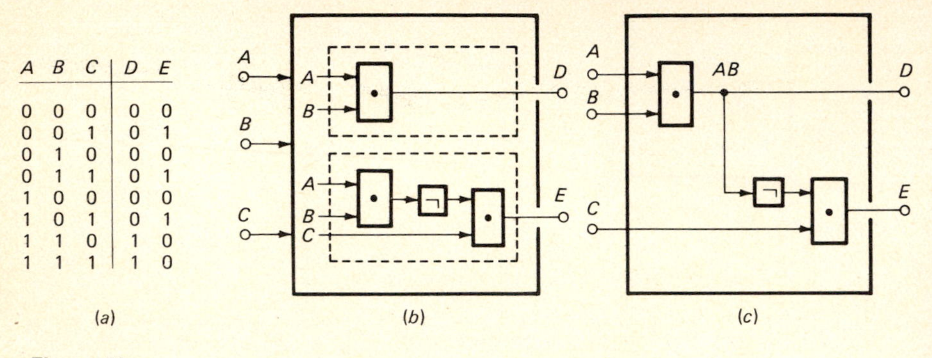

A	B	C	D	E
0	0	0	0	0
0	0	1	0	1
0	1	0	0	0
0	1	1	0	1
1	0	0	0	0
1	0	1	0	1
1	1	0	1	0
1	1	1	1	0

Figure 54

The required two-output system can be constructed as two separate systems within one box, as in Figure 54*b*. However, by using the *AB* term which is common to both *D* and *E* we can share some components to achieve the simpler network of Figure 54*c*.

Mechanical Level Networks Mechanical networks can also act as level networks. The values of the algebra could correspond to the position of movable rods:

A position to the left indicates the value 0.
A position to the right indicates the value 1.

Some possible mechanical level operators are given in Figure 55. The springs serve to return the inputs to the 0 position when the inputs are not applied. The circled dots indicate pivots. The behavior of the NOT and OR elements should be self-explanatory; the behavior of the AND element is shown in Figure 56 for all four positions. This shows that the output is 1 (at right) only when both inputs *X* and *Y* are 1 (at right).

A mechanical majority network patterned after the electronic level network would have the form of Figure 57. Of course it could be left in the simpler form of Figure 48.

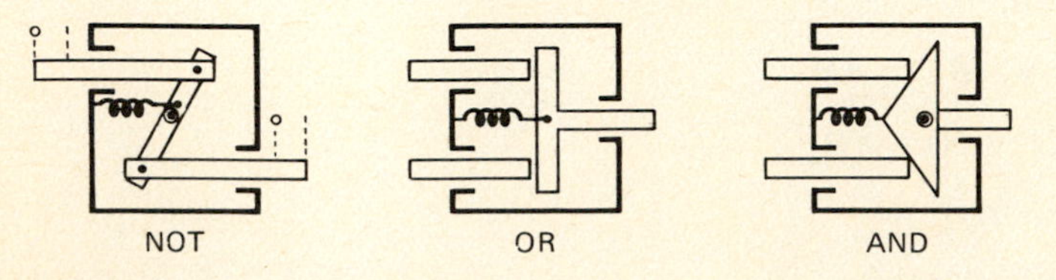

Figure 55

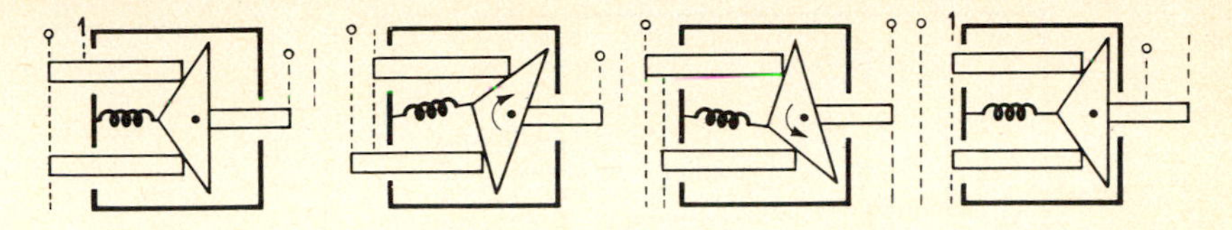

Figure 56

EXERCISE SET 6

1. Optimize the given hydraulic-flow network of Figure 58.

2. Construct a mechanical network equivalent in behavior to Figure 59 but having a minimum number of AND, OR, and NOT components. *Ans.*: $\overline{A \cdot B}$

3. Write a switching algebraic expression for the level network of Figure 60 and optimize the number of input arrows. Draw the simplified network. It should have half the complexity of the original one.

4. Draw level networks corresponding to the following expressions; optimize, draw the simplified networks, and compare

(*a*) $\bar{A}(\bar{B} \vee B\bar{C})$ try for $C = 5$ arrows

(*b*) $(\bar{A} \vee BC \vee \bar{D})(\bar{E} \vee BC)$ try for $C = 9$ arrows

(*c*) $\bar{A}\bar{B} \vee C(A \vee B)$ try for $C = 5$ arrows

5. Synthesize a *comparator system* (also called a coincidence gate, biconditional network, or all-or-none system) whose output Z has value 1 only when both inputs X and Y have the same value. Construct the network from AND, OR, and NOT level components, optimizing the number of input arrows. Construct it again using only the EXCLUSIVE-OR and the NOT.

6. Since mechanical OR and NOT components are simpler than the AND, construct the 3-input MAJORITY network using the fewest number of OR and NOT components only.

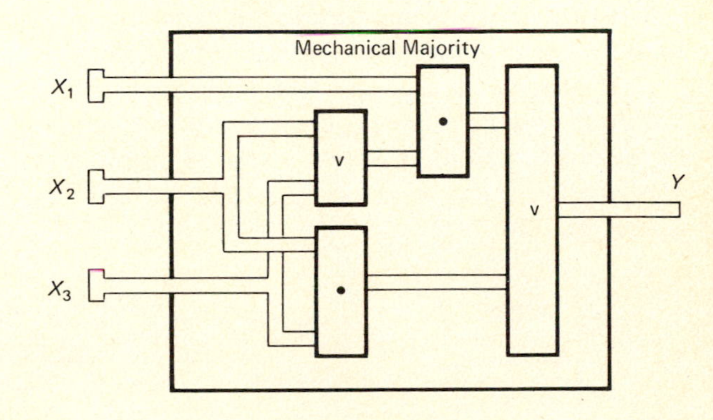

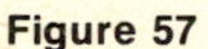

Figure 57

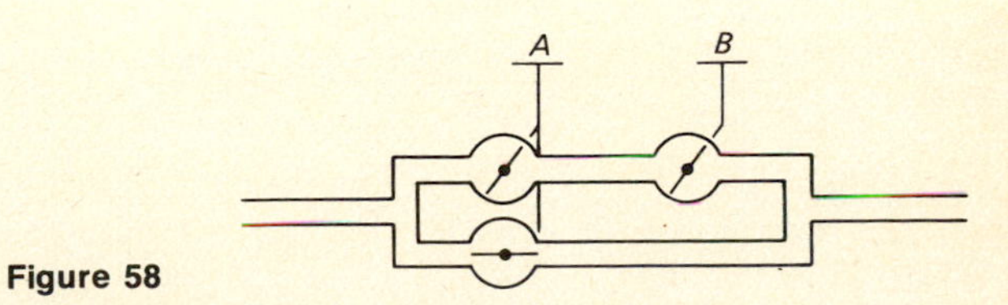

Figure 58

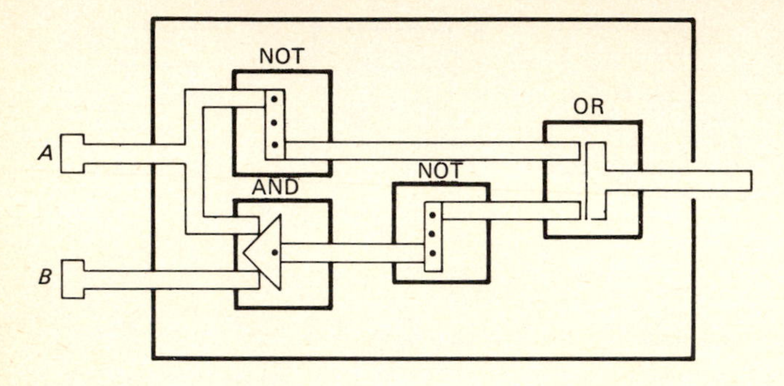

Figure 59

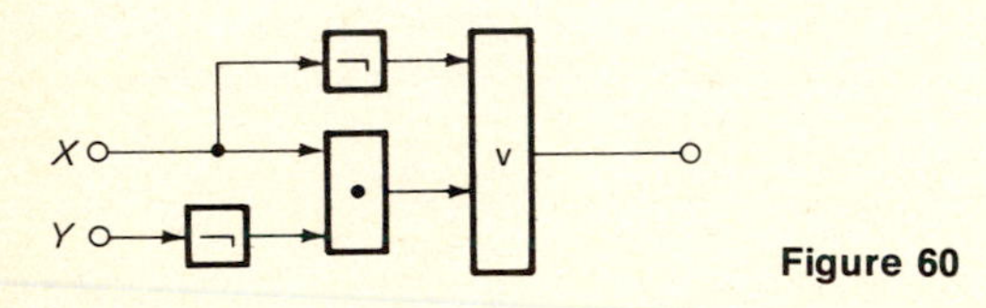

Figure 60

8. IDENTIFICATION

One significant type of problem arising in the study of systems is the identification problem, i.e., determining the characteristics or structure of a system from its input-output behavior.

This is also known as the black-box or unknown problem. It restricts knowledge of the "inside" or structure of a system, thus forcing us to experiment in order to determine the behavior.

Consider, for example, the level system of Figure 61. Since there are three input variables, the experiment requires at least 2^3 or 8 different combinations of levels. The experiment is conveniently done by applying the input waveforms A, B, C of Figure 61b and observing the output waveform D.

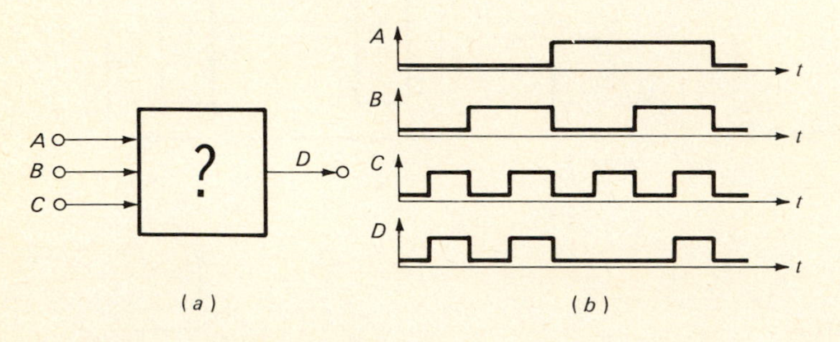

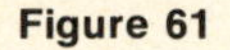

Figure 61

Notice that these waveforms correspond to a table-of-combinations (laid on its side). The problem now becomes one of synthesis from this table-of-combinations. The corresponding algebraic expression is

$$D = \bar{A}\bar{B}C \vee \bar{A}BC \vee ABC$$

which can be simplified to

$$\begin{aligned} D &= (\bar{A}\bar{B} \vee B(\bar{A} \vee A))C \\ &= (\bar{A}\bar{B} \vee B)C \\ &= (\bar{A} \vee B)C \end{aligned}$$

The systems corresponding to the first and last of these expressions are shown in Figure 62. Notice the use of ANDS and ORS having three inputs.

Although the behavior of the given system is unique, the structure that could yield this behavior is not unique since either of the two networks of Figure 62, as well as many others, could produce the same waveforms.

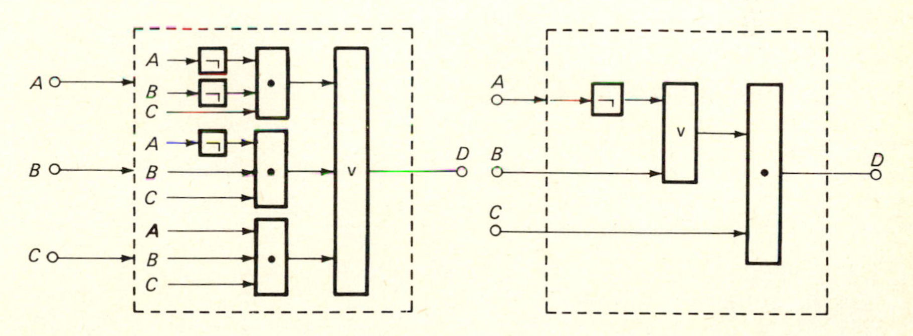

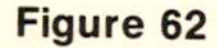
Figure 62

EXERCISE SET 7

1. A hydraulic network consists of some inaccessible pipe (buried underground in concrete) and some valves which are operated by three rods. To identify the network some experiments are performed, and the results are indicated in Figure 63. The values D, U, F, N represent down, up, flow, and no flow respectively. Indicate at least two networks which could yield these data.

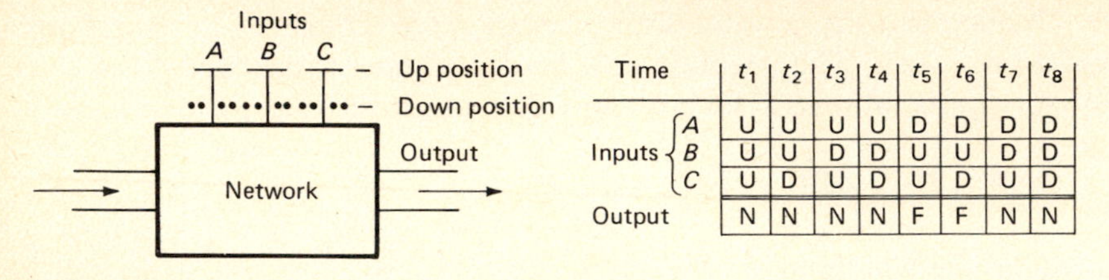

Time		t_1	t_2	t_3	t_4	t_5	t_6	t_7	t_8
Inputs	A	U	U	U	U	D	D	D	D
	B	U	U	D	D	U	U	D	D
	C	U	D	U	D	U	D	U	D
Output		N	N	N	N	F	F	N	N

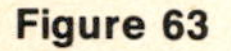
Figure 63

2. How many different (nonequivalent) systems could yield the following data? Find the simplest flow system by inspection (knowing it has two springs).

Inputs:	A	1	1	0	1	0	0	1	0
	B	0	1	0	0	1	0	0	0
	C	1	0	1	0	0	1	1	1
Output:	D	1	0	1	1	0	1	1	1

3. A mechanical level component (Figure 55) behaves like an OR with the convention that any position at the left indicates value 0 and the position to the right indicates value 1. If this convention is reversed, so that the position at the left indicates value 1, the behavior cannot be described as an OR. Why not? How would you describe the behavior?

4. Comment on the nature of the digital system of Figure 64. What could, or could not, be in the black box?

9. APPLICATION TO COMPUTATION

A significant application of two-valued switching systems is in numerical computations. Human beings compute with the digits 0, 1, 2, 3, 4, 5, 6, 7, 8, 9 (partly because they have 10 fingers), whereas most computers use the two digits 0, 1 (partly because they have two values or, effectively, "two fingers"). These number systems are called the *decimal* and *binary* number systems respectively.

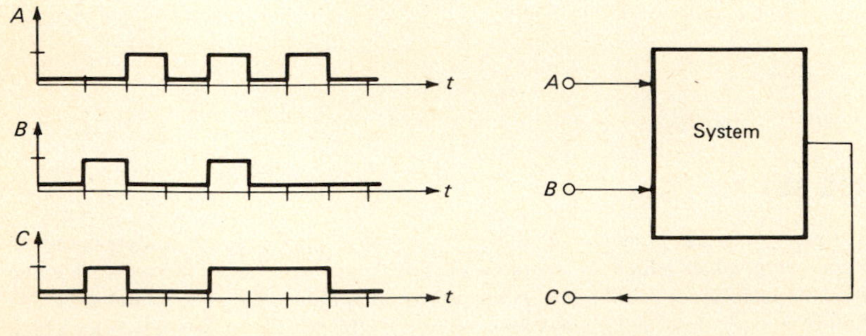

Figure 64

Numbers, Positional Notation

In the decimal number system a number such as 121 means

$$121 = 1*100 + 2*10 + 1*1$$

Notice that the digit 1 appears twice but has a different value depending on its position (hence the name, positional notation). This number can be expanded in powers of 10 as

$$121 = 1*10^2 + 2*10^1 + 1*10^0$$

The number 10, which appears taken to some power in every term, is called the *base* or radix of the number system. Similarly, in the binary number system the base is 2, and a number such as 1111001 means

$$\begin{aligned} 1111001 &= 1*2^6 + 1*2^5 + 1*2^4 + 1*2^3 + 0*2^2 + 0*2^1 + 1*2^0 \\ &= 64 + 32 + 16 + 8 + 0 + 0 + 1 \\ &= 121 \end{aligned}$$

Notice that the same number has been expressed in the decimal and binary systems, and that it requires more digits when expressed in the binary system. To avoid confusion between numbers expressed in different bases the numbers are often put in parenthesis with the base appearing as a subscript outside the parenthesis. For example,

$$(1111001)_2 = (121)_{10} \qquad (11)_2 = (3)_{10} \qquad (11)_{10} = (1011)_2$$

In general, a sequence of coefficients c_i representing an integer in base b is denoted and defined as

$$\begin{aligned} (c_n c_{n-1} \cdots c_2 c_1 c_0)_b &= c_n b^n + \cdots + c_2 b^2 + c_1 b^1 + c_0 b^0 \\ &= \sum_{i=0}^{n} c_i b^i \end{aligned}$$

Any decimal number can be converted to a binary number by successively dividing the decimal number by 2 and recording each remainder. When the decimal number has finally been reduced to 0, the remainders, taken in reverse order, constitute the required binary representation. For example, the binary representation of the decimal number 11 is computed as follows:

$$11/2 = 5 + r = 1$$
$$5/2 = 2 + r = 1$$
$$2/2 = 1 + r = 0$$
$$1/2 = 0 + r = 1$$

Thus

$$(11)_{10} = (1011)_2$$

This can also be done by expanding the number into a sum of powers of 2 as

$$(11)_{10} = 8 + 3 = 8 + 2 + 1$$
$$= 1*2^3 + 0*2^2 + 1*2^1 + 1*2^0$$
$$= (1011)_2$$

The binary representation of some decimal numbers is given in Figure 65. A table-of-combinations is also shown to indicate that the order of arranging the combinations of values is precisely the order of the decimal integers (except that some zeros may be added in front of any number, such as 10 = 010 = 0010).

We have thus used the binary representation of decimal numbers for ordering in the table-of-combinations perhaps without being aware of it. This is also the reason for starting counting at 0 rather than 1 in previously labeling the table-of-combinations.

Decimal Number	Binary Number	Table-of-Combinations
0	0	0000
1	1	0001
2	10	0010
3	11	0011
4	100	0100
5	101	0101
6	110	0110
7	111	0111
8	1000	1000
9	1001	1001
10	1010	1010
11	1011	1011
12	1100	1100
13	1101	1101
14	1110	1110
15	1111	1111

Figure 65

A \ B	0	1	2	3	4	5	6	7	8	9
0	0	1	2	3	4	5	6	7	8	9
1	1	2	3	4	5	6	7	8	9	10
2	2	3	4	5	6	7	8	9	10	11
3	3	4	5	6	7	8	9	10	11	12
4	4	5	6	7	8	9	10	11	12	13
5	5	6	7	8	9	10	11	12	13	14
6	6	7	8	9	10	11	12	13	14	15
7	7	8	9	10	11	12	13	14	15	16
8	8	9	10	11	12	13	14	15	16	17
9	9	10	11	12	13	14	15	16	17	18

Figure 66

A \ B	0	1
0	0	1
1	1	10

$0 + 0 = 0$
$0 + 1 = 1$
$1 + 0 = 1$
$1 + 1 = 10$

A	B	C	S
0	0	0	0
0	1	0	1
1	0	0	1
1	1	1	0

Figure 67

Addition of Digits

One of the most common arithmetic operations on numbers is the sum operation, used so often in computations. The sum operation is defined for the decimal number system by the table of Figure 66. However, in the binary number system, the sum operation is extremely simple, consisting of only four combinations as shown in Figure 67.

A corresponding sum network could be constructed from this table-of-combinations. One output of this network yields the sum digit S, and the other yields the carry digit C. This network is called a *half adder* for reasons which will soon be apparent. It is synthesized as follows and shown in Figure 68. Notice that the sum is the EXCLUSIVE-OR system of Figure 53.

$$C = A \cdot B$$

$$S = A \veebar B$$

$$S = \bar{A}B \veebar A\bar{B}$$

$$= (A \veebar B)(\overline{AB})$$

Addition of Numbers

The addition of two numbers usually proceeds one column at a time, beginning with the least significant digit and proceeding to the left. This

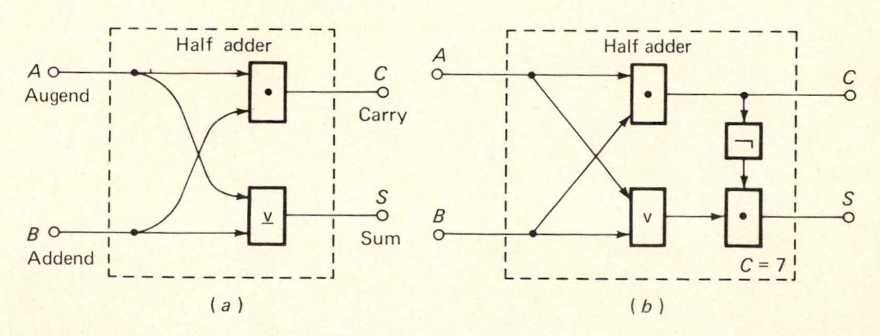

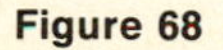
Figure 68

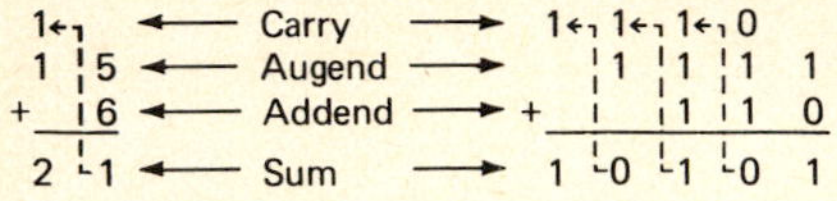

Decimal		Binary
1	← Carry →	1 1 1 0
1 5	← Augend →	1 1 1 1
+ 6	← Addend →	+ 1 1 0
2 1	← Sum →	1 0 1 0 1

Figure 69

method is illustrated in Figure 69 by adding the numbers 15 (called the *augend*) and 6 (called the *addend*) using both the decimal and binary number systems.

Design of an Adder

This process of adding two numbers can also be accomplished by switching networks. For each column there is an adder which adds the addend digit, the augend digit, and the carry digit from the previous column. It then produces at its output a sum digit and a carry digit. The carry digit serves as input to the next column. The sequence of sum digits (such as 10101) is the final required sum. A diagram showing a *parallel adder* is shown in Figure 70.

Notice that the previously designed *half adder* could be used to add only the least significant digits since it has only two inputs. The remainder of the adder components must have a third input corresponding to a carry digit. These three-input adders, called *full adders*, can be designed from the table-of-combinations of Figure 71*a*. The sum S can be synthesized from the table as

$$
\begin{aligned}
S &= \bar{A}\bar{B}C \vee \bar{A}B\bar{C} \vee A\bar{B}\bar{C} \vee ABC \\
&= (\bar{A}B\bar{C} \vee A\bar{B}\bar{C}) \vee (\bar{A}\bar{B}C \vee ABC) \\
&= (\bar{A}B \vee A\bar{B})\bar{C} \vee (\bar{A}\bar{B} \vee AB)C \\
&= (A \underline{\vee} B)C \vee (\bar{A}\bar{B} \underline{\vee} AB)C && \text{by definition of EXCLUSIVE-OR} \\
&= (A \underline{\vee} B)C \vee \overline{(A \underline{\vee} B)}C && \text{see Problem 3}a \text{ of Exercise Set 2} \\
&= (A \underline{\vee} B) \underline{\vee} C
\end{aligned}
$$

The carry C' is similarly determined from the tables as

$$
\begin{aligned}
C' &= \bar{A}BC \vee A\bar{B}C \vee AB\bar{C} \vee ABC \\
&= (\bar{A}B \vee A\bar{B})C \vee AB(\bar{C} \vee C) \\
&= (A \underline{\vee} B)C \vee AB
\end{aligned}
$$

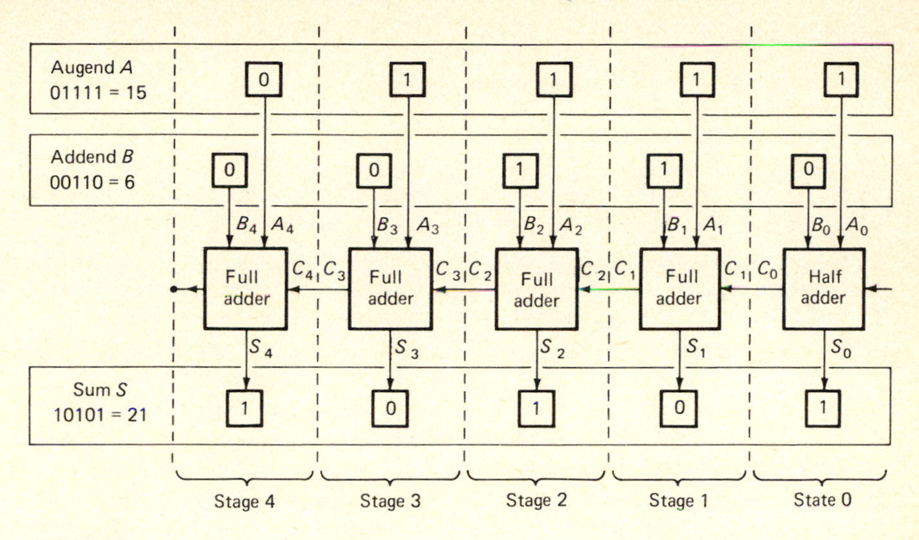

Figure 70

Notice that the EXCLUSIVE-OR ($A \veebar B$) is common to both the sum and carry expressions. The full adder can be drawn as shown in Figure 71*b*. It consists of two half adders (and an OR). The EXCLUSIVE-OR network could be replaced by the network of ANDs, ORs, and NOTs of Figure 68*b*, yielding a total complexity of 16 arrows.

If a digital computer has such a parallel adder consisting of 20 full adders, it could add integers up to 2^{20}, which is slightly over 1 million; 30 such full adders could add up to 1 billion. Some computers have 64 full adders, allowing addition of numbers greater than the square of 1 billion!

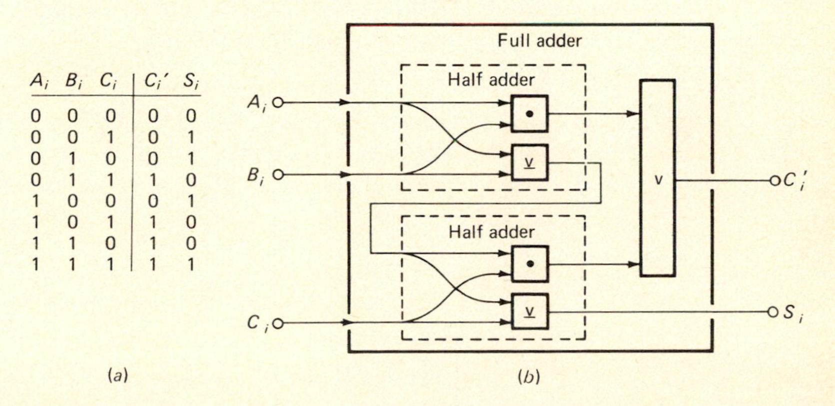

A_i	B_i	C_i	C_i'	S_i
0	0	0	0	0
0	0	1	0	1
0	1	0	0	1
0	1	1	1	0
1	0	0	0	1
1	0	1	1	0
1	1	0	1	0
1	1	1	1	1

(*a*) (*b*)

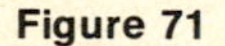

Figure 71

EXERCISE SET 8

1. Represent the following numbers in base 10:
(*a*) $(123)_4$ (*b*) $(210)_3$
(*c*) $(202)_7$ (*d*) $(1111111)_2$

2. Convert the following decimal numbers into binary numbers:
(*a*) 52 (*b*) 117
(*c*) 1984 (*d*) 100,000

3. The *octal*-number system has a base of 8, and is very useful in computer applications. Convert the decimal numbers, 8, 44, and 121 into their binary and octal equivalents, and notice how blocks of three binary digits are related to the corresponding octal digits.

4. What is the largest number that can be represented in base 2 if n binary digits can be used? *Ans.*: $2^n - 1$

5. Convert the numbers 15 and 6 into base 3 and do the addition in base 3.

6. Suppose you are informed of inheriting 1000000 dollars from some relative. Knowing that this relative expresses all amounts in base 2, how much money did you inherit?

7. The table-of-combinations given in Figure 72 describes a binary subtractor. Synthesize its simplest level implementation using OR, AND, and NOT components only. Try for 16 arrows. Note that m denotes minuend, s denotes subtrahend, b denotes borrow, and d denotes the difference.

10. EXTENDING THE SYSTEM

In this section we briefly show how the deterministic systems of this chapter can be extended to more operators and more values.

Other Binary Systems

We have concentrated on the three operators of AND, OR, and NOT, but there are other possible level operators such as the EXCLUSIVE-OR, NOR, NAND, and INHIBIT-AND. All the previous methods of analysis apply to such operators, but these operators may have different properties than the previous algebra.

The EXCLUSIVE-OR Operator The EXCLUSIVE-OR system of Figure 53 can be considered an operator as defined in Figure 73. Properties of such an operator can be studied by the method of complete induc-

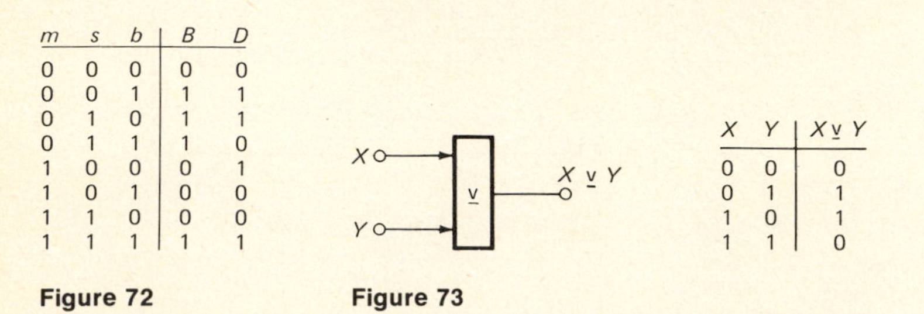

m	*s*	*b*	*B*	*D*
0	0	0	0	0
0	0	1	1	1
0	1	0	1	1
0	1	1	1	0
1	0	0	0	1
1	0	1	0	0
1	1	0	0	0
1	1	1	1	1

Figure 72

X	*Y*	*X* ⊻ *Y*
0	0	0
0	1	1
1	0	1
1	1	0

Figure 73

Figure 74

X	$\underline{\vee}$	X	$\neq$	X
0	0	0	T	0
1	0	1	F	1

X	$\underline{\vee}$	$\bar{X}$	$=$	1
0	1	1	T	1
1	1	0	T	1

tion, using tables-of-combinations. For example the tables of Figure 74 show that the property of idempotence does not hold but that complementarity does hold. Similarly, while attempting to study the property of dualization, the following similar property is discovered:

$$\bar{X} \,\underline{\vee}\, \bar{Y} = X \,\underline{\vee}\, Y$$

Figure 75*a* shows the proof of this property, and Figure 75*b* indicates its usefulness in optimization.

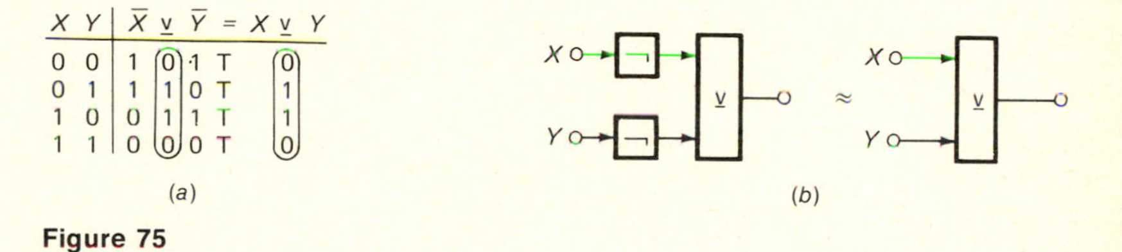

X	Y	$\bar{X}$	$\underline{\vee}$	$\bar{Y}$	$=$	$X \,\underline{\vee}\, Y$
0	0	1	0	1	T	0
0	1	1	1	0	T	1
1	0	0	1	1	T	1
1	1	0	0	0	T	0

Figure 75

A summary of some other properties of the EXCLUSIVE-OR operator is given in Figure 76. This operator is often used in communication systems for error detecting and correcting.

EXCLUSIVE – OR Properties

a. $X \,\underline{\vee}\, X = 0$

b. $X \,\underline{\vee}\, \bar{X} = 1$

c. $\bar{X} \,\underline{\vee}\, \bar{Y} = X \,\underline{\vee}\, Y$

d. $1 \,\underline{\vee}\, X = \bar{X}$

e. $0 \,\underline{\vee}\, X = X$

f. $X \,\underline{\vee}\, \bar{X}Y = X \vee Y$

g. $(X \,\underline{\vee}\, Y) \,\underline{\vee}\, Z = X \,\underline{\vee}\, (Y \,\underline{\vee}\, Z)$

h. $X \,\underline{\vee}\, XY = X\bar{Y}$

i. $(X \,\underline{\vee}\, Y)(X \,\underline{\vee}\, Z) = (X \,\underline{\vee}\, Z)\, Y$

j. $X \vee Y = X \,\underline{\vee}\, Y \,\underline{\vee}\, XY$

k. If $X \,\underline{\vee}\, Y = X \,\underline{\vee}\, Z$ then $Y = Z$

l. If $X \,\underline{\vee}\, Y = Z$ then $X = Y \,\underline{\vee}\, Z$

Figure 76

The NOR Operator Another commonly used switching operator is the NOR. Its defining table, level diagram, and one possible electronic network are shown in Figure 77. Other electronic NOR networks can be found in the Appendix. Notice that this operator is simply the NOT of an OR, which leads to the name NOT-OR or NOR operator. The operator symbol is formed from the v of the OR with a vertical bar through it.

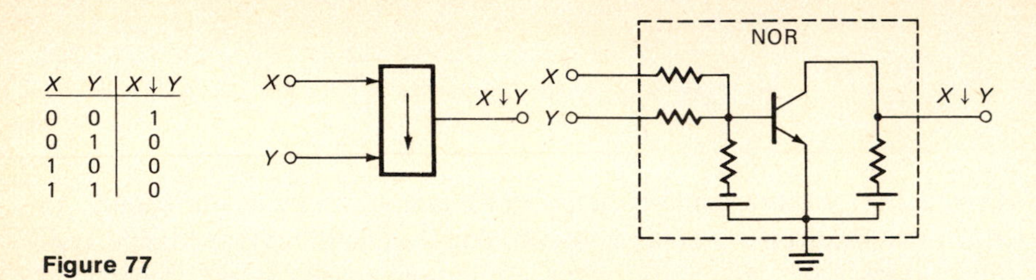

Figure 77

The NOR can be written in terms of the more common operators as

$$X \downarrow Y = \overline{X \vee Y} = \bar{X} \cdot \bar{Y}$$

The following three properties are significant:

$$X \downarrow X = \bar{X}$$

$$(X \downarrow Y) \downarrow (X \downarrow Y) = X \vee Y$$

$$(X \downarrow X) \downarrow (Y \downarrow Y) = X \cdot Y$$

The corresponding diagrams are shown in Figure 78. These three properties and diagrams show that NOR elements alone can be interconnected to make NOT, OR, and AND elements. In other words, there is no need for NOT, OR, and AND elements if we have NOR elements available; NOR elements alone could be used to synthesize any system! Many ex-

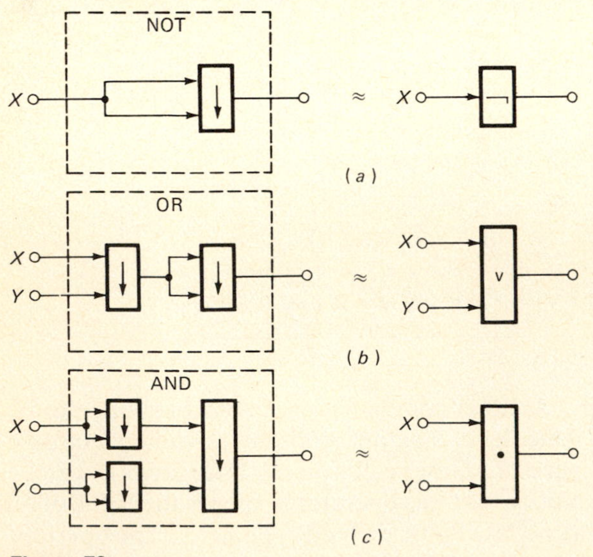

Figure 78

isting systems are in fact constructed with only NOR elements as the fundamental building blocks.

A set of operators is defined to be *functionally complete* (or f-complete) if any system can be synthesized using interconnections of these operators only. Thus the set of three operators AND, OR, and NOT is f-complete. Also the single operator NOR is f-complete.

For example, we could write any expression such as

$$D = \bar{A}(B \vee \bar{C})$$

in terms of NOR elements alone. Of course we could do this by replacing each operator by its equivalent NOR network of Figure 78, but this would lead to a complex network. Instead we could make use of the algebraic properties

$$X \vee Y = \overline{X \downarrow Y} \qquad \text{from } X \downarrow Y = \overline{X \vee Y}$$

$$X \cdot Y = \bar{X} \downarrow \bar{Y} \qquad \text{from } X \downarrow Y = \bar{X} \cdot \bar{Y}$$

$$\bar{X} = X \downarrow X$$

Applying these properties to the above expression yields

$$\begin{aligned} D &= \bar{A}(B \vee \bar{C}) \\ &= \bar{\bar{A}} \downarrow \overline{(B \vee \bar{C})} \\ &= A \downarrow (B \downarrow \bar{C}) \\ &= A \downarrow (B \downarrow (C \downarrow C)) \end{aligned}$$

The original and final networks are shown in Figure 79. Notice that the network involving NORs has fewer components.

Example. Prove that the two components, OR and EXCLUSIVE-OR, are f-complete if some terminals can be connected to a fixed value of 0 or 1.

Solution Recall from Figure 76*d* that $X \underline{\vee} 1 = \bar{X}$. This indicates that a NOT element can be formed by keeping one terminal of the EXCLUSIVE-OR constantly connected to

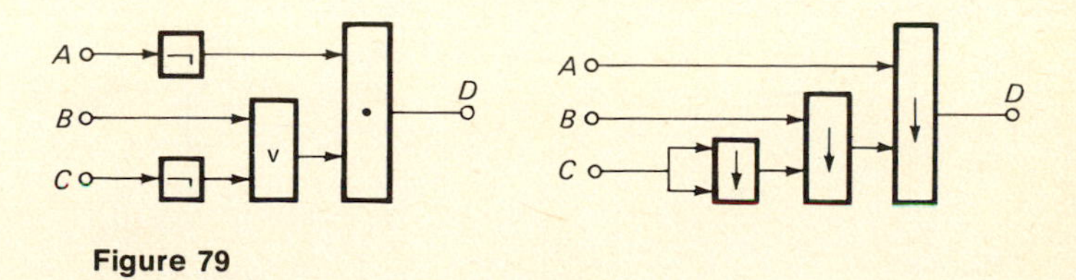

Figure 79

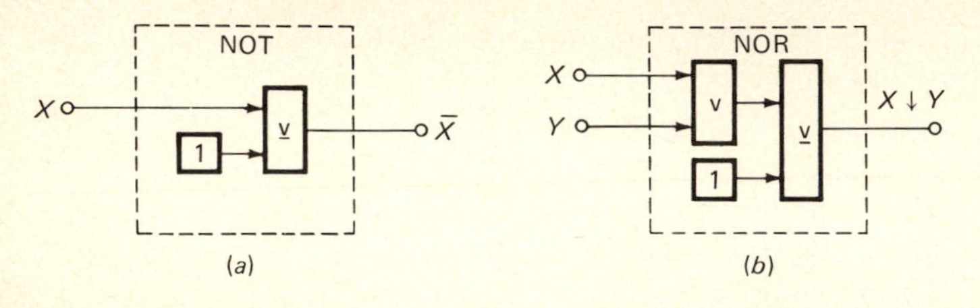

Figure 80

the value of 1 as in Figure 80*a*. From this expression we can construct a NOR

$$X \downarrow Y = \overline{X \vee Y} = (X \vee Y) \underline{\vee} 1$$

which is shown in Figure 80*b*. We have shown that the OR and the EXCLUSIVE-OR (and the constant 1) can be interconnected to form a NOR. And since NORs can be used to synthesize any two-valued system, these two operators can also synthesize any system.

The NAND and INHIBIT-AND. There are more binary operators which may be useful. Two of these, the NAND and INHIBIT-AND, are defined in Figure 81. Their analysis will be left as an exercise.

EXERCISE SET 9

1. Construct a network and table-of-combinations for each of the expressions

$$(A \downarrow \bar{B}) \underline{\vee} \bar{B} \qquad \text{and} \qquad A \downarrow (\bar{B} \underline{\vee} \bar{B})$$

Construct the equivalent level networks using the fewest AND, OR, and NOT components:

(*a*) From the table-of-combinations

(*b*) By algebraic methods (after substituting the equivalent AND, OR, and NOT operators).

2. Prove that the following *cancellation property*

$$\text{If} \quad X * Y = X * Z \qquad \text{then} \qquad Y = Z$$

holds when * is the EXCLUSIVE-OR but not the INCLUSIVE-OR.

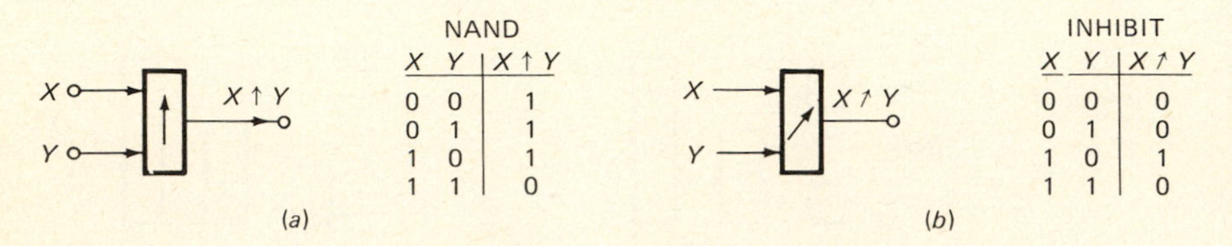

Figure 81

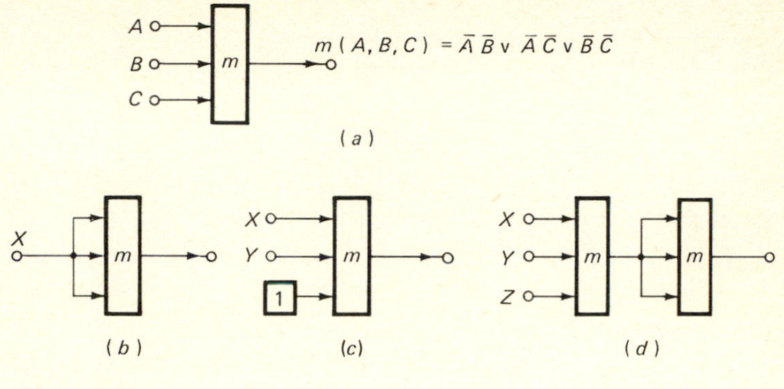

Figure 82

3. Prove that the NAND operator is functionally complete, and construct the network and table-of-combinations for the expression

$$(A \uparrow (A \uparrow B)) \uparrow (B \uparrow (A \uparrow B))$$

4. The MINORITY network of Figure 82*a* defined as

$$m(A, B, C) = \bar{A}\bar{B} \vee \bar{A}\bar{C} \vee \bar{B}\bar{C}$$

can be used as a three-input operator. For the remaining networks of Figure 82, construct the simplest equivalent AND, OR, and NOT networks. Prove that this MINORITY operator is functionally complete. Is the MAJORITY operator also functionally complete?

*Nonbinary Systems

All the operators considered so far have been two-valued, with the values designated as 0 to 1. However, many physical systems appear more naturally to be nonbinary, having three or more values. For example,

Voltage levels may be high, low, or intermediate.
Polarities may be positive, negative, or zero.
Positions may be in, out, or midway.
Rotations may be clockwise, counterclockwise, or still.
Directions may be north, south, east, or west.
Grades may be assigned as A, B, C, D, and F.

The general methods of this chapter can also be extended to such many-valued systems. For example, let us consider the three mechanical networks of Figure 83. If the rods occupy only three positions, to the left,

*Sections marked with a star are suitable for self-study. They may be omitted without loss of continuity.

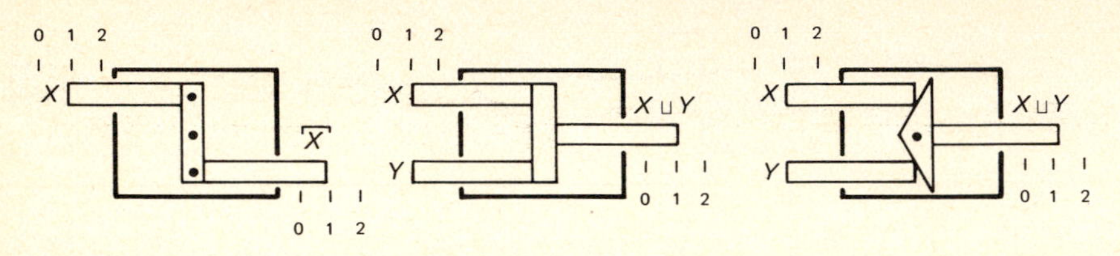

Figure 83

right, and midway, we can assign the positions any three values such as 0, 1, 2 or −1, 0, +1. Let us choose the following convention:

The position of the rods to the left has value 0.
The position of the rods midway is given value 1.
The position of the rods to the right is value 2.

The behavior of these components could be described by table-of-combinations as in Figure 84.

Notice that the first operator simply inverts the input, and so it could be called an inverter. The output of the second operator indicates whatever input is maximum (rightmost) and so could be called the *max operator*. Similarly the third operator could be called the *min operator*.

Again it is possible to determine properties of these operators by complete induction. For example let us check idempotence and complementarity for the *max operator*. The table-of-combinations of Figure 85 shows that idempotence holds but complementarity does not hold. The property of dualization, however, does hold, as shown in Figure 86. To prove a property such as associativity involving three variables would require a table-of-combinations of 27 rows. Try it!

This three-valued system is unfortunately not very useful because it is not f-complete. With these three types of operators we cannot synthesize all three-valued systems. However the max, min, and the three oper-

X	$\overline{X}$
0	2
1	1
2	0

X	Y	$X \sqcup Y$
0	0	0
0	1	1
0	2	2
1	0	1
1	1	1
1	2	2
2	0	2
2	1	2
2	2	2

X	Y	$X \sqcap Y$
0	0	0
0	1	0
0	2	0
1	0	0
1	1	1
1	2	1
2	0	0
2	1	1
2	2	2

Figure 84

X	$X \sqcup X$	=	X
0	0 0 0	T	0
1	1 1 1	T	1
2	2 2 2	T	2

X	$X \sqcup \overline{X}$	$\overset{?}{=}$	1
0	0 2 2	F	1
1	1 1 1	T	1
2	2 2 0	F	1

Figure 85

ators of Figure 87 are f-complete, but we will not go into any further detail about three-valued systems.

The significance of this section is to show that the principles of this chapter are fundamental and apply to many other systems. In essence it shows how to construct and study a mathematical system corresponding to any physical system of interest.

EXERCISE SET 10

1. If the value convention is interchanged so that 0 denotes the rightmost position and 2 denotes the leftmost position, how would you rename the operators of Figure 83?

2. Construct the table-of-combinations for the max, min, and inverter if four positions 0, 1, 2, and 3 are possible. The lowest number (0) denotes the leftmost position, and the largest (3) denotes the rightmost position.

3. Prove the following properties concerning the three-valued systems of Figure 84 and Figure 87:

(*a*) $2 \sqcup X = 2$ (*b*) $2 \sqcap X = X$

(*c*) $\bar{X}_0 \sqcup \bar{X}^1 \sqcup \bar{X}^2 = 2$ (*d*) $(1 \sqcap \bar{X}^1) \sqcup (2 \sqcap \bar{X}^2) = X$

(*e*) $\bar{X}^2 \sqcup \bar{Y}^2 = (\overline{X \sqcup Y})^2$

4. Construct the table-of-combinations for the network of Figure 88*a* and use it to find a synthesis procedure. Show by the table-of-combinations that it is equivalent to Figure 88*b*. (The operators with one input are defined in Figure 87.)

5. Construct a ternary (base 3) half adder. Construct a table-of-combinations for a ternary full adder. Note that only 18 rows are required. Show that a simple interconnection of ternary half adders can behave as a ternary full adder.

X	Y	$\overline{X \sqcup Y}$	=	$\overline{X} \sqcap \overline{Y}$
0	0	0 0 0 2	T	2 2 2
0	1	0 1 1 1	T	2 1 1
0	2	0 2 2 0	T	2 0 0
1	0	1 1 0 1	T	1 1 2
1	1	1 1 1 1	T	1 1 1
1	2	1 2 2 0	T	1 0 0
2	0	2 2 0 0	T	0 0 2
2	1	2 2 1 0	T	0 0 1
2	2	2 2 2 0	T	0 0 0

Figure 86

X	$\bar{X}^0$	$\bar{X}^1$	$\bar{X}^2$
0	2	0	0
1	0	2	0
2	0	0	2

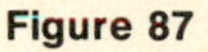

Figure 87

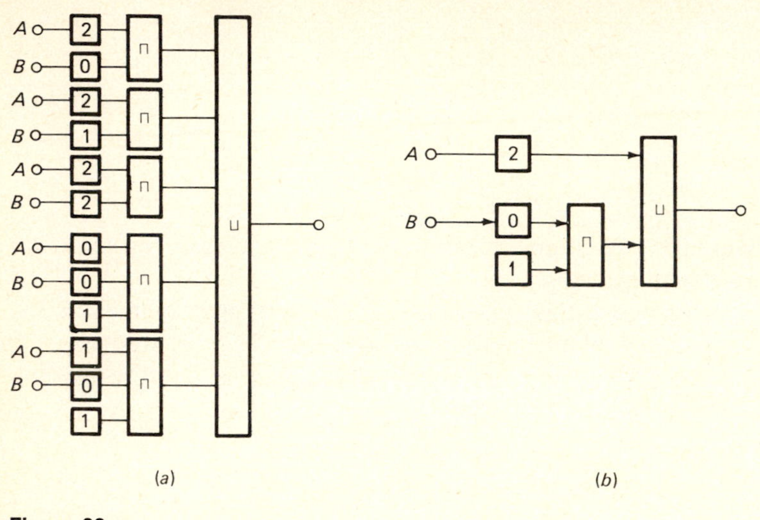

Figure 88

SUMMARY

This chapter provides an example of the complete study of one type of system, the digital, deterministic, instantaneous system. The summary will not emphasize the mathematical details but will cover the underlying philosophical and system-theoretic aspects, which may have been overshadowed by the mathematics. It will show how you could develop the mathematics corresponding to whatever particular devices are of interest to you. There are 16 two-valued operators such as AND, OR, and NOT and 19,683 three-valued operators such as max, min, and inverter; there are therefore many possible systems and corresponding algebras that can be studied by the methods of this chapter.

In general a *system* consists of some sets and relations. In this chapter we could consider the system as a set of contacts or valves (in the flow networks) related by their interconnections. We could also consider the system as a set of electric or mechanical black boxes (in the level networks) connected together. More abstractly, we could consider the system as a set of variables related by the operations such as OR, AND, and NOT. All these systems are essentially equivalent mathematically.

The systems approach enables the physical considerations to be separated from the logical and mathematical considerations so that an electronically oriented person may view a black box in terms of diodes, a mechanically oriented person may view this same box in terms of levers, a chemically oriented person may view the box in terms of transport processes, a mathematically oriented person may view the box as an abstraction without physical significance but with beautiful structure. Thus the systems approach gives a *unity* to systems, despite the variety of physical components. A system is everything to everyone!

The systems approach has two aspects, called the *behavioral* and *structural*, corresponding to an external or internal emphasis, respectively. The behavioral aspect considers the system as a black box which can be studied only through its external input and output terminals. The structural aspect is concerned with the internal interconnections between the input and output terminals. The difference between the behavioral and structural view is analogous to the difference between psychology and anatomy. It is also analogous to the difference between the point of view of an automobile driver and a mechanic (or between any operator and repairman).

General Systems Processes

Some general processes can be studied for any system, namely those of modeling, representation, analysis, synthesis, realization; optimization, identification, and evaluation. These processes are of course motivated by, and dependent upon, a goal, purpose, or application. Most of the applications in this book will be related to control, communication, and computation. These systems processes and their relation to the systems of this chapter are described below.

Modeling is the process of abstracting relevant attributes from a physical system to yield a simplified version of this reality which is similar to the original physical system in important respects. It proceeds by observation and experimentation and involves making assumptions to gain simplicity. In this chapter systems have been assumed:

Instantaneous (as opposed to delayed)
Two-valued (as opposed to many-valued)
Deterministic (as opposed to probabilistic)

Some other assumptions may have crept in unintentionally because of restrictions of viewpoint. For example, in this chapter we did not consider interconnecting systems in which the outputs may affect the inputs; i.e., there have been no feedback loops. In later chapters some of these assumptions will be removed.

Representation is the process of describing a system in a convenient, useful, powerful, and insightful way. This may take various diagrammatic, tabular, algebraic, or symbolic forms. In this chapter we have used:

Structural diagrams to aid in understanding the interconnections
Switching algebra to aid in simplifying the structure
Tables-of-combinations to aid in analysis and synthesis

There are other representations (such as Karnaugh maps, matrices, n-dimensional hypercubes) which we did not consider.

Analysis is the process of determining the properties of interconnected systems from the properties of the components and their interconnection. In this chapter it is a straightforward process making use of the table-of-combinations and an algebra.

Synthesis is the process of starting from a given behavioral goal (or input-output specification) and producing a structure consisting of interconnected smaller components (such as OR, AND, and NOT). The two methods described here involve a direct verbal translation into an algebra and the transformation of a table-of-combinations.

Implementation or realization is the process of converting an abstract structure into a physical form. The AND, OR, and NOT blocks are replaced by networks of electric, mechanical, electromechanical, hydraulic, or other components. This realization also depends upon the goal or application since, for example, a relay network arcing in an explosive gaseous environment is a potential problem.

Optimization of structure, or simplification, or minimization, is the process of simplifying a structure while maintaining the same behavior. It makes use of the results of analysis, the concept of equivalence of behavior, and the costs of the physical components in the realization. In later chapters we will also optimize behavior.

Design is the creative process of synthesis, implementation, and optimization to accomplish some goal.

Identification is the process of determining a structure from input-output behavior. It assumes that a certain model holds, and it attempts to show what classes of equivalent structures could yield this same input-output behavior. The structure is not unique.

Evaluation is the process of determining, appraising, or estimating how good a system is and to what extent the goals have been satisfied. It proceeds by comparing or testing. Although this evaluation aspect of systems has not been discussed in great detail in this chapter, it is nevertheless important. It appears particularly important when it is recalled that many assumptions have been made in modeling, possible wrong interpretations could have been made in synthesis, poor choices made in realization, and some possibilities unforeseen in optimization. Unfortunately, sometimes the only method of evaluation involves constructing some systems and comparing them in actual use. Other methods involving evaluation of reliability will be studied in a later chapter on probabilistic systems.

These system-theoretic aspects of modeling, representation, analysis, synthesis, implementation, optimization, identification, and evaluation will be applied later to the many other types of systems such as probabilistic, sequential, and stochastic. Before meeting any other types of systems we will first consider some mathematical concepts of sets and relations motivated by these system-theoretic concepts. After all, any system consists of sets and relations.

PROBLEMS

1. Optimize the number of springs in the given network.

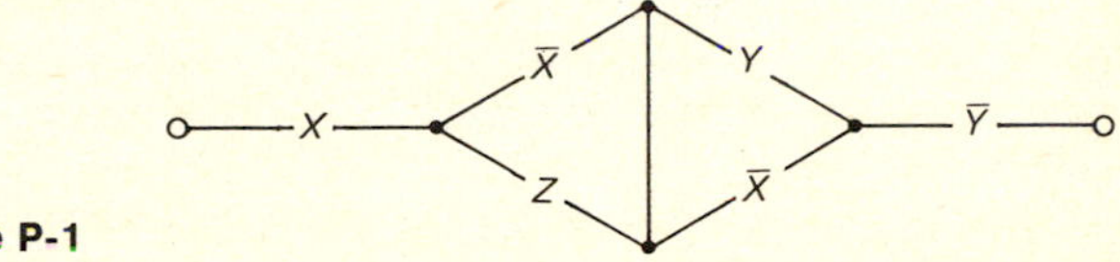

Figure P-1

2. All the following expressions can be optimized by inspection in *one* step, or at most two (if the proper identity is recognized). Write the optimized expression, indicating the appropriate justification.

(*a*) $AB \vee (\bar{A} \vee \bar{B})C$

(*b*) $(B\bar{C} \vee D)(\bar{A} \vee B\bar{C})$ try 8 springs

(*c*) $(A\bar{B})(\overline{C \vee \bar{A}B})(\bar{A} \vee B)$

(*d*) $(B \vee C \vee D\bar{E}F)C$ try 2 springs

3. Optimize the following networks:

(*a*) $A \vee B(C \vee DE \vee \bar{E} \vee \bar{D})$ *Ans.*: $A \vee B$

(*b*) $A \vee B(C \vee DE \vee \bar{E} \vee B)$

(*c*) $A \vee (B\bar{C} \vee C(\bar{A} \vee B)(B \vee C)$

(*d*) $(AB \vee C)(A \vee (\bar{A} \vee D)\overline{(B\bar{C} \vee \bar{A})}$ *Ans.*: $A(B \vee C)$

(*e*) $(AB \vee C)(A \vee (\bar{A} \vee D)\overline{(B\bar{C} \vee \bar{A})}$ *Ans.*: AC

4. Even though the following networks are not of the series-parallel type, write an algebraic expression for each by following all paths through the network.

Ans.: *a*, $AC \vee BD \vee AED \vee BEC$

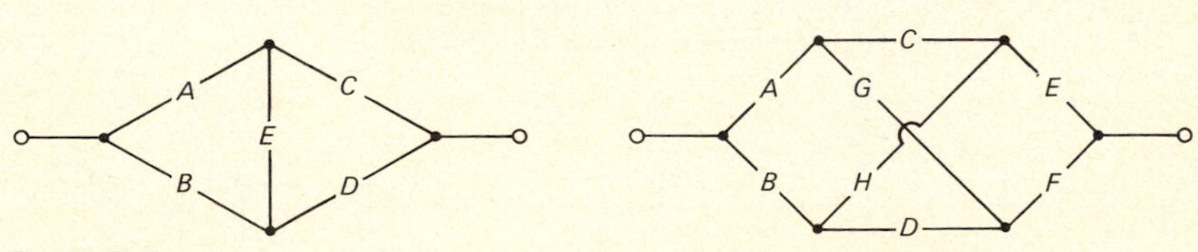

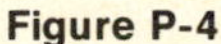

Figure P-4

5. For the following switching algebraic expressions, optimize the number of springs in the relay implementation and optimize the number of arrows in the level implementation (using AND, OR, and NOT elements only). Draw the diagrams in both cases.

(*a*) $(\bar{A} \vee \bar{B} \vee \bar{D})(\bar{A} \vee \bar{C} \vee \bar{D})$

(*b*) $\bar{A}\bar{B} \vee AC \vee A\bar{B}\bar{D} \vee \bar{A}B$ *Ans.*: $\bar{A} \vee \bar{B}\bar{D} \vee C, \overline{A(B \vee D)} \vee C$

(*c*) $ABC \vee DC \vee B\bar{D} \vee ABD \vee \bar{A}B\bar{C}$

(d) $(A \vee B)(A \vee C)(\bar{A} \vee \bar{B} \vee \bar{C})$ *Ans.*: 10 springs, 9 arrows
(e) $(A \vee B \vee C)(A \vee C \vee D)(A \vee \bar{B} \vee \bar{D})$
(f) $(A \vee \bar{B})(\bar{A} \vee C)$ try for 7 springs, 7 arrows

6. Write an algebraic expression describing the following network and construct a behaviorally equivalent level network using the smallest number of AND, OR, and NOT components. Use both table-of-combinations and algebraic methods.

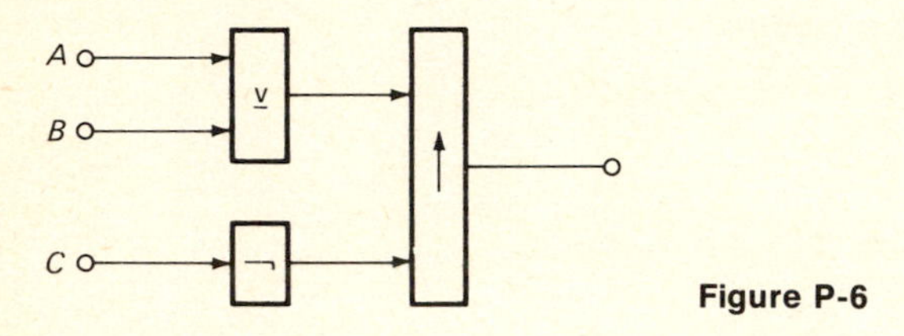

Figure P-6

7. The switching expression $N = \bar{X} \vee (\overline{Y \cdot \bar{Z}})$ is to be realized using only AND, OR, and NOT level components. Sketch the simplest level network if the complexity (cost) is proportional to:

(a) The total number of components
(b) The total number of input arrows
(c) The total sum when the cost of NOT components is 2 units, the cost of OR components is 3 units, and the cost of AND components is 6 units.

8. The given level component is called an INHIBIT element and read X "is inhibited by" Y or X "but not" Y. The output is X if not inhibited by applying input Y.

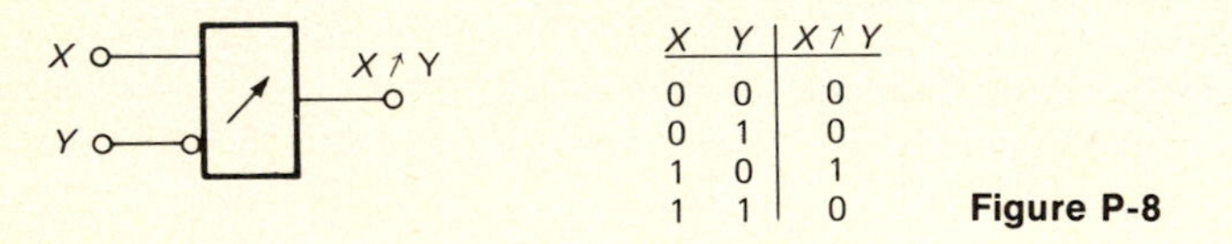

X	Y	X ↗ Y
0	0	0
0	1	0
1	0	1
1	1	0

Figure P-8

(a) Describe it in terms of AND, OR, and NOT.
(b) Check the commutative property.
(c) Draw the simplest interconnection $(A \nearrow B) \nearrow (A \nearrow B)$ in terms of INHIBIT elements.
(d) Construct the table-of-combinations and network for $(1 \nearrow X) \nearrow (1 \nearrow (1 \nearrow Y))$
(e) Show that it is f-complete if some terminals can be connected to a constant value 1.

9. Sketch all input waveforms X which can combine with the input waveform Y to produce the given output waveform Z. What name could describe the operation of this network? *Ans.*: $X \leq Y$

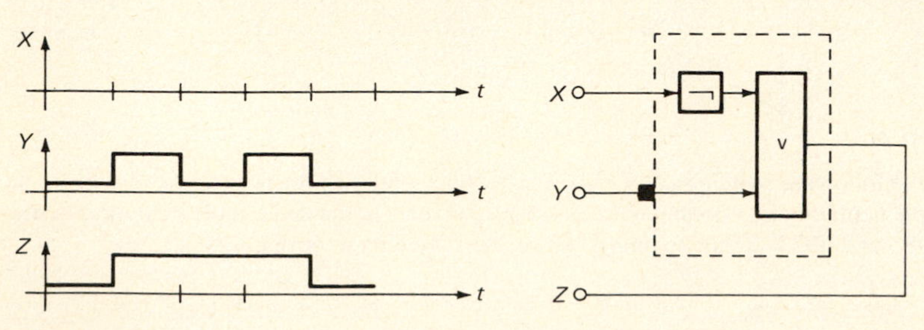

Figure P-9

10. The given component is constructed using one relay having two windings. Only two values of voltage (0 volts and E volts) are applied to the windings. Describe by table-of-combinations all possible two-valued functions this component could behave as. *Hint*: There are many.

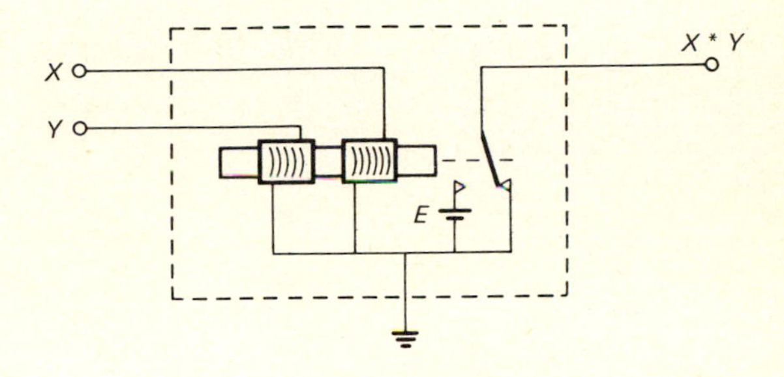

Figure P-10

11. Convert all the following numbers in the given bases to numbers in base 10:

(*a*) 201 in base 3

(*b*) 21.2 in base 3 *Ans.*: 7.6666· · ·

(*c*) 110 in base −2

(*d*) 10.2 in base −3

(*e*) 100.1 in base $\frac{2}{3}$ *Ans.*: approximately 2

12. Construct a half-adder using the smallest number of NORs only. Try for five NORs.

13. Prove that the three ordinary arithmetic operators of sum, difference, and product are functionally complete by:

(*a*) Writing the AND, OR, and NOT operators in terms of the three ordinary algebraic operations of +, −, and *. *Hint*: $X \vee Y = (X + Y) - (X * Y)$.

(*b*) Writing the one operator of NOR in terms of these three operators in the ordinary algebra.

14. A fluid reservoir has three transducers X_1, X_2, X_3 mounted at various heights which have value 1 when the fluid is at their level or above. A network is to indicate (by having value 1) whenever the fluid level is below X_1 or between X_2 and X_3.

(*a*) Construct a table-of-combinations for such a network.

(*b*) How many different networks are possible? *Ans.*: 16

(*c*) Show that one possible network is $\bar{X}_1 \vee X_2\bar{X}_3$.

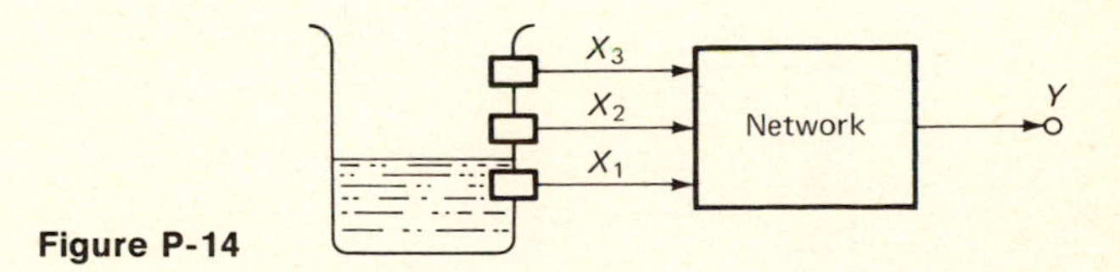

Figure P-14

15. The following is a description of a simple electrical lock-and-key mechanism. The key can be marked in any way with magnetic ink (as indicated by dots). This marking can be hidden (for security) by an opaque nonmagnetic material. The lock consists of four magnetic sensors which sense a particular four areas on any key and operate a corresponding relay if magnetic ink is detected.

(*a*) Design and optimize a network which recognizes keys with areas 1, 3, and 4 only inked in, or with areas 1 and 4 only inked in.

(*b*) Design this network for a symmetrical key (as above) so that it operates the lock even when it is inverted (interchanging 1 with 4 and 2 with 3). (Try for eight springs.)

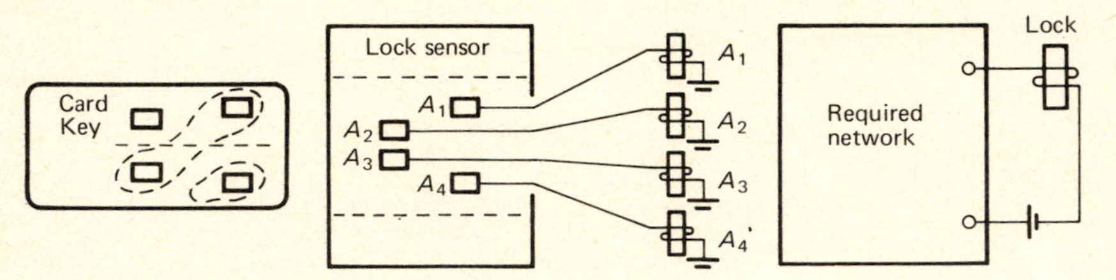

Figure P-15

16. A weather-alarm system receives the following binary signals:

T having value 1 if the *t*emperature is below freezing.
R having value 1 if the *r*ainfall is above a critical level.
W having value 1 if the *w*ind velocity is greater than 5 mph.
F having value 1 if the *f*og causes poor visibility.

This system sounds a minor emergency alarm (value 1) if there is excessive rain with either high winds or low temperature and sounds a major alarm if there is low temperature, excessive rain, and at least one of the remaining two dangerous conditions.

(*a*) Write a switching algebraic expression for each of the two alarm conditions.
(*b*) Construct a level network, optimizing the number of AND, OR, and NOT components.
(*c*) If it is specified that only one alarm (the most urgent one) can sound at any one time, is the design made any easier? Is the resulting system any simpler?

17. A threshold network is a model of a nerve cell. It consists of a number of inputs X_i each having weight W_i, a single output Y, and a variable threshold T. The output has value 1 if and only if the algebraic sum of the (ordinary) products $W_i * X_i$ is greater than or equal to the threshold T. That is,

$$Y = 1 \quad \text{when} \quad (W_1 * X_1 + W_2 * X_2 + \cdots + W_n * X_n) \geq T$$

For the two-input threshold network, with $W_1 = 2$ and $W_2 = 3$ construct the equivalent binary level network for the following threshold values of $T = 0, 1, 2, 3, 4, 5, 6$.

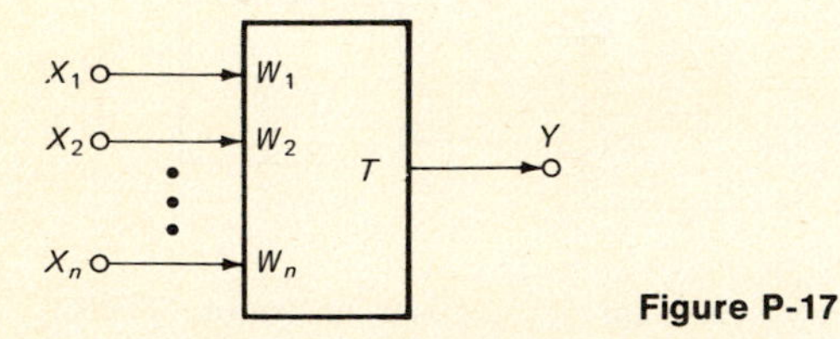

Figure P-17

18. *Project* (*major investigation*). The properties of the AND, OR, and NOT are summarized in Figure 26; the properties of the EXCLUSIVE-OR are summarized in Figure 76. In a similar way analyze the following systems and summarize them:

(*a*) The NAND, of Figure 81
(*b*) The NOR, of Figure 77

(*c*) The INHIBIT, of Figure P-4
(*d*) The unnamed operator of Figure P-9
(*e*) The equivalence operator of Figure P-18

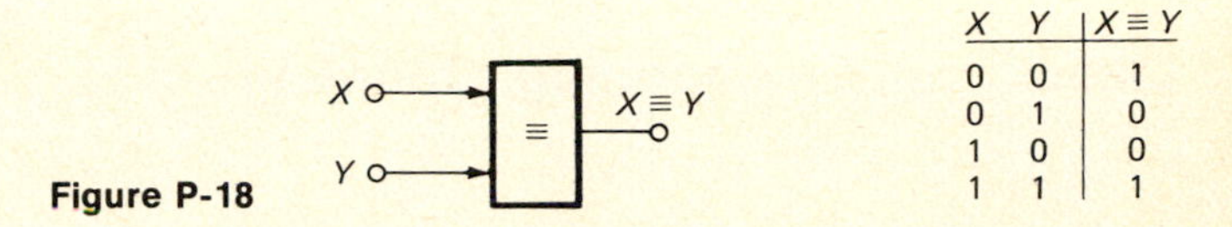

X	Y	X ≡ Y
0	0	1
0	1	0
1	0	0
1	1	1

Figure P-18

19. *Project.* The given diagram describes the modulo system having operators mod-sum, mod-product, and mod-inverse. It is similar in many respects to the ordinary algebra and also to the EXCLUSIVE-OR. Study this system in detail noticing some properties such as

$$X \oplus X = X'$$

$$X' \oplus Y' = (X \oplus Y)'$$

$$X \oplus (Y \otimes Z) \neq (X \oplus Y) \otimes (X \oplus Z)$$

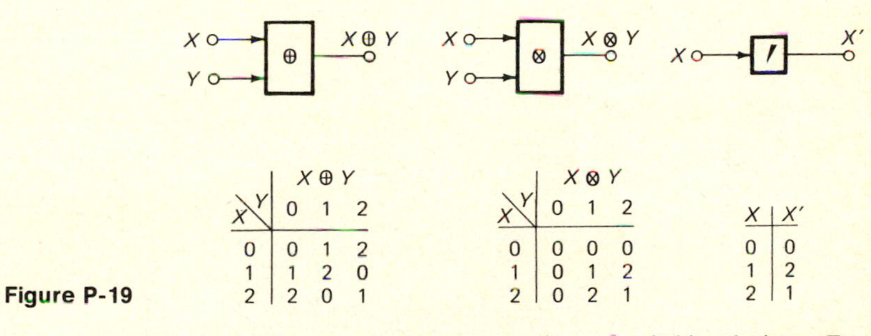

X \ Y	0	1	2
0	0	1	2
1	1	2	0
2	2	0	1

($X \oplus Y$)

X \ Y	0	1	2
0	0	0	0
1	0	1	2
2	0	2	1

($X \otimes Y$)

X	X′
0	0
1	2
2	1

Figure P-19

20. *Project.* Investigate in detail some other implementations of switching devices. For example you may choose fluidic devices, rotational devices, or devices operated by light.

Chapter 2

Sets and n-Tuples

Classifying, Enumerating, Counting

Man appears to have an inborn urge to compare, relate, order, and classify. He classifies people as male or female; ages as young or old; weather as good, bad, or fair; systems as stable or unstable; sizes as large, medium, or small; foods as edible or inedible; qualities as grades A, B, or C.

In many cases there are two alternatives, and an item must be classified into one or the other of these two categories. In such cases the concept of a set is convenient and useful.

A set is a simple concept but a powerful one. It is both an abstraction from everyday experience and the fundamental basis for all of mathematics. Indeed, virtually all of present-day mathematics can be derived from the theory of sets. Thus the concept of a set is both practical and theoretical. As usual here, we will start with the practical view.

1. SETS: REPRESENTATION AND ANALYSIS

A *set* (intuitively) is any collection, aggregate, grouping, gang, bunch, ensemble, compilation, or class of elements that has the properties of:

Definiteness, making it possible to determine whether an element is within a set or not, and

Distinctness, making it possible to determine whether all elements are different or not.

Example Some sets are:
The 10 decimal digits
All human beings living on earth now
All the positive integers

The first two sets are finite, whereas the third set is infinite. We will tend to concentrate on finite sets.

Example Some collections which are not sets are:
All young people (not definite)
The five most important books on earth (not definite)
The four digits 1, 0, 1, 2 (not distinct)

The criterion for definiteness is not as important as the fact that there is a criterion. For example, you may wish to classify as young anyone who is under the age of thirty. Or perhaps you would rather choose the "legal age" of twenty-one, or the advertiser's youth of seventeen, or sociologist's category of anyone who behaves in a certain way. Of course, once a criterion is established, it must be consistently adhered to.

The property of distinctness simply prevents any repetition of elements within a set. This property is important in counting the number of elements in sets.

Representation of Sets

Sets can be represented in many ways: by extension, by intension, and by set diagrams.

Representation by Extension Specification of a set by extension involves listing all the elements of the set between braces in the form

$$X = \{x_1, x_2, x_3, \ldots, x_n\}$$

The common letter x indicates that all elements have something in common (by virtue of their being in the set), and the different subscripts indicate that all elements are different.

Examples

$A = \{0, 1, 2, 3, 4, 5, 6, 7, 8, 9\}$

$B = \{0, 1\}$

$C = \{a, b, c, d, f\}$

$D = \{\text{AND, OR, NOT}\}$

$E = \{10, +, \text{ego, Sunday}\}$

The relation of membership in a set denoted $e \in S$ is read "element e is a member of set S." Similarly $e \notin S$ is read "element e is not a member of S." Thus for example in the previous sets we may write

$0 \in A \qquad 0 \notin C \qquad \text{ego} \in E$

Representation by Intension Specification of a set by intension involves giving a property which is common to any typical element of the set. It takes the form

$X = \{x\colon\ P(x)\}$

which is read "set X is the set of all elements x that have property P."

Examples

$A = \{x\colon\ x \text{ is a digit of the decimal number system}\}$

$B = \{x\colon\ x * (x - 1) = 0\} \qquad = \qquad \{y\colon y * y - 1) = 0\}$

$C = \{g\colon\ g \text{ is a letter grade at some school}\}$

$F = \{x\colon\ 0 < x \leq 1\}$

$H = \{h\colon\ h \text{ is a living human being}\}$

Notice that some sets are most conveniently specified by extension (set E); whereas some are most conveniently specified by intension (set H); and some are equally easily specified by either method (sets A, B, C). Although it seems difficult to find a property common to all elements of the set

$E = \{10, +, \text{ego, Sunday}\}$

it can easily be specified by intension as

$E = \{x\colon\ (x \text{ is "10"}) \text{ or } (x \text{ is "+"}) \text{ or } (x \text{ is "ego"}) \text{ or } (x \text{ is "Sunday"})\}$

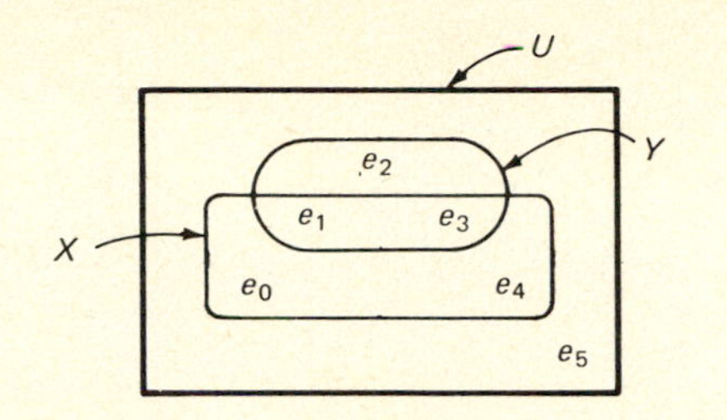

$U = \{e_0, e_1, e_2, e_3, e_4, e_5\}$
$X = \{e_0, e_1, e_3, e_4\}$
$Y = \{e_1, e_2, e_3\}$

Figure 1

Representation by Set Diagram It is useful to represent sets graphically by means of a set diagram. Variations of set diagrams are known as Venn diagrams, Euler diagrams, Veitch maps, Mahoney maps, or Karnaugh maps. On a set diagram a set is represented as a closed curve surrounding the elements within the set. Some sets and their corresponding set diagram are shown in Figure 1.

For larger sets the elements are not usually shown on the set diagram. Instead, the number of elements within a set may be shown. For example, consider a set (universe) of 100 people. Any such set can be classified in many ways. The people could be classified according to age, as in Figure 2*a*. A second classification according to sex could partition this universe into the four sets shown in Figure 2*b*. This partition could be further refined according to the opinion (like or dislike) of a certain product or program, yielding eight partitions of Figure 2*c*.

Set Diagrams in Karnaugh Map Form The sets shown on set diagrams could be drawn in any shape as illustrated in Figure 3, for the case of three sets. From these diagrams it is apparent that some methods of drawing the sets may be preferable to others. A most convenient set diagram results when the sets are drawn as rectangles and extended to the

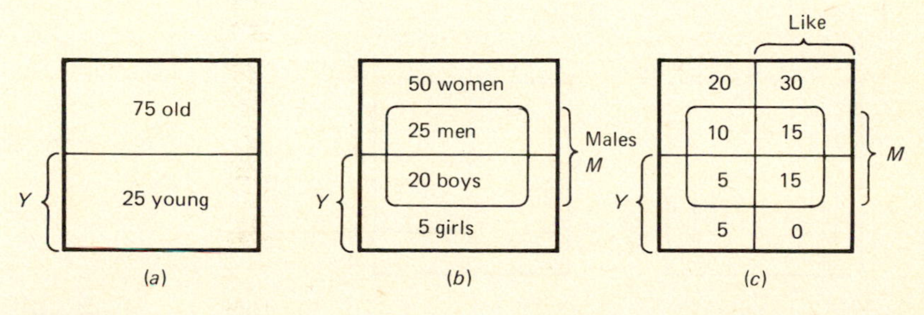

Figure 2

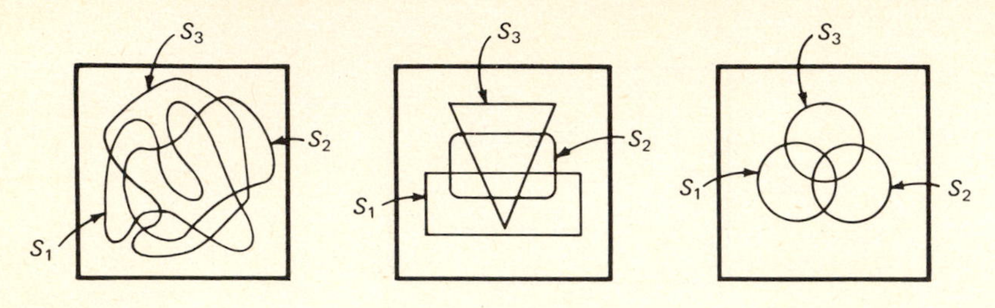

Figure 3

boundaries of the diagram as in Figure 4. This yields a tabular form often known as a Karnaugh map. It is simple to draw and useful to work with. These set diagrams are illustrated in Figure 5 for one, two, three, and four sets. Diagrams for more than four sets are rather awkward but will be considered later. There are other ways of labeling these diagrams, but we will often use the labeling of Figure 5 for the convenience which comes from consistency.

Fundamental Concepts of Sets

A *universe*, denoted U, is the set of all elements under consideration. Although the universe is often understood and therefore not explicitly stated, it is still beneficial to clarify at the outset what is being studied; is it the set of all people, or all American people, or a sample of American consumers of one type?

A set S_1 is a *subset* of set S_2, denoted $S_1 \subseteq S_2$, if all elements of S_1 are also elements of S_2. For example,

$$\{e_1, e_3\} \subseteq \{e_1, e_2, e_3\}$$

Also, the set of boys is a subset of the set of all human males. The subset relation is shown in Figure 6*b*. Notice that all sets are subsets of a universe.

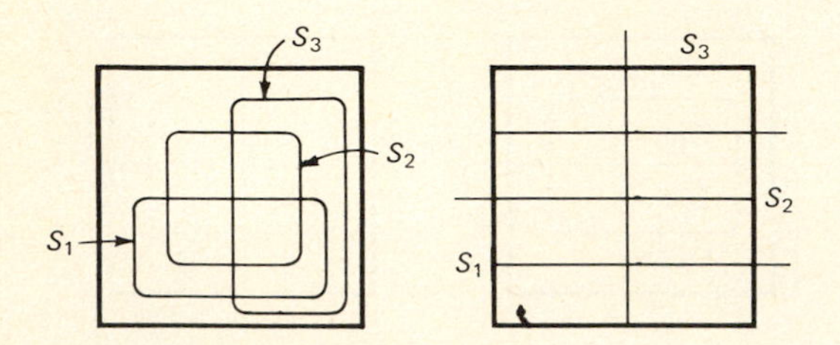

Figure 4

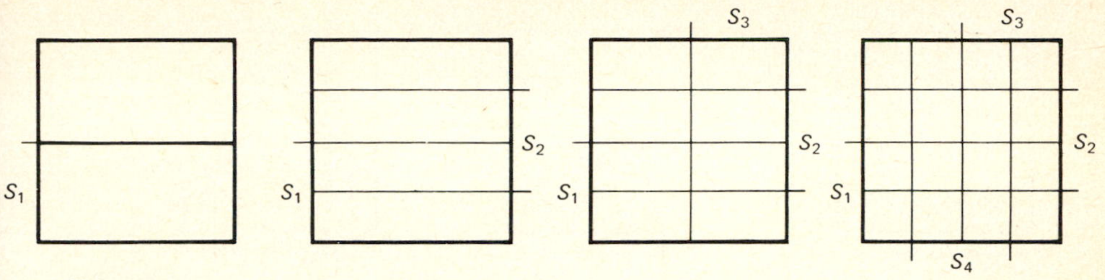

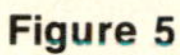
Figure 5

Operations on Sets There are three main operations on sets, as shown in Figure 7.

The *complement* of a set S, denoted $\tilde{S}$, is the set of all elements which are not in set S (but are still in the universe). This can be written symbolically as:

$$\tilde{S} = \{x : x \notin S\}$$

Example If

$$X = \{e_0, e_1, e_3, e_4\}$$

and

$$U = \{e_0, e_1, e_2, e_3, e_4\}$$

then

$$\tilde{X} = \{e_2\}$$

Also the set of all human males is the complement of the set of all human females (in the universe of all humans) and is written

$$M = \tilde{F}$$

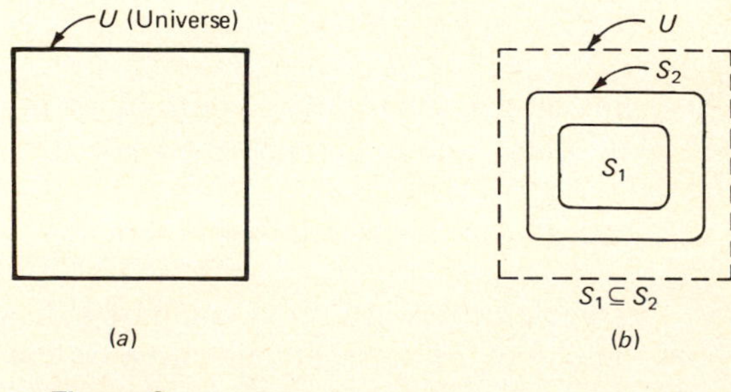

Figure 6

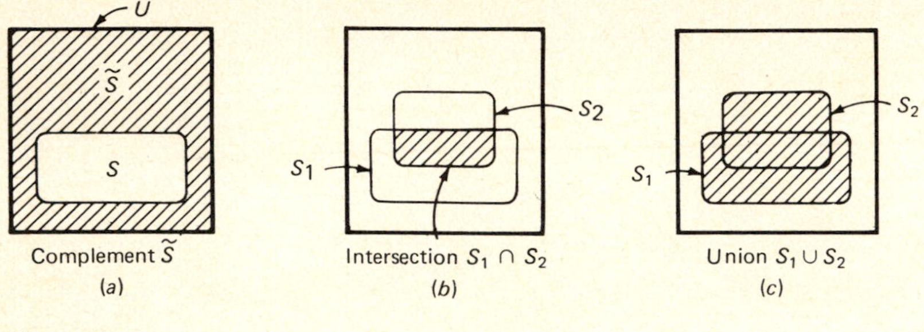

Figure 7

The *intersection* of sets S_1 and S_2, denoted $S_1 \cap S_2$ or more simply as $S_1 S_2$, is the set of elements which are in both S_1 and S_2. Symbolically this is written

$$S_1 \cap S_2 = \{x : x \in S_1 \text{ and } x \in S_2\}$$

For example,

$$\{e_0, e_1, e_3, e_4\} \cap \{e_1, e_2, e_3\} = \{e_1, e_3\}$$

Also, the intersection of the set of all youths Y and the set of all males M is the set of all young males or boys, i.e.,

$$Y \cap M = Y \cdot M = \{x\text{: } x \text{ is young and } x \text{ is male}\}$$

The *union* of sets S_1 and S_2, denoted $S_1 \cup S_2$, is the set of all elements which are in set S_1 or set S_2 (or both). Symbolically

$$S_1 \cup S_2 = \{x : x \in S_1 \text{ or } x \in S_2\}$$

For example,

$$\{e_0, e_1, e_3, e_4\} \cup \{e_1, e_2, e_3\} = \{e_0, e_1, e_2, e_3, e_4\}$$

Also the union of the set of all youths Y and the set of all females F is the set of all those who are young or female, i.e.,

$$Y \cup F = \{x\text{: is young or } x \text{ is female}\}$$

If an element is in a set, it could be indicated by marking (shading or cross-hatching) the corresponding area on the set diagram. The example

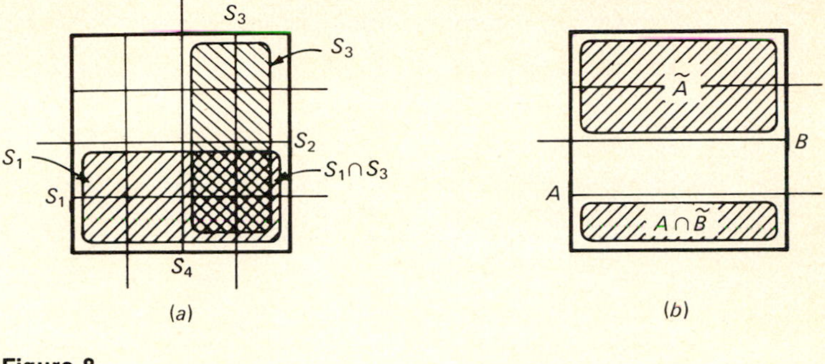

Figure 8

in Figure 8*a* indicates that an element is in set S_1 and that an element is in set S_3. The doubly hatched area indicates an element in the intersection $S_1 \cap S_3$.

If two sets do not intersect, we say that they are *mutually exclusive* or disjoint. Another way of viewing this is to say that the intersection is the empty set or *null set* ϕ, the set containing no elements. The example in Figure 8*b* shows that sets $(A \cap B)$ and $\tilde{A}$ are mutually exclusive, i.e.,

$$(A \cap \tilde{B}) \cap \tilde{A} = \phi$$

A *partition* of a set S is a class of subsets of S which are mutually exclusive (have no elements in common) and exhaustive (include all elements of S). For example, in Figure 9 the set of males M and the set of youths Y partition the universe of human beings into the four subsets corresponding to boys, girls, men, and women. If each set splits (partitions) all the existing sets in two, the total number of partitions possible with n sets is 2^n. Thus for three sets there are eight partitions, for 10 sets there are 2^{10} or 1,024 partitions, and for 20 sets there are over 1 million partitions. Of course partitions may also be created by splitting sets into three

$\tilde{Y} \cdot \tilde{M}$ Women
$\tilde{Y} \cdot M$ Men
$Y \cdot M$ Boys
$Y \cdot \tilde{M}$ Girls
M
Y
(*a*)

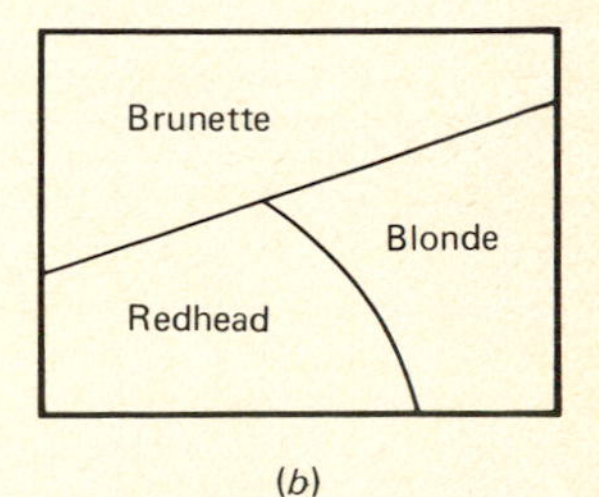

(*b*)

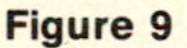

Figure 9

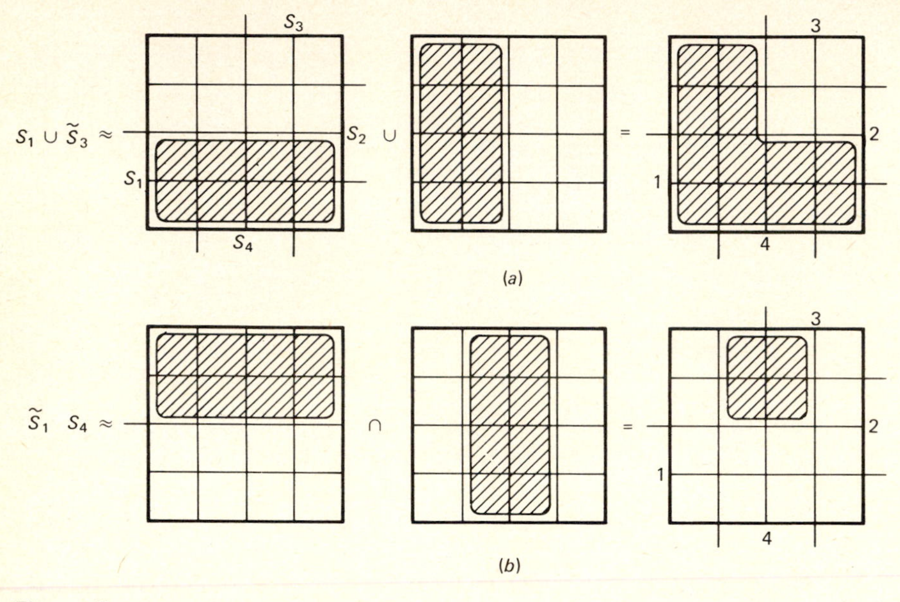

Figure 10

or more mutually exclusive and exhaustive subsets, as shown in Figure 9*b*, which describes the coloring of human hair.

Operations on Set Diagrams The set operations of union, intersection, and complement may be viewed as operations on set diagrams, as the two examples of Figure 10 illustrate. These operations can also be extended, step by step, to any number of sets. Figure 11 illustrates how the representation of a complex expression can be constructed from its simpler parts (which were done separately in Figure 10). After some experience with such diagrammatic manipulations it is often possible to perform all the manipulations in one diagram. This is illustrated in Figure 12

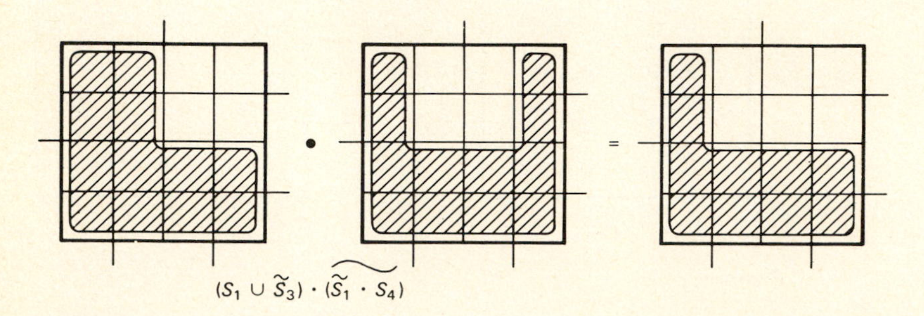

Figure 11

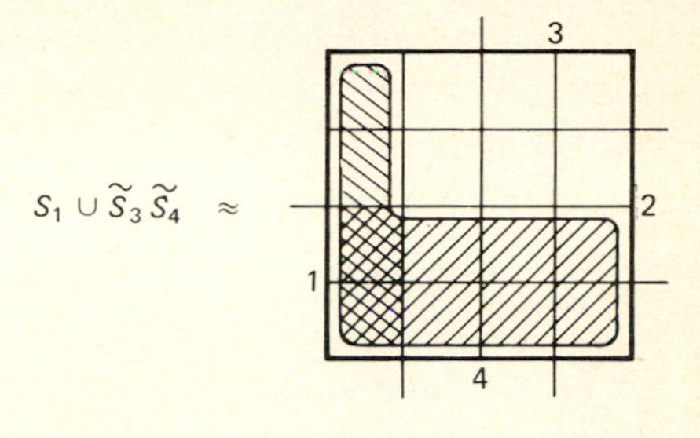

Figure 12

for the simpler set

$$S_1 \cup \tilde{S}_3 \tilde{S}_4$$

Upon comparing the last two set diagrams we see that they are identical. Thus we conclude that the corresponding set algebraic expressions are equal, i.e.,

$$(S_1 \cup \tilde{S}_3)(\tilde{S}_1 S_4) = S_1 \cup \tilde{S}_3 \tilde{S}_4$$

Algebra of Sets By using set diagrams many equalities in the set algebra can be obtained. A summary of some of these is given in Figure 13.

1. Idempotence	$X \cap X = X$	$X \cup X = X$
2. Operation of unit	$U \cap X = X$	$U \cup X = U$
3. Operation of zero	$\phi \cap X = \phi$	$\phi \cup X = X$
4. Complementarity	$X \cap \tilde{X} = \phi$	$X \cup \tilde{X} = U$
5. Commutativity	$X \cap Y = Y \cap X$	$X \cup Y = Y \cup X$
6. Dualization	$\widetilde{X \cap Y} = \tilde{X} \cup \tilde{Y}$	$\widetilde{X \cup Y} = \tilde{X} \cap \tilde{Y}$
7. Absorption	$X \cap (X \cup Y) = X$	$X \cup X \cap Y = X$
8. Associativity	$X \cap (Y \cap Z) = (X \cap Y) \cap Z$	$X \cup (Y \cup Z) = (X \cup Y) \cup Z$
9. Distributivity	$X \cap (Y \cup Z) = X \cap Y \cup X \cap Z$	$X \cup Y \cup Z = (X \cup Y) \cap (X \cup Z)$

Figure 13

On examining these equalities it will be discovered that this set algebra is indeed identical to the switching algebra of the previous chapter. You may have suspected this from similarities much earlier. The two-valued nature of sets comes from the definiteness property that insists that an element must be in a set or not in it. The other correspondences are indicated in Figure 14.

Algebra of sets	$\sim$	$\cap$	$\cup$	ϕ	U
Algebra of switches	—	$\cdot$	v	0	1

Figure 14

Such a similarity of structures is known as an *isomorphism*. The isomorphism between the set algebra and the switching algebra is not only interesting but also very useful.

For example the previous equality of sets

$$(S_1 \cup \tilde{S}_3)(\widetilde{\tilde{S}_1 S_4}) = S_1 \cup \tilde{S}_3 \tilde{S}_4$$

which was determined by set diagrams, can also be determined by algebra:

$$\begin{aligned} (S_1 \cup \tilde{S}_3)(\widetilde{\tilde{S}_1 \cap S_4}) &= (S_1 \cup \tilde{S}_3)(\tilde{\tilde{S}}_1 \cup \tilde{S}_4) && \text{by dualization} \\ &= (S_1 \cup \tilde{S}_3)(S_1 \cup S_4) && \text{by involution} \\ &= S_1 \cup \tilde{S}_3 \tilde{S}_4 && \text{by distributivity} \end{aligned}$$

Thus we have seen how the switching algebra applies to sets. In the next section we will see how sets can apply to switching systems.

EXERCISE SET 1

1. Represent each of the following sets by intension, and then represent them all on one set diagram:

$A = \{1, 3, 5, 7, 9\}$
$B = \{0, 1, 2, 3\}$
$C = \{3, 6, 9\}$
$U = \{0, 1, 2, 3, 4, 5, 6, 7, 8, 9\}$

2. For the sets of Problem 1, determine the following sets:

(*a*) $A \cup B\tilde{C}$ (*c*) $\tilde{A} \cup \tilde{B} \cup C$
(*b*) $(\widetilde{A \cup B})C$ (*d*) $\widetilde{AB} \cup \tilde{B}C$

3. Construct set diagrams to check the following equalities:

(*a*) $X \cup \tilde{X} = U$
(*b*) $\tilde{X} \cup \tilde{Y} = \widetilde{X \cap Y}$
(*c*) $X \cup (Y \cap Z) = (X \cup Y) \cap (X \cup Z)$

2. APPLICATIONS OF SETS

Set Diagrams for Optimizing Switching Networks

Consider the switching network of Figure 15*a*, which is described by the expression

$$N = X_1 X_2 X_3 X_4 \vee \bar{X}_1 X_2 X_4 \vee X_1 X_3 \bar{X}_4 \vee X_1 \bar{X}_3 \vee X_1 \bar{X}_2$$

Because of the isomorphism between the switching algebra and the set algebra this expression can be drawn on a set diagram, as shown in Figure

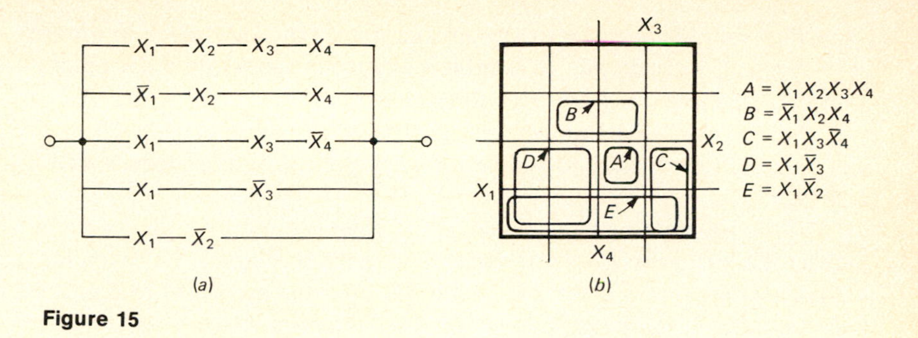

(a) (b)

Figure 15

15*b*. From this example we should observe a very significant fact: *a long term in the algebra corresponds to a small area on the set diagram, and a short term in the algebra corresponds to a large area on the set diagram.* For example, term $X_1X_2X_3X_4$ corresponds to only one block whereas the term X_1 corresponds to eight blocks. (Note also that the number of blocks in every term is some power of 2 such as 1, 2, 4, 8, or 16.)

This observation suggests that networks can be simplified by regrouping blocks to form large areas. One possible regrouping follows in Figure 16, with its much simpler algebraic expression and network. Notice that the shaded areas are represented by 1s.

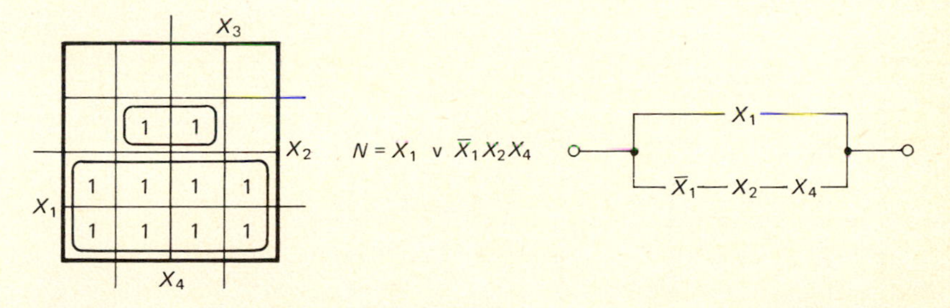

Figure 16

However a second alternative regrouping of two larger areas results in an even simpler expression of Figure 17.

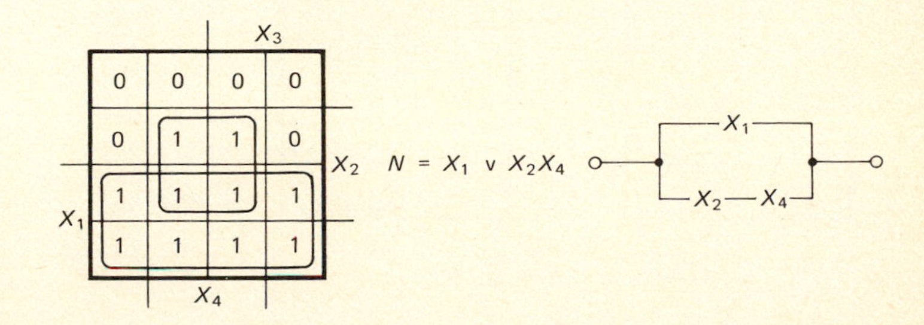

Figure 17

The simplification basically makes use of the fact that the identity $Ax \vee A\bar{x} = A$ corresponds to combining adjacent areas on a set diagram. The two examples of Figure 18 illustrate this.

Notice also that the top and bottom of the diagram as well as the sides should be considered adjacent for simplification purposes as in Figure 19.

If the top and bottom as well as the sides were drawn physically adjacent, the diagram would become a doughnut-shaped figure called a *toroid*, as shown in Figure 20. Since the toroid is awkward to draw, we will continue with the two-dimensional set diagram, always remembering that the edges are also adjacent.

Examples Simplify the expressions corresponding to the given set diagrams of Figure 21. The solutions are given in Figure 22. Each illustrates a useful point. The first example illustrates that further algebraic simplification may be necessary. The second example illustrates the fact that some large blocks ($C\bar{D}$) are not necessary if their areas are already covered by other blocks. The third example shows that blocks of 0s may be more convenient to visualize, leading to the NOT of the required expression.

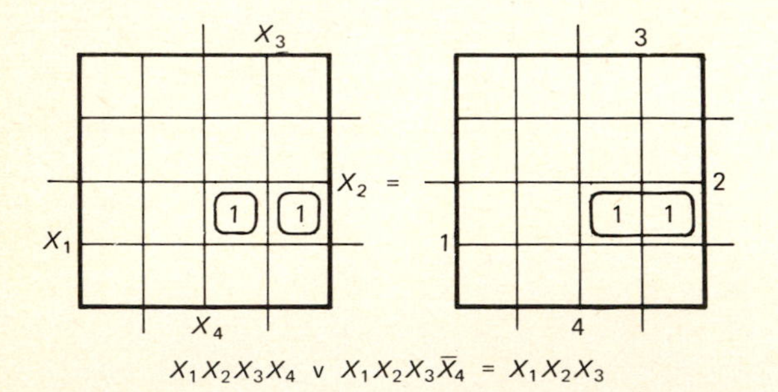

$X_1X_2X_3X_4 \vee X_1X_2X_3\bar{X}_4 = X_1X_2X_3$

(*a*)

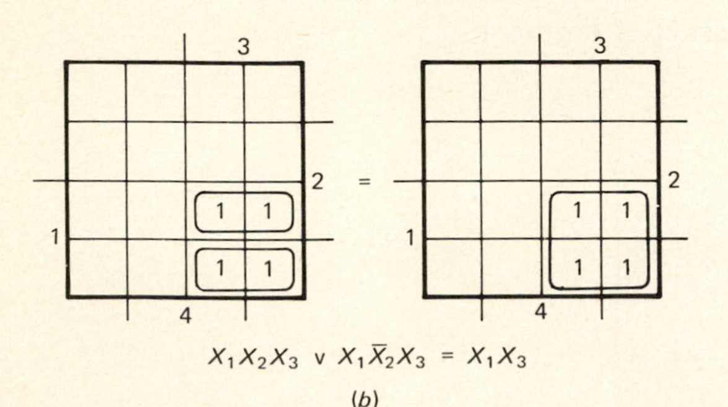

$X_1X_2X_3 \vee X_1\bar{X}_2X_3 = X_1X_3$

(*b*)

Figure 18

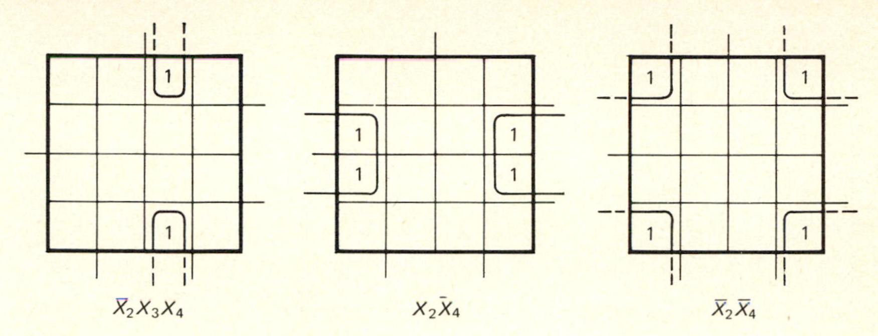

Figure 19

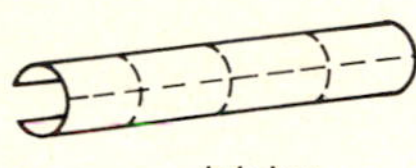
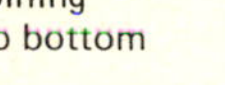

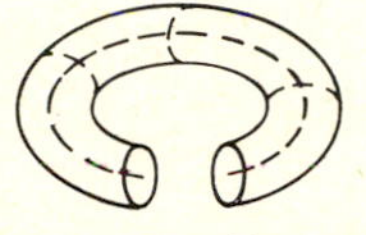

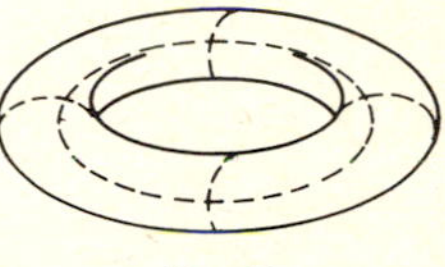
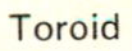

Figure 20

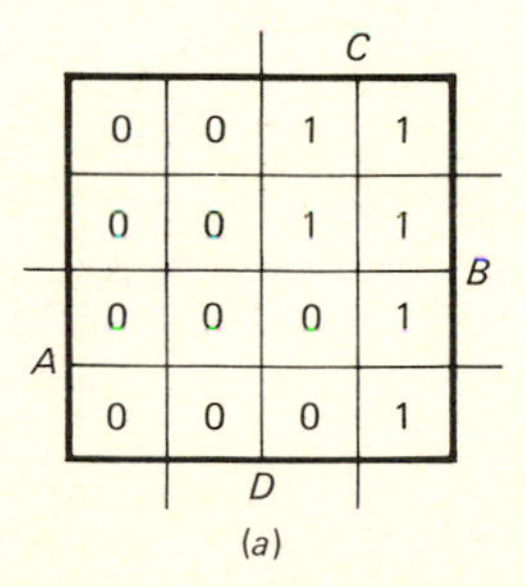

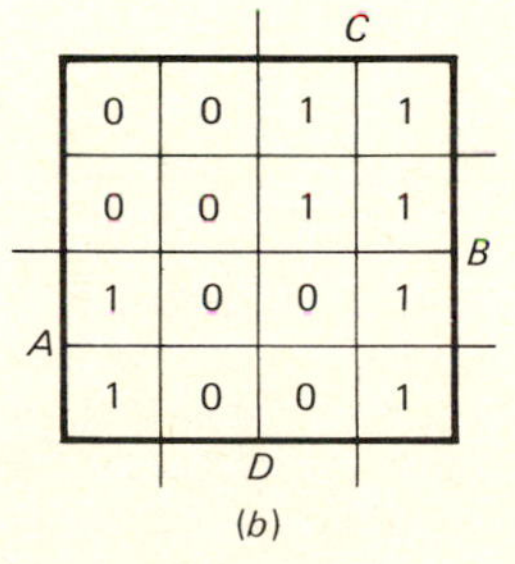

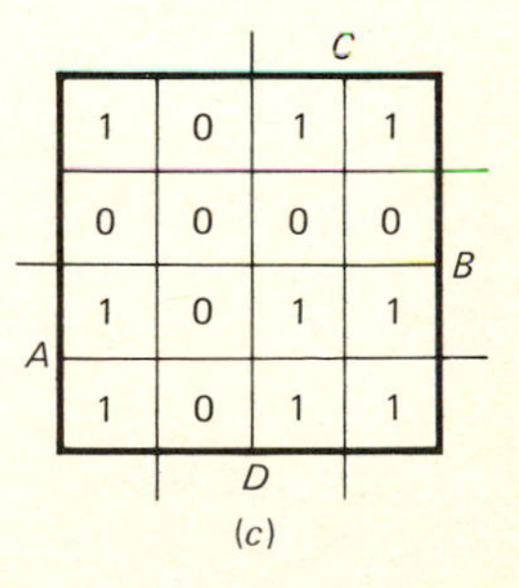

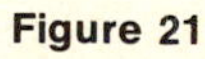
Figure 21

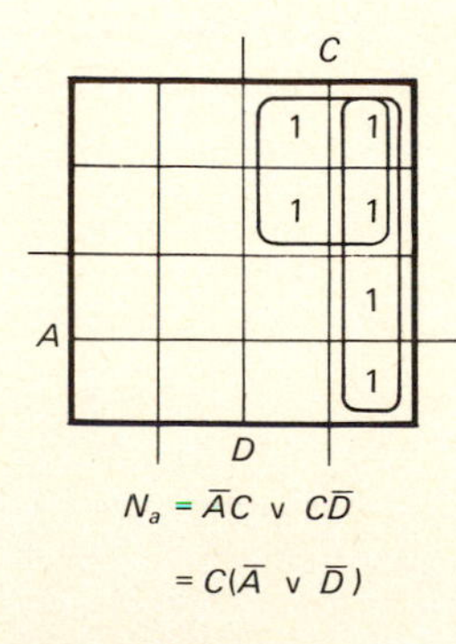

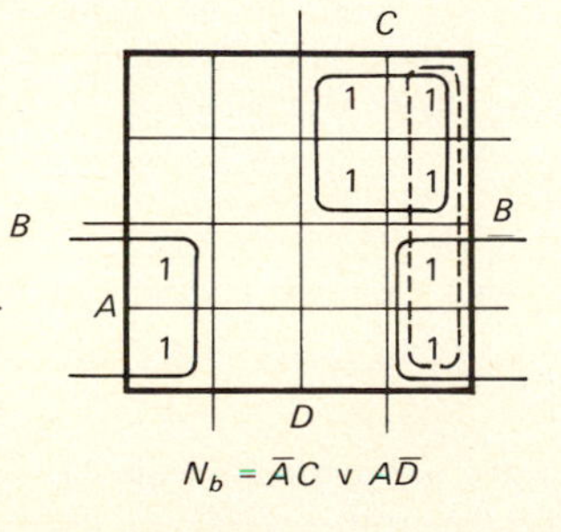

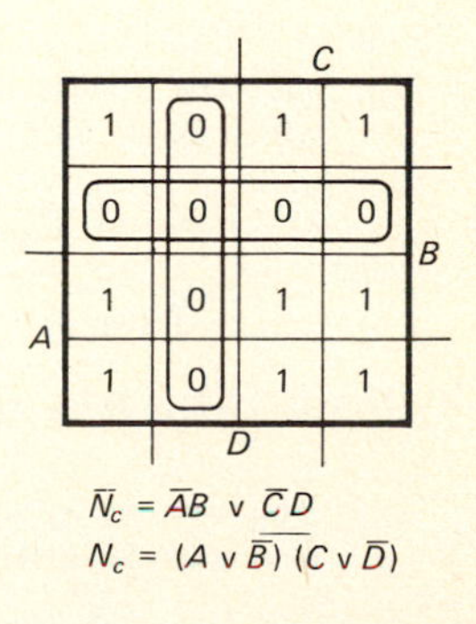

Figure 22

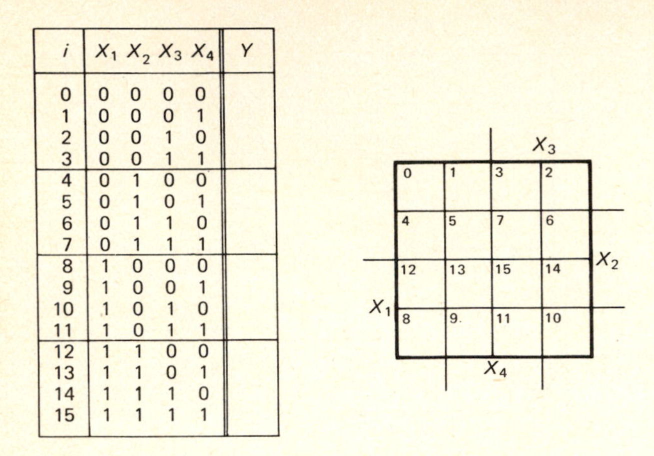

i	X_1	X_2	X_3	X_4	Y
0	0	0	0	0	
1	0	0	0	1	
2	0	0	1	0	
3	0	0	1	1	
4	0	1	0	0	
5	0	1	0	1	
6	0	1	1	0	
7	0	1	1	1	
8	1	0	0	0	
9	1	0	0	1	
10	1	0	1	0	
11	1	0	1	1	
12	1	1	0	0	
13	1	1	0	1	
14	1	1	1	0	
15	1	1	1	1	

Figure 23

Set Diagrams and Tables-of-Combinations. There is a correspondence between areas on a set diagram and rows in a table-of-combinations, as shown in Figure 23. For example the tenth combination 1010 corresponds to $X_1 \bar{X}_2 X_3 \bar{X}_4$, which is represented as the rightmost bottom block on the set diagram. You may wish to check out some other combinations.

This correspondence is convenient for the synthesis and simplification of switching systems, as the following example shows.

Example: Binary Coded Decimal (BCD) Representation. The ten decimal digits 0, 1, 2, . . . 9 can be represented by a binary coded decimal method using four binary variables A, B, C, D. For example the combination $ABCD = 0101$ corresponds to the decimal digit 5, whereas the combination $ABCD = 1101$ does not correspond to any decimal digit. Design a level network which checks whether a combination of four binary digits corresponds to the BCD code; and if not, it indicates that an error has occurred by having an output value of 1. Optimize the number of arrows in the AND, OR, and NOT network.

Solution A table-of-combinations is drawn in Figure 24*a*. The combinations corresponding to digits 10 through 15 have an output value of 1. The output function E, which can be represented by its combinations

$$E = V(10, 11, 12, 13, 14, 15)$$

can be drawn directly onto a set diagram as in Figure 24*b*. It can be simplified to

$$E = AB \vee AC = A(B \vee C)$$

The resulting optimal level network is shown in Figure 24*c*. Notice that input D is not necessary for detecting the error.

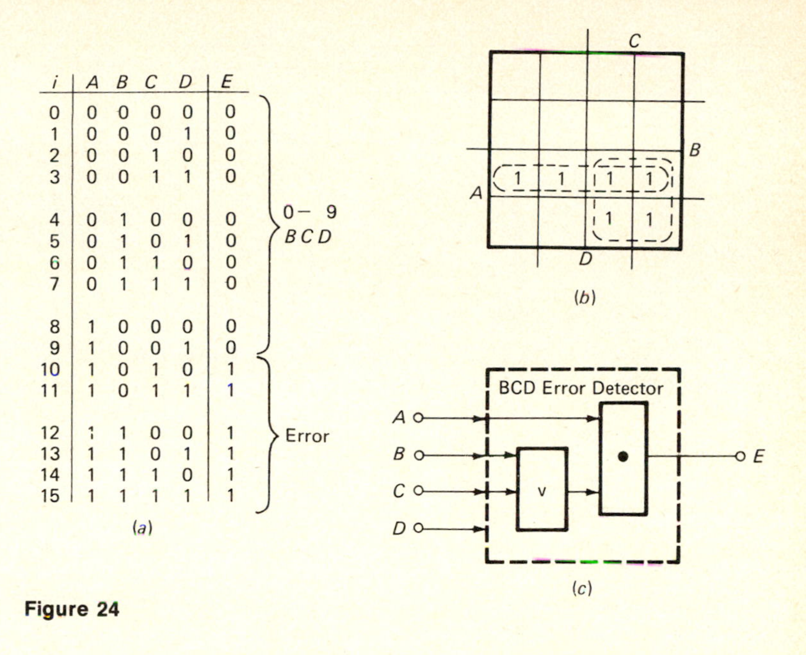

i	A	B	C	D	E
0	0	0	0	0	0
1	0	0	0	1	0
2	0	0	1	0	0
3	0	0	1	1	0
4	0	1	0	0	0
5	0	1	0	1	0
6	0	1	1	0	0
7	0	1	1	1	0
8	1	0	0	0	0
9	1	0	0	1	0
10	1	0	1	0	1
11	1	0	1	1	1
12	1	1	0	0	1
13	1	1	0	1	1
14	1	1	1	0	1
15	1	1	1	1	1

Figure 24

***Optimizing Incompletely Specified Systems.**

Set diagrams are also extremely convenient for assigning don't-care conditions by inspection. A don't-care condition d is chosen as value 1 if it aids in constructing a large block and is chosen as value 0 otherwise.

Consider for example the three-variable system of Figure 25a and its set diagram representation of Figure 25b. This is actually the simple elevator door of Chapter 1, Figure 38. By inspection we can immediately see not only one but two assignments of Figure 25c and d which lead to simple networks. This method is far superior to the previous exhaustive method of algebraically simplifying for each possible condition.

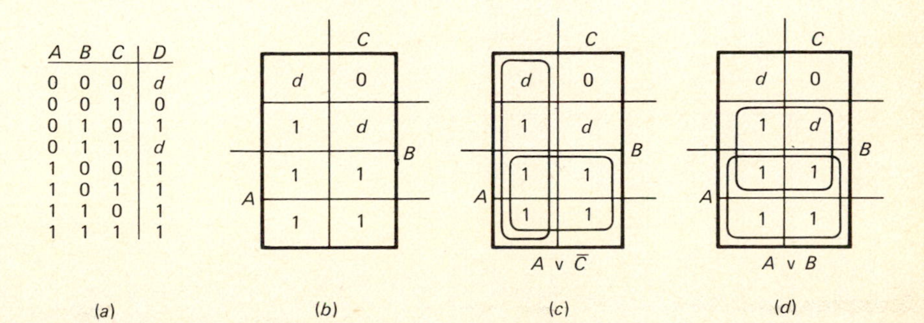

A	B	C	D
0	0	0	d
0	0	1	0
0	1	0	1
0	1	1	d
1	0	0	1
1	0	1	1
1	1	0	1
1	1	1	1

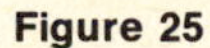

Figure 25

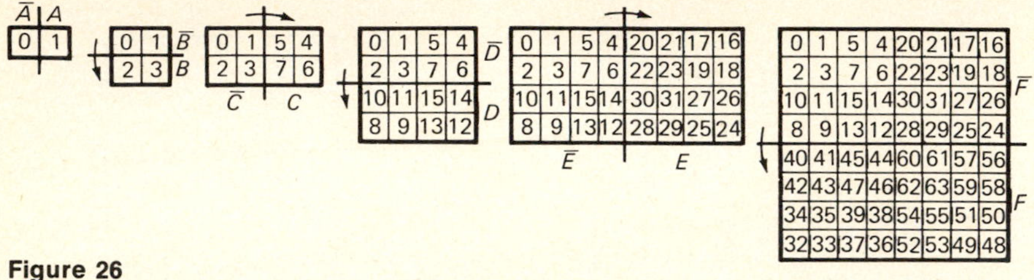

Figure 26

***Extending Set (Karnaugh) Diagrams**

There is a systematic way of constructing two-dimensional set diagrams for any number of variables. The method, indicated in Figure 26, consists simply of unfolding a set diagram alternately about a horizontal and then a vertical axis, with each unfolding creating a reflected mirror image. Even the numbers, which correspond to a table-of-combinations, are generated by this reflecting manner.

An example of optimization is shown in Figure 27. Notice that on the table-of-combinations the least significant digit corresponds to the first variable A, to match the numbering on the set diagram. Optimization

E	D	C	B	A	i	F	E	D	C	B	A	i	F
0	0	0	0	0	0	0	1	0	0	0	0	16	0
0	0	0	0	1	1	0	1	0	0	0	1	17	0
0	0	0	1	0	2	0	1	0	0	1	0	18	0
0	0	0	1	1	3	0	1	0	0	1	1	19	0
0	0	1	0	0	4	0	1	0	1	0	0	20	0
0	0	1	0	1	5	1	1	0	1	0	1	21	1
0	0	1	1	0	6	0	1	0	1	1	0	22	0
0	0	1	1	1	7	1	1	0	1	1	1	23	0
0	1	0	0	0	8	0	1	1	0	0	0	24	0
0	1	0	0	1	9	1	1	1	0	0	1	25	1
0	1	0	1	0	10	0	1	1	0	1	0	26	0
0	1	0	1	1	11	1	1	1	0	1	1	27	1
0	1	1	0	0	12	0	1	1	1	0	0	28	0
0	1	1	0	1	13	1	1	1	1	0	1	29	1
0	1	1	1	0	14	0	1	1	1	1	0	30	0
0	1	1	1	1	15	1	1	1	1	1	1	31	1

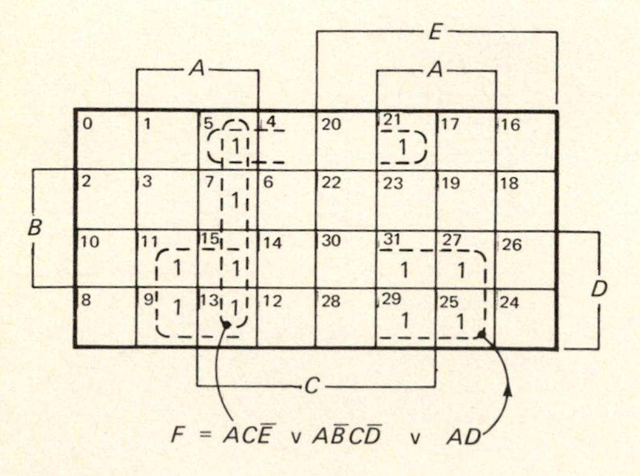

Figure 27

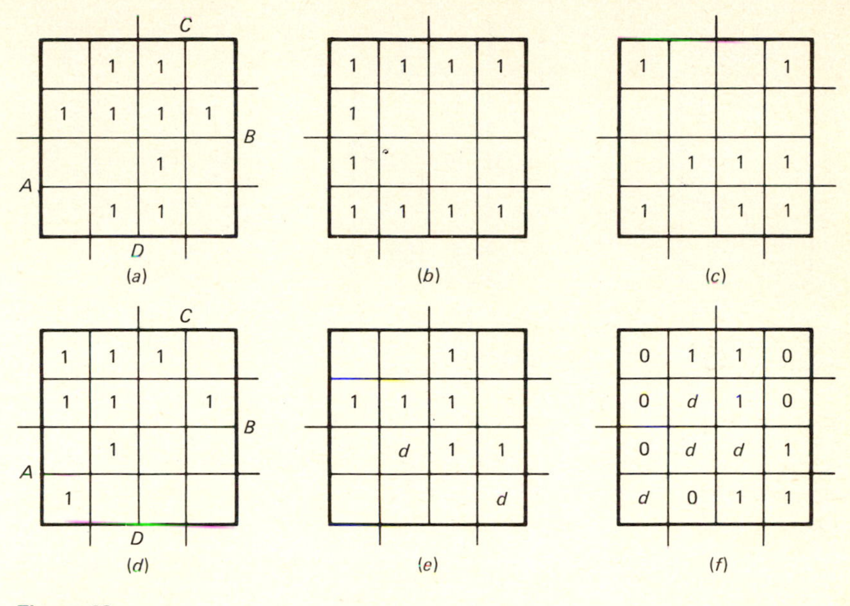

Figure 28

is indicated by the usual adjacencies, as well as any adjacencies which occur upon folding. Thus symmetry about a folding axis is related to optimality. Notice that the result of Figure 27 can be further optimized algebraically.

EXERCISE SET 2

1. Draw set diagrams and optimize:

(a) $A \vee B\bar{C} \vee \bar{A}BC \vee ABCD$ (b) $ABCD \vee \overline{ABCD}$

(c) $(A\bar{B} \vee \bar{A}B)(C\bar{D} \vee \bar{C}D)$ (d) $\bar{B}D \vee B\bar{C}E \vee \bar{A}\bar{B}CD \vee ABC\bar{D}\bar{E}$

2. Optimize the networks corresponding to the diagrams of Figure 28.

3. Show all possible largest blocks which could cover the given set diagram of Figure 29. Select the fewest blocks for the simplest level network. Comment!

4. Synthesize a MAJORITY network having four inputs A, B, C, D. Optimize this network using the set diagram (Karnaugh map). In the case of a tie (no majority) make the most economical choice in terms of the number of springs in a flow-type implementation. How many different optimal assignments can you find?

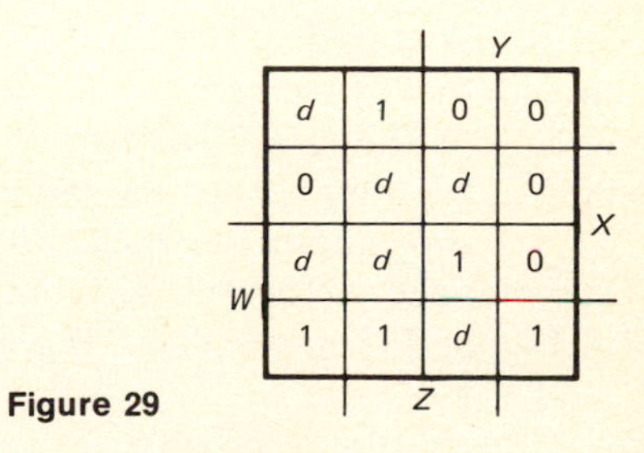

Figure 29

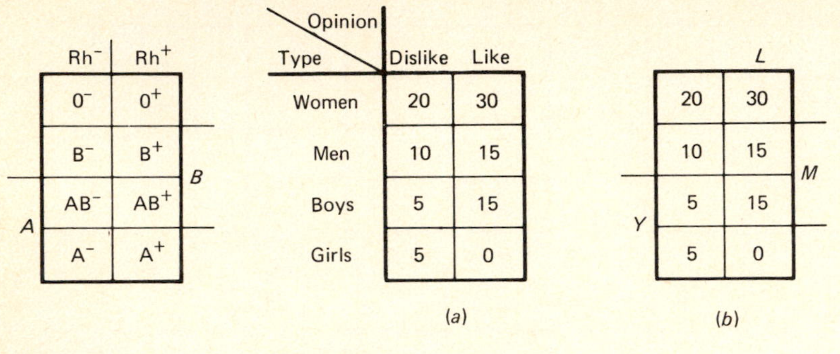

Figure 30 **Figure 31**

Applications of Sets to Classifying, Counting, Checking

This section illustrates through some examples how sets and their diagrammatic representations can be used to classify types of objects, to present data for counting purposes, and to detect inconsistencies in data.

Classifying Blood Human blood is often classified as type A, B, O, or AB depending on which (or none, or both) of the A, B antigens it contains. It is also classified as RH positive or negative depending on the presence or absence of the Rhesus antigen. A transfusion is safe if and only if the recipient's blood has all antigens of the donor's blood (or more). Describe this blood classification in terms of sets.

It is important to distinguish between type A blood and blood having the A antigen, since type A blood has only A (and not B) type antigens. The set diagram of Figure 30 with its eight blood types should clarify any confusion resulting from this labeling convention.

The condition for a safe transfusion in terms of sets is simply

$$D \subseteq R$$

where D is the set of antigens in the blood of the donor and R is the set of antigens in the blood of the recipient. For example, people of type AB^- can receive blood of type A^-, B^-, AB^-, or O^- since the corresponding donor sets $\{\mathrm{A}\}$, $\{\mathrm{B}\}$, $\{\mathrm{A,B}\}$, and ϕ are subsets of the recipient set $R = \{\mathrm{A,B}\}$. Notice also that people having blood type AB^+ are *universal recipients* since all donors are subsets of this recipient set $R = \{A, B, +\}$.

Size of Sets. The *size* (or cardinality) of a set S, denoted $n[S]$, is the number of elements within the set S. In general the size may be infinite, as is the size of the set of all positive integers. But in all our examples the number will be finite and given as some integer.

For example, the results of a survey about a product or program are summarized in Table 31a. Notice that this table is simply the set diagram of Figure 31b in disguise. It was also encountered in Figure 2.

From this set diagram we could compute the sizes of some sets such as:

$$n[Y \cap M] = 5 + 15 = 20$$

$$n[\widetilde{\tilde{Y} \cap \tilde{L}}] = 100 - 30 = 70$$

$$n[\tilde{M} \cap \tilde{Y} \cap L] = 0$$

$$n[Y \cup M] = 100 - 20 - 30 = 50$$

Notice in particular that

$$n[Y \cup M] \neq n[Y] + n[M]$$

that is,

$$50 \neq 25 + 45$$

since the intersection $n[Y \cap M]$ has been included twice. This expression can be changed by subtracting one of these contributions, yielding

$$n[Y \cup M] = n[Y] + n[M] - n[Y \cap M]$$

This leads to the general result

$$n[S_1 \cup S_2] = n[S_1] + n[S_2] - n[S_1 \cap S_2]$$

Using the isomorphism between sets and switching expressions we can express the number of people who are young or else male as

$$n[Y \underline{\vee} M] = 5 + 10 + 15 = 30$$

In general we include no intersections (or subtract both contributions) to get

$$n[S_1 \underline{\vee} S_2] = n[S_1] + n[S_2] - 2n[S_1 \cap S_2]$$

Similarly we could express the number of people that are neither young nor male as

$$n[Y \downarrow M] = n[\widetilde{Y \cup M}] = 20 + 30 = 50$$

Conclusions from Sizes From such a survey we can attempt to draw some conclusions. For example, we may want to find a relation between the opinion and the sex of the surveyed people.

The number of males who liked the program is

$$n[M \cap L] = 15 + 15 = 30$$

Similarly, the number of females who liked the program is

$$n[\tilde{M} \cap L] = 30 + 0 = 30$$

Because the last two numbers are the same, we should not conclude that males and females have the same opinion: this is because more females were surveyed. A better indication of the relation between opinion and sex is given by considering ratios or proportions. For example, the portion of males who liked the program is

$$\frac{[M \cap L]}{n[M]} = \frac{30}{45} = 0.67$$

whereas the portion of females who liked the program is

$$\frac{n[\tilde{M} \cap L]}{n[\tilde{M}]} = \frac{30}{55} = 0.55$$

Thus we conclude that males in general like the program more than females.

We could also inquire about the relation between opinion and age. We would find that opinion is not dependent on age since the same portion of young and old (nonyoung) like the program, i.e.,

$$\frac{n[L \cap Y]}{n[Y]} = 0.6 \qquad \text{and} \qquad \frac{n[L \cap \tilde{Y}]}{n[\tilde{Y}]} = 0.6$$

EXERCISE SET 3

1. A survey indicates that:

40 percent of the people have blood labeled as type A.
10 percent of the people have blood labeled as type B.
5 percent of the people have blood labeled as type AB.

(*a*) What percent of the people have blood with A antigens? *Ans.* 45 percent

(*b*) What percent have type O blood?

2. A universal donor has a blood type which can be given to anyone. What type is this? *Ans.*: 0^-

3. A survey of 1,000 people indicates that 300 people listen to radio station A, 800 people listen to station B, and 400 listen to both. What is the matter with this survey?

4. In a certain country one language can be spoken by 40 percent of the people and another language spoken by 70 percent of the people. What can be said about the percent of the people who speak both languages?

5. The survey of this section could be refined by the introduction of another set

$E = \{x\colon x \text{ reacted extremely}\}$ as shown in Figure 32

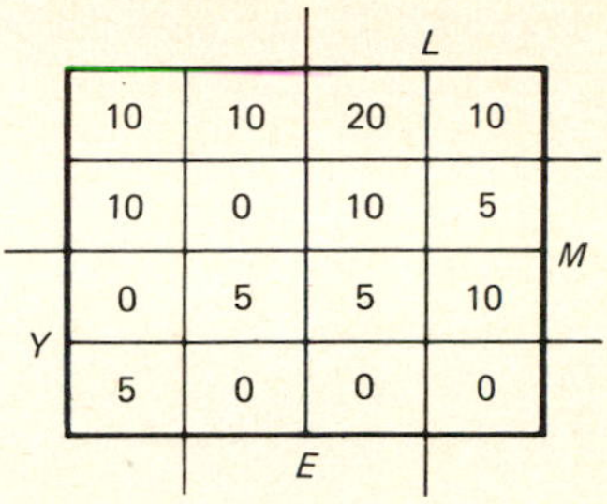

10	10	20	10
10	0	10	5
0	5	5	10
5	0	0	0

L M Y E

Figure 32

This set E (extreme) along with the set L (like) yields four degrees of opinion: excellent, good, fair, and poor corresponding to the set intersections: $L \cap E$, $L \cap \tilde{E}$, $L \cap \tilde{E}$, and $\tilde{L} \cap E$ respectively.

Determine the portion of the population in the following sets. Describe each set in words.

(*a*) $\tilde{M} \cup \tilde{E}L$

(*b*) $(\widetilde{LE}) \cup (\tilde{L} \cup Y)$

(*c*) $(EL) \cup (\widetilde{\tilde{M}L})$

3. *n*-TUPLES (ORDERED ARRANGEMENTS)

Recall that within a set the elements are not listed in any particular order. Thus all the following sets are equal.

$$\{a, b, c\} = \{b, a, c\} = \{c, a, b\} = \{b, c, a\} = \{a, c, b\} = \{c, b, a\}$$

However, in some cases the order of elements may be significant, indicating a progression in size, quality, time, position, or magnitude. In such cases the concept of an *n*-tuple (or vector) will replace the concept of a set.

An *n-tuple* is any ordered listing of *n* elements. It is represented by placing angular brackets around the list, such as

$$\langle c_1, c_2, \ldots, c_n \rangle$$

The ordered elements c_i, which need not be distinct, are called the *coordinates* or *terms* of the *n*-tuple. Thus c_1 is the first term, c_2 is the second, and c_n is the last term. Sometimes 2-tuples are called *ordered pairs* and 3-tuples are called *ordered triplets.*

Example The success (S) or failure (F) of some system on five consecutive days could be given as a 5-tuple such as

$$\langle S, S, F, F, S \rangle \quad \text{or} \quad \langle 1, 1, 0, 0, 1 \rangle$$

Example A vector indicating the position of a point in the usual rectangular coordinates is

$\langle x, y, z \rangle$

Example The decimal number 2001 could be written more formally as

$\langle 2, 0, 0, 1 \rangle$

Example A very common 2-tuple is the *input-output pair*

$\langle x(t), y(t) \rangle$

when x represents the input to a system at one moment t and y represents the corresponding output.

Example An interesting 3-tuple requiring no further explanation is

$\langle 38, 26, 36 \rangle$

The order is most important!

Equality of *n*-tuples

Two *n*-tuples are equal when their corresponding terms are all equal. Symbolically, if $A = \langle a_1, a_2, \ldots, a_n \rangle$ and $B = \langle b_1, b_2, \ldots, b_n \rangle$, then $A = B$ if and only if $a_1 = b_1$, and $a_2 = b_2$, and . . . , $a_n = b_n$. For example, the equality

$$\langle t, P(t), T(t) \rangle = \langle 2, 10, 20 \rangle$$

could indicate that at the second time instant the pressure P has value 10 units and the temperature T has value 20 units.

Also

$$\langle a, b, c \rangle \neq \langle b, a, c \rangle \neq \langle c, a, b \rangle \neq \langle b, c, a \rangle$$

and so on.

Sets of n-tuples and n-tuples of sets. Because of the order of terms and the nondistinctness property, *n*-tuples should be distinguished from sets. However, *n*-tuples may be elements of sets. For example, the outcome of the throw of two dice that yield a sum of 10 can be listed in the set

$$\{\langle 4, 6 \rangle, \langle 5, 5 \rangle, \langle 6, 4 \rangle\}$$

Also, a set may be a term of an *n*-tuple. For example, three sets could be listed in decreasing order of size as follows:

$$\langle \{a, b, d, e\}, \{a, 3\}, \{b\} \rangle$$

Set Product

An important operation on sets is the set product, denoted $S_1 \times S_2$, which must be distinguished from the set intersection $S_1 \cap S_2$.

The *set product* of two sets S_1 and S_2, denoted $S_1 \times S_2$, is the set of all ordered pairs with the first term of this pair taken from set S_1 and the second term from S_2. Formally

$$S_1 \times S_2 = \{ \langle x, y \rangle : x \in S_1 \text{ and } y \in S_2 \}$$

For example, the product of sets $\{0, 1\}$ and $\{0, a, b\}$ is

$$\{0, 1\} \times \{0, a, b\} = \{ \langle 0, 0 \rangle, \langle 0, a \rangle, \langle 0, b \rangle, \langle 1, 0 \rangle, \langle 1, a \rangle, \langle 1, b \rangle \}$$

For convenience in writing, the angular brackets and the commas within the n-tuples can be removed, provided that it is always understood that these combinations are n-tuples. For example,

$$\{0, a, b\} \times \{0, 1\} = \{00, 01, a0, a1, b1, b0\}$$

Comparing the above two examples demonstrates the simplicity of notation and it also shows that in general the product is not commutative, i.e.,

$$S_1 \times S_2 \neq S_2 \times S_1$$

Notice also that the order of the n-tuples within the set is not significant, i.e. n-tuple $b1$ can be interchanged with $b0$ but cannot be commuted to $1b$.

Applications of Set Product

The set product is a convenient way of exhaustively listing or enumerating all combinations of elements from two sets. This exhaustive enumeration will be most useful in the study of probability, as well as in the study of relations.

Example Given the sets

$A = \{\text{large, medium, small}\}$ and

$B = \{\text{male, female}\}$

in the universe of dogs, the set product $A \times B$ enumerates all combinations of sizes and sex of dogs.

$$A \times B = \left\{ \begin{array}{l} \text{large male, medium male, small male,} \\ \text{large female, medium female, small female} \end{array} \right\}$$

Example If the outcomes on the throws of a thick coin are heads, tails, and edge, represented by the set {H, T, E}, then all possible outcomes on two throws of this coin are {H, T, E} × {H, T, E} which can be denoted $\{H, T, E\}^2$, and enumerated as

$$\{H, T, E\} \times \{H, T, E\} = \{HH, HT, HE, TH, TT, TE, EH, ET, EE\}$$

Similarly the set of all outcomes on the throws of two dice is

$$\{1, 2, 3, 4, 5, 6\}^2 = \{1, 2, 3, 4, 5, 6\} \times \{1, 2, 3, 4, 5, 6\}$$

Example The set product $\{0, 1\}^2$ is a concise way of indicating all possible combinations of zeros and ones as might appear in a table-of-combinations for two switching variables, i.e.,

$$\{0, 1\}^2 = \{0, 1\} \times \{0, 1\} = \{00, 01, 10, 11\}$$

Extending the Set Product

Although the set product is most useful when defined on two sets, it may be extended to any number of sets.

The *n*-ary set product of sets $S_1, S_2, \ldots, S_n$ is denoted by $S_1 \times S_2 \times \cdots \times S_n$ (or sometimes as $\mathop{\times}\limits_{i=1}^{n} S_i$) and defined as

$$\mathop{\times}\limits_{i=1}^{n} S_n = \{\langle x_1, x_2, \ldots, x_n \rangle : x_i \in S_i \text{ for } 1 \leq i \leq n\}$$

Example The product $\{0, 1\}^3$ is a concise description of all combinations of three switching variables, i.e.,

$$\{0, 1\}^3 = \{000, 001, 010, 011, 100, 101, 110, 111\}$$

Representation of Set Products

A set product may be represented diagrammatically in many ways: by a point-set diagram, a tree, and a table-of-combinations. These should be self-explanatory from Figure 33, which describes the toss of two thick coins.

The point-set diagram is graphical and most convenient for the product of two sets; the tree can be readily extended to more sets; and the table-of-combinations is most general for any number of sets.

These representations also illustrate the simple counting result

$$n[S_1 \times S_2] = n[S_1] * n[S_2]$$

which relates the number of elements in the product of two sets to the number of elements in each set. In Figure 33 each set has three elements, and so the product set has $3 * 3 = 9$ elements.

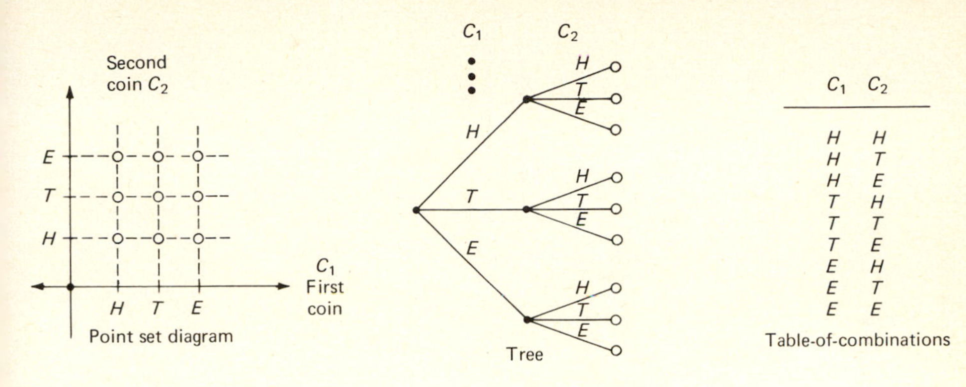

Figure 33

EXERCISE SET 4

1. Compute $\{0, -1, -2\} \times \{0, 2\}$
2. Describe all possible outcomes on the toss of three (thin) coins, four coins, and five coins by the point-set graph, the tree, and the table-of-combinations. Comment.
3. Represent in various ways all possible combinations of sex (male, female), age (young, intermediate, old), and size (large, medium, small).
4. Express the concept of n-tuple in terms of sets.
5. From your own experience, select three meaningful examples of n-tuples and three examples of useful set products.

4. COUNTING AND COMMUNICATING

Combinatorial analysis (or combinatorics) is the study of counting, arranging, selecting, ordering, coding, or enumerating. In this section we will consider some of the simple but fundamental aspects of this theory with a motivation to communication of digital information.

In communication, each message (letter, word, or statement) is often coded as some ordered and distinguishable arrangement of objects, such as flags, voltage levels, or sequences of pulses. Any such ordered distinguishable arrangement, called a *permutation*, corresponds to an n-tuple. For example, the letter A may be coded as

$$A = \langle 1, 1, 0, 0, 1 \rangle$$

and communicated by an ordering of flags, or a combination of lights at one instant, or a sequence of voltage levels, as shown in Figure 34.

Since many systems involve a finite number of messages or symbols (such as the English language with 26 letters and a space) we are interested in counting them for comparing various methods of communication.

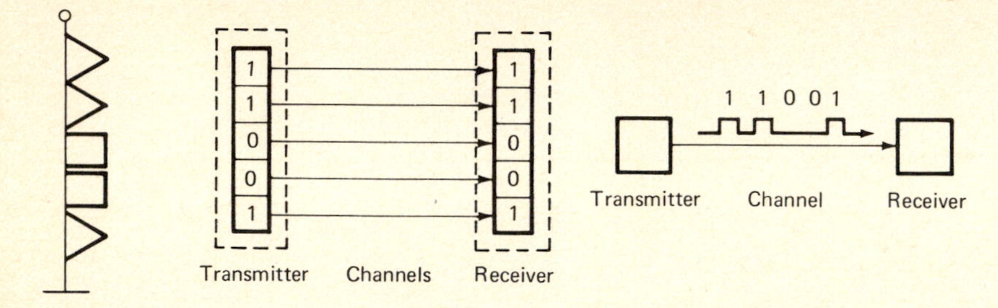

Figure 34

A most fundamental principle in counting such n-tuples (messages) is the *product principle*:

> If there are n_1 ways of performing one operation and then n_2 ways of performing a second operation, there are $n_1 n_2$ ways of performing both operations.

This principle is in essence an application of the product of two sets, since $n[S_1 \times S_2] = n[S_1] * n[S_2]$. The product principle can be extended to any number of operations. Some examples follow.

Example: Communication with Flags There are 6 different colored flags available, but each message consists of only 2 flags showing at any time. How many different messages are possible?

Solution We consider each message as resulting from two operations of choosing flags. For the first position we have 6 choices; after making this choice we have only 5 choices for the second position. Thus, from the product principle, there are $6 * 5 = 30$ messages possible, still sufficient for communicating all 27 symbols of the English alphabet.

Similarly, for 4 distinct flags, with only 3 showing for each message, there are $4 * 3 * 2 = 24$ possible messages (from the extended product principle).

In general, the number of distinguishable ordered arrangements (permutations) of n distinct objects in r positions is

$$D_r^n = (n)(n-1)(n-2) \cdots (n-r+1)$$

For the particular case where $n = r$, this number of distinguishable arrangements of n distinct objects in n positions is

$$D_n^n = (n)(n-1)(n-2) \cdots (3)(2)(1)$$

This expression is usually read as "*n factorial*" and denoted

$$n! = (n)(n-1)(n-2) \cdots (3)(2)(1)$$

Example: Communication with Parallel Channels Given five communication channels, each of which can transmit only two levels, find the number of messages possible.

Solution Electric voltage levels, unlike flags, are "repeatable"; having chosen one of the two values to send on the first communication channel still leaves two possible levels for the second channel. Extending the product principle for five steps yields

$$2 * 2 * 2 * 2 * 2 = 32 \text{ possible messages}$$

which allows communication of the 27-symbol English alphabet

Notice that if a three-valued system is used (with positive, zero, and negative voltages), only three lines are necessary since $3^3 \geq 27$.

Example Communication with Lanterns Given five lanterns, two of which are red and three are green, how many distinguishable combinations (messages) are possible if all five are arranged horizontally?

Solution The total number of ways of arranging the five lanterns is 5!, which is 120. However some of these arrangements are not distinguishable from others. For example, in each of the 120 arrangements, the two red lanterns can be interchanged (permuted) thus decreasing the total by a factor of 2. Similarly the three green lanterns can be permuted (interchanged among themselves) 3! or 6 times, decreasing the number of arrangements by a factor of 6. Thus the total of 120 arrangements must be reduced by both a factor of 2 and 6 yielding only 10 distinguishable arrangements, i.e.,

$$\frac{(2+3)!}{2!3!} = \frac{5!}{2!3!} = \frac{120}{2 * 6} = 10$$

As a check these 10 arrangements (messages) are enumerated below.

RRGGG GRRGG GGRRG GGGRR RGRGG

GRGRG GGRGR RGGRG GRGGR RGGGR

In general, if there are n objects, r_1 of one type T_1, and r_2 of a second type T_2, up to r_k of a type T_k, then the total number of arrangements $n!$ must be reduced by a factor $(r_1)!(r_2)!(r_3)\cdots(r_k)!$ since each object of type T_i can be permuted $(r_i)!$ times. The total number of distinguishable arrangements is

$$\frac{n!}{(r_1)!(r_2)!\cdots(r_k)!} = \frac{(r_1 + r_2 + \cdots + r_k)!}{(r_1)!(r_2)!\cdots(r_k)!}$$

which is abbreviated in either of the following ways:

$$D[r_1, r_2, \ldots, r_k] \quad \text{or} \quad \binom{n}{r_1, r_2, r_3, \ldots, r_k}$$

A particular sort of arrangement or permutation, which is sometimes called a combination, is simply a restricted type of permutation, where the objects are of two types only, say r of one type and consequently $n - r$ of another type. The number of permutations of r such objects from n objects is then

$$D[r, n-r] = \frac{[r + (n-r)]!}{(r)(n-r)!} = \frac{(n)!}{(r)!\,(n-r)!}$$

This is often known as the "number of *combinations* of n things taken r at a time" and is denoted in various ways such as

$$nCr \qquad C_r^n \qquad \binom{n}{r}$$

This quantity can also be interpreted as the number of ways of selecting (rather than arranging in order) r objects from a total of n objects (and rejecting $n - r$ objects).

Counting by Enumeration

Unfortunately many counting problems cannot be solved by the above methods. Often the only method of solution is by complete enumeration, listing all possible arrangements.

Example A communication system used at night has a transmitter consisting of four identical lights strung equally spaced vertically on a pole. The receiver, on some moving vehicle, distinguishes only the presence or absence of the light; he is unable to tell the absolute distance between lights (not knowing his distance from the lights). List all messages that can be communicated with this system.

Solution All 2^4 or 16 combinations of lights may be listed on a table-of-combinations as in Figure 35. Each combination is either a new distinguishable message or is indistinguishable from the previously listed messages. All the distinguishable arrangements are shown in Figure 36. Notice that combinations 3, 5, and 9 are indistinguishable because of the physical constraints or restrictions. There are only six distinguishable arrangements from the 16 possible combinations; not including combination 0 with no lights. In some cases this could be a message, but it could also be the absence of any message.

Lights \ Combinations	0	1	2	3	4	5	6	7	8	9	10	11	12	13	14	15
Highest	0	0	0	0	0	0	0	0	1	1	1	1	1	1	1	1
Second	0	0	0	0	1	1	1	1	0	0	0	0	1	1	1	1
Third	0	0	1	1	0	0	1	1	0	0	1	1	0	0	1	1
Lowest	0	1	0	1	0	1	0	1	0	1	0	1	0	1	0	1

Figure 35

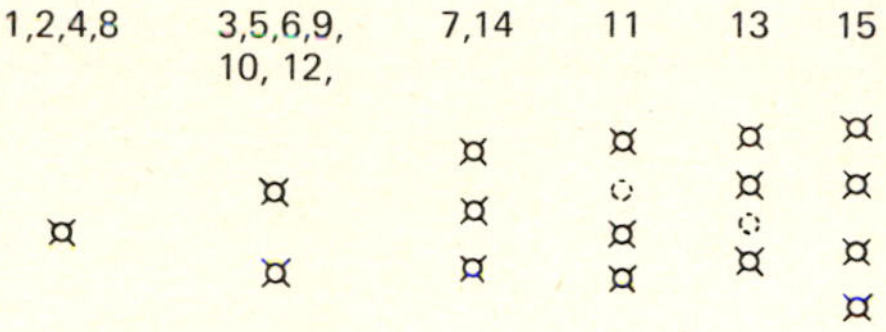

1,2,4,8 3,5,6,9, 10, 12, 7,14 11 13 15

Figure 36

c b a	Sets
0 0 0	ϕ
0 0 1	$\{a\}$
0 1 0	$\{b\}$
0 1 1	$\{b, a\}$
1 0 0	$\{c\}$
1 0 1	$\{c, a\}$
1 1 0	$\{c, b\}$
1 1 1	$\{c, b, a\}$

Figure 37

In a similar way we can enumerate all the subsets of a set $\{a, b, c\}$. This can be done systematically by first constructing a table-of-combinations as in Figure 37. The columns are labeled by the elements of the set, and the rows indicate which of these elements are to be accepted (value 1) or rejected (value 0) to make each subset. Thus the table gives us all 2^3 ways of selecting and rejecting the elements of the set to make subsets as shown at the right of Figure 37.

The set of all subsets of a set is often called the *power set* and is defined

$$\Pi(S) = \{x \colon x \subseteq S\}$$

in this case

$$\Pi[\{a, b, c\}] = \{\phi, \{a\}, \{b\}, \{c\}, \{a, b\}, \{b, c\}, \{a, c\}, \{a, b, c\}\}$$

This procedure shows that for a set having n elements there are 2^n subsets. It further shows how to generate all these subsets in a systematic manner.

EXERCISE SET 5

1. An automobile license number could consist of

(*a*) One alphabetic letter followed by five integers or

(*b*) Two alphabetic letters followed by three integers or

(*c*) Three alphabetic letters followed by one integer.

Compare the total number of licenses possible with each method. What nonmathematical aspects should be considered before adopting one of these methods?

2. Given four differently colored flags, any number of which can be showing on a flagpole at one time, find the total number of messages possible if spacing is ignored. *Ans.*: 64

3. What is wrong with the following solution to Problem 2? In the first position there are five possibilities (four flags or none), in the second there are four possibilities, etc., so by the product principle

$$5 * 4 * 3 * 2 = 120 \text{ messages}$$

4. Consider the previous lantern problem with two red lanterns and three green ones. If this arrangement can be approached from two directions, i.e., if strung over a road, how many distinguishable messages are possible now?

5. Consider a previous nighttime communication system with five instead of four lights mounted on a pole. How many messages are possible?

6. If a nighttime communication system consists of four identical lights arranged in a square, how many messages are possible?

5. OPTIMIZING COMMUNICATION

In this section we will consider a system which illustrates most of the significant concepts of this chapter, including sets, counting, and communication. It will also introduce "crummy," or *nonideal*, systems and a way of improving them.

Consider a two-valued communication system having three lines or channels, as in Figure 38. It may be used to communicate 2^3 or 8 messages. Unfortunately the channels may be crummy; i.e., occasionally a channel fails to transmit a proper value, as in Figure 38, where a value of 1 has been received as a 0.

One method of detecting such errors is illustrated in Figure 39. A *parity encoder* labeled *E* monitors all the lines at the transmitter, and it transmits a 1 only if there is an odd number of 1s sampled. Thus the total number of 1s sent from the transmitter (on the four lines) is always even. At the receiver the total number of 1s should then also be even. A second *parity-checking network* labeled *F* simply checks whether the even number of 1s has been received.

If, however, an odd number of 1s is received, this network indicates a value 1 which could in turn sound an alarm or cause the message to be repeated.

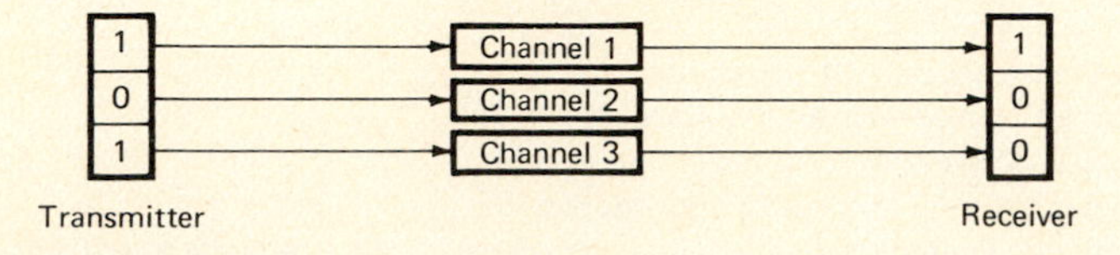

Figure 38

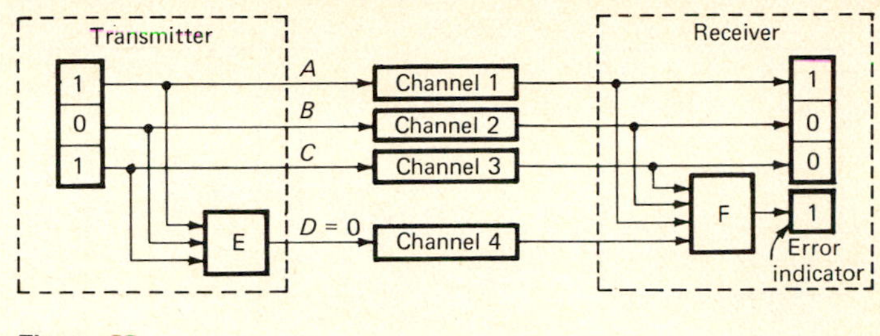

Figure 39

The two switching networks at the transmitter and receiver could be synthesized and optimized by either algebraic or set-diagrammatic methods.

The parity encoder at the transmitting end can be synthesized from Figure 40a and optimized algebraically as follows:

$$
\begin{aligned}
N &= \bar{A}\bar{B}C \vee \bar{A}B\bar{C} \vee A\bar{B}\bar{C} \vee ABC \\
&= \bar{B}(\bar{A}C \vee A\bar{C}) \vee B(\bar{A}\bar{C} \vee AC) \\
&= \bar{B}(A \,\underline{\vee}\, C) \vee B(\overline{A \,\underline{\vee}\, C}) \\
&= B \,\underline{\vee}\, (A \,\underline{\vee}\, C) \\
&= A \,\underline{\vee}\, (B \,\underline{\vee}\, C)
\end{aligned}
$$

The resulting network is shown in Figure 40c.

Similarly the table-of-combinations, set diagram, and optimized network could be obtained for the parity-checking network at the receiver. The set diagram and resulting optimal network are given in Figure 41.

Notice that although the set diagrams show no adjacent terms, considerable optimizing was still possible! This illustrates one of the serious

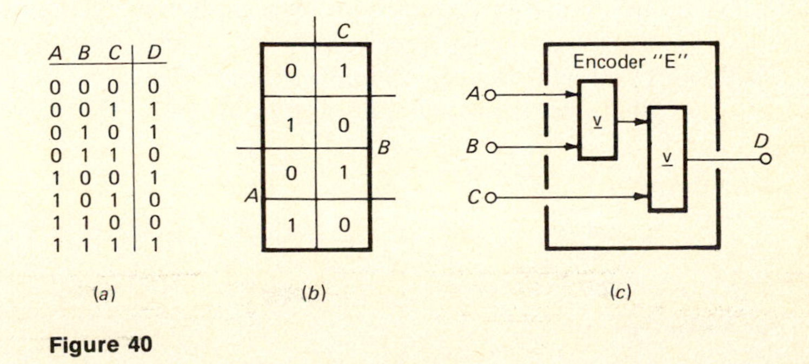

A	B	C	D
0	0	0	0
0	0	1	1
0	1	0	1
0	1	1	0
1	0	0	1
1	0	1	0
1	1	0	0
1	1	1	1

(a) (b) (c)

Figure 40

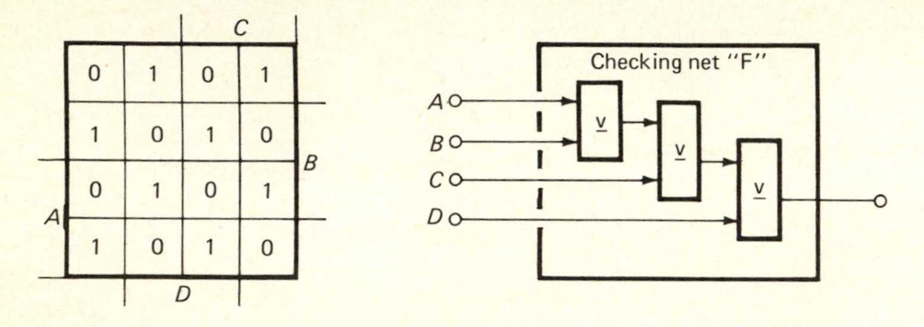

Figure 41

shortcomings of the set-diagrammatic optimization; it may be convenient for AND, OR, and NOT realizations but not for others such as this EXCLUSIVE-OR realization.

It should also be understood that this new communication system will not detect all errors. For example, if errors occur at the same instant on two channels, they are not detected since an even number of 1s is still received. However, since double errors usually occur much less frequently than single errors, such systems are still useful. A more detailed probabilistic analysis of this system will follow in Chapter 4.

This system can be extended equally well to any number of channels simply by increasing the number of EXCLUSIVE-OR elements in the encoding and decoding (or checking) networks.

SUMMARY

This has been a short introduction to sets and n-tuples with an emphasis on finite or digital systems, especially in communication.

Sets were found to have an algebra isomorphic with the previous switching algebra. This resulted in a set diagram (Karnaugh) representation which is very convenient for optimizing systems.

n-tuples were introduced when the order of elements in a list was significant. Counting of the number of ordered arrangements often corresponded to counting the number of messages for communication.

Communication has been briefly introduced, both in the ideal case with flags and lights and in a nonideal case with channels.

All these concepts will continue to be useful throughout this book.

PROBLEMS

1. For the given data

$$n[X] = 300 \qquad n[Y] = 400 \qquad n[XY] = 100 \qquad n[U] = 1{,}000$$

determine the following:

(*a*) $n[X \cup Y]$ (*c*) $n[X\tilde{Y} \cup \tilde{X}Y]$

(*b*) $n[XY \cup X\tilde{Y}]$ (*d*) $n[\widetilde{X \cup \tilde{Y}}]$

2. In a certain school of 1,000 students there are

600 students enrolled in engineering courses
450 students enrolled in mathematics courses
300 students enrolled in physics courses
300 students in both engineering and mathematics
150 students in both engineering and physics
200 students in both physics and mathematics
100 students in all three areas

How many students are not enrolled in any of the three areas of engineering, mathematics, or physics? *Ans.*: 200

3. It is given that

$n[AB]=50 \quad n[A]=50 \quad n[AC]=40$
$n[C]=40 \quad n[B]=60 \quad n[\tilde{A}\cdot\tilde{B}\cdot\tilde{C}]=50$

Determine $n[U]$.

4. In a magazine survey it was found that:

30 percent read magazine *A*.
30 percent read magazine *B*.
40 percent read magazine *C*.
20 percent read magazine *B* only.
10 percent read all three.
20 percent read *A* and *C*.

What percent did not read any of the three magazines? *Ans.*: 30%

5. Three switches are observed to behave in the following manner:

Switch *A* is closed 66 percent of the time.
Switch *B* is closed 83 percent of the time.
Switches *A* and *B* are both closed together for 55 percent of the time.
Switch *A* alone is closed 1 percent of the time.
Switches *A* and *B* are open together for 6 percent of the time.
Switches *B* and *C* are both closed together 75 percent of the time.
Switch *A* is closed when switch *C* is open for 6 percent of the time.
Switch *C* is closed when switch *A* is open for 30 percent of the time.

Find the percent of the time that none of the switches are closed. *Ans.*: 1%

6. Suppose an insurance company has acquired the following data (by surveying 1000 people) in the hope of relating accident-proneness *A* to sex and the habits of smoking *S* and drinking *D*.

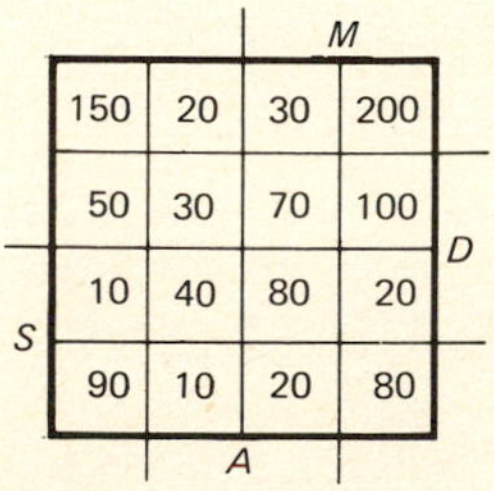

M

150	20	30	200
50	30	70	100
10	40	80	20
90	10	20	80

D, S, A

$S = \{x:$ x smokes (every day)$\}$
$D = \{x:$ x drinks (every day)$\}$
$M = \{x:$ x is a male$\}$
$A = \{x:$ x is accident prone involved in more than 3 accidents in the last 2 years$\}$

Figure P-6

(*a*) Are males more accident-prone than females?
(*b*) Are smoking males more accident-prone than drinking females?
(*c*) What portion of the smokers drink? *Ans.*: 15/35
(*d*) What portion of the drinkers smoke? *Ans.*: 15/40
(*e*) What portion of the (surveyed) population are accident-prone?
(*f*) What portion of those who smoke and drink are accident-prone?
(*g*) What portion of those who are accident-prone drink or smoke?
(*h*) Comment on the accident-proneness of drinking men as compared to drinking women.
(*i*) Of the eight possible combinations of sex, drinking, and smoking which combination is most accident-prone?
(*j*) What are some of the conclusions you can draw from such a survey?

7. Consider the following 10-year study performed in a retirement home to relate the effect of diet on the health of individuals. Four hundred individuals were given a special diet whereas 600 were given a normal diet.

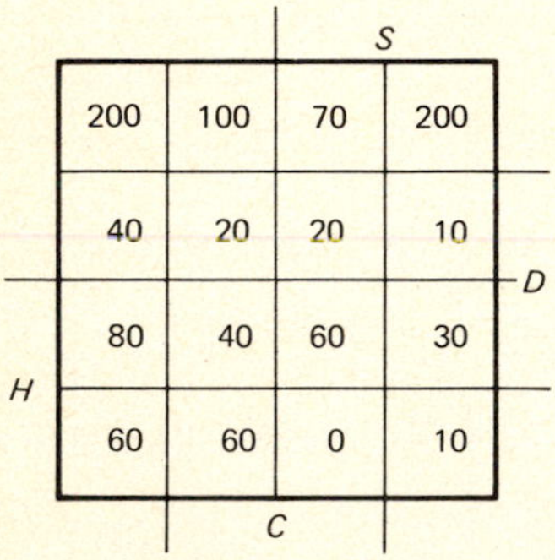

200	100	70	200
40	20	20	10
80	40	60	30
60	60	0	10

S, D, H, C

$C = \{x: x$ has cancer$\}$

$D = \{x: x$ dies (in the 10 year period)$\}$

$H = \{x: x$ suffers heart attacks$\}$

$S = \{x: x$ is on a special diet$\}$

Figure P-7

(*a*) What portion of those on the diet suffered heart attacks?
(*b*) What portion of those not on the diet suffered heart attacks?
(*c*) What portion of those on the diet died having cancer?
(*d*) What portion of those not on the diet died having cancer?
(*e*) Did the diet influence the death rate? *Ans.*: no
(*f*) What can be said about those who were on the diet and had both heart attacks and cancer?
(*g*) What other conclusions and interpretations can you find?

8. The temperature in five areas (of a city, or a lake, or a building) is sampled each day and indicated as being high (value 1) or low (value 0), with the corresponding percentages indicated on the given set diagram.

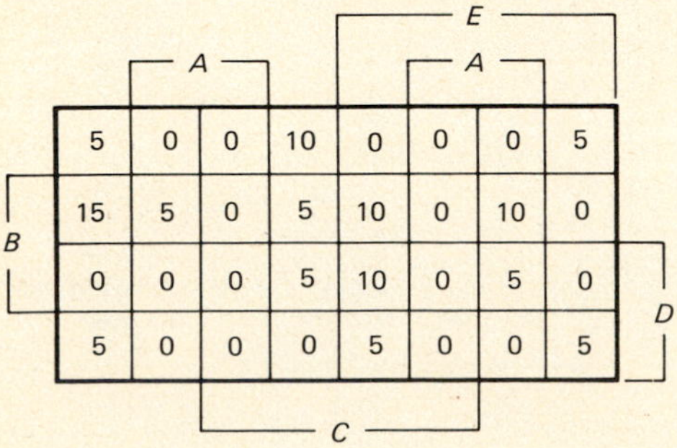

5	0	0	10	0	0	0	5
15	5	0	5	10	0	10	0
0	0	0	5	10	0	5	0
5	0	0	0	5	0	0	5

E, A, A, B, D, C

Figure P-8

(*a*) Which area is the warmest (for the largest portion of time)?
(*b*) What percent of the time is the temperature low in all areas?
(*c*) When the temperature at A is high, what can be said about the temperature at C?
Ans.: It is always low.
(*d*) If the temperature at B is low, what can be said about the temperature at D?
(*e*) If the temperature at A is high and at B is low, what can we say about the temperature at D? How often does this occur?
(*f*) If the temperature at D is high, what percent of the time is the temperature at B also high?
(*g*) What percent of the time does exactly one area have high temperature?

9. Construct a survey which concludes that the minority of beautiful women are able but the majority of able women are beautiful.

10. Enumerate systematically all 16 possible binary operators corresponding to the given table. Identify the AND, OR, NOT, NOR, and NAND. How many binary operators of three variables (like the MAJORITY) are possible? *Ans.*: 256

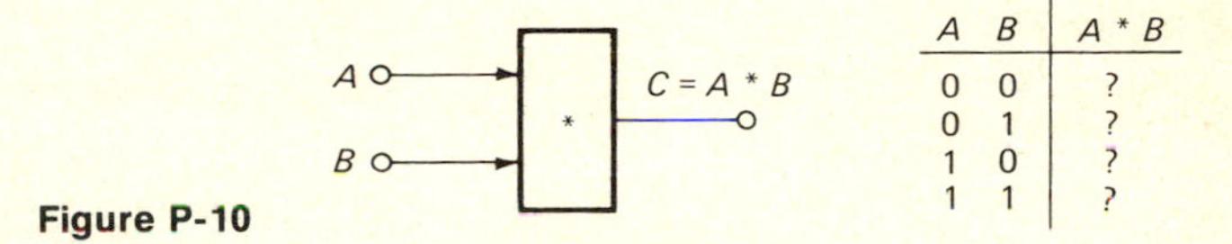

Figure P-10

11. In a three-valued number system, if only four positions are used, then 81 numbers can be represented in some weighted positional notation

$$(C_3, C_2, C_1, C_0)_b = C_0 b^0 + C_1 b^1 + C_2 b^2 + C_3 b^3$$

What is the range of these 81 numbers if:
(*a*) $C_i \in \{0, 1, 2\}$ and $b = 3$
(*b*) $C_i \in \{-1, 0, +1\}$ and $b = 3$ *Ans.*: $-39 \leq x \leq +39$
(*c*) $C_i \in \{0, 1, 2\}$ and $b = -3$
(*d*) $C_i \in \{-1, 0, +1\}$ and $b = -\frac{1}{3}$

12. Suppose you had six flags but only three different colors to dye them. How could you color the flags to obtain the maximum number of messages if all six flags must be showing at any time? What result do you think holds in the general case of m flags and n colors?

***13.** A parallel communication system consists of c channels each transmitting n values (which correspond to n discrete voltage levels) for a total of $m = n^c$ messages. If the cost of such a system is proportional to the product of c and n, show that a three-valued system has minimal cost. (A knowledge of calculus is needed here.)

14. On a set diagram (Karnaugh map) draw and optimize the expressions for:
(*a*) $\bar{A}\bar{B}\bar{C}\bar{D}\bar{E} \vee A\bar{B}CD \vee AD$
(*b*) $V(3, 6, 7, 8, 11, 12, 13, 14, 15, 22, 24, 30)$
(*c*) $V(0, 5, 7, 8, 12, 13, 15, 16, 21, 23, 24, 28, 29, 31)$

15. Many of the problems of Chapter 1 can be optimized more easily by the map methods of this chapter. You may wish to go back and redo some of the problems that caused you difficulty. Redo also the following problems:
(*a*) Figure 37
(*b*) Figure $P14$
(*c*) Problem 4, Exercise Set 5.

16. Design a control circuit for a vehicle which is to follow a line along a roadway. The line is straddled by three photocells, A, B, C, which have an output of 1 if they detect the line. If the line (which may be detected by at most two photocells) is not at the center of the photocells, a signal is sent indicating which direction the vehicle is to turn (right R or left L).

(*a*) Construct a table-of-combinations for this three-input, two-output network indicating reasons for the nonobvious combinations.

(*b*) Construct a simple level network to serve this control purpose.

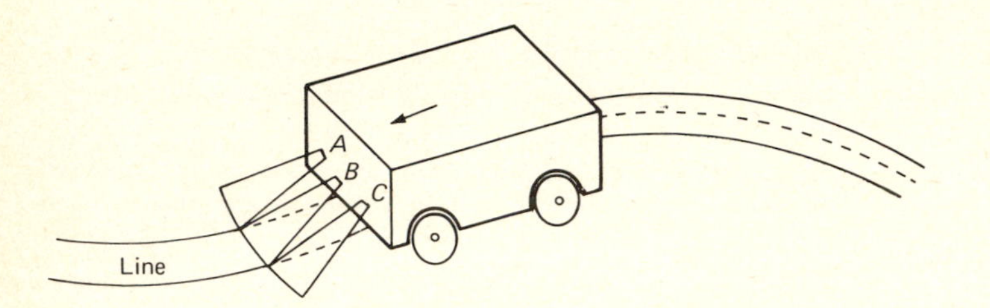

Figure P-16

17. The signal light at a busy intersection is to be controlled by the traffic at that intersection to relieve and avoid congestion. On each of the four lanes (northbound, southbound, eastbound, and westbound) a device senses the traffic and indicates 0 for light traffic and 1 for heavy traffic. Synthesize a system with the above four directional sensors N, S, E, W as inputs. This system then controls a timer which extends the green-light duration in the east-west direction when it receives a signal of value 1 and decreases this duration when it receives a signal of value 0. The speed limit in the east-west direction is 40, whereas in the north-south direction it is 30.

(*a*) Construct a table-of-combinations, indicating reasons for any arbitrary choices.

(*b*) Optimize by the map method. (Try for a complexity of eight springs, or seven arrows, or four level components.)

(*c*) Comment on your design, assumptions, further problems, implementation, etc.

18. Another method of optimizing communication is to use *redundancy*, as in the figure. The same signal is transmitted over five channels, and at the receiver the majority of the five received signals is assumed correct. Represent the MAJORITY network on a Karnaugh map and optimize it.

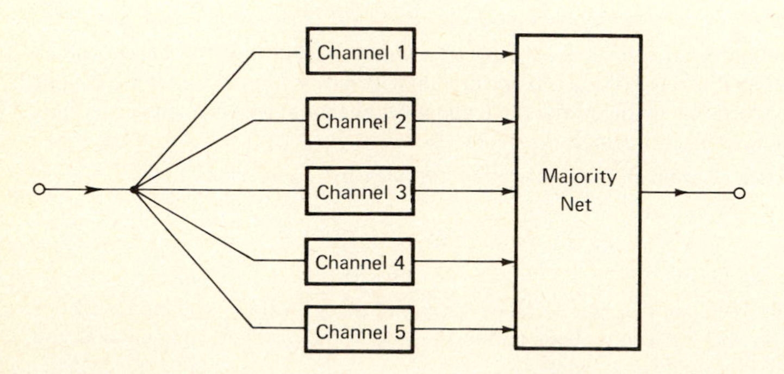

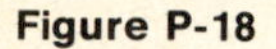

Figure P-18

19. *Project.* Conduct a survey involving three or four binary variables (age, sex, marital status, overweight, grades, opinions, activities, habits, etc.) and attempt to find relations between these variables using set diagrams and other methods of this chapter. Later, when studying probabilistic systems, you can analyze this further.

20. *Project.* Investigate the isomorphism between the set algebra (or switching algebra) and the symbolic logic (having values true or false, and connectives of conjunction, disjunction, negation, implication, and biconditional).

Chapter 3

Interaction

Relations, Functions, Algorithms

Systems have been defined as sets of objects which are related by some interaction. We have already considered sets; now we will consider relations. This will be a brief introduction to relations and may appear somewhat more abstract than the previous chapters. Perhaps it is best read quickly, then returned to at later times.

1. RELATIONS

The concept of relation occurs frequently in everyday experience. It indicates some sort of connection, dependence, association, correlation, or correspondence between things. Some examples of common relations follow:

1. Number x is less than y.
2. Man x is skilled at task y.
3. Set x is a subset of set y.
4. Transmitted signal x was received as signal y.
5. Student x is registered for course y.

6. Freeway x connects to freeway y.
7. At time x the value is y.
8. Town x is to the north of town y.
9. Event x is more likely to occur than event y.
10. Man x lives within 10 miles of y.
11. Function x is the derivative of function y.
12. Output x follows input y.
13. Woman x is the mother of person y.
14. Terminal x is electrically connected to terminal y.
15. Person x loves person y.

From these examples it is seen that relations are very general and that a theory of relations could be useful. Such intuitive relations will now be formalized and abstracted to more convenient and powerful forms.

All the above relations have much in common. First, they all have the same symbolic form "xRy," where R represents such statements as "is less than" or "loves" or "is registered in." The elements x, y are from two sets, say S_1 and S_2 (in some cases $S_1 = S_2$). Such relations between two sets are called *binary* relations. Notice that the correspondence between elements in the two sets is directed. This direction, or order, must be respected since, for example, x may love y but y may not love x.

To emphasize the two sets and the direction, a relation is often symbolized by R: $S_1 \rightarrow S_2$ instead of just simply R. This is read as "relation R, from S_1 to S_2".

Representation of Relations

Representation of a relation as a sagittal diagram. It may be useful or convenient to represent a relation diagrammatically. One method indicates the two sets by closed curves surrounding the elements and indicates the relation by an arrow joining the elements of the two sets. This representation is known as a *sagittal diagram* or sometimes as a bipartite graph. A formal example is shown in Figure 1, illustrating that not all elements of set S_1 have a corresponding element in set S_2 and that some elements of S_1 may have many corresponding elements in set S_2. Some less

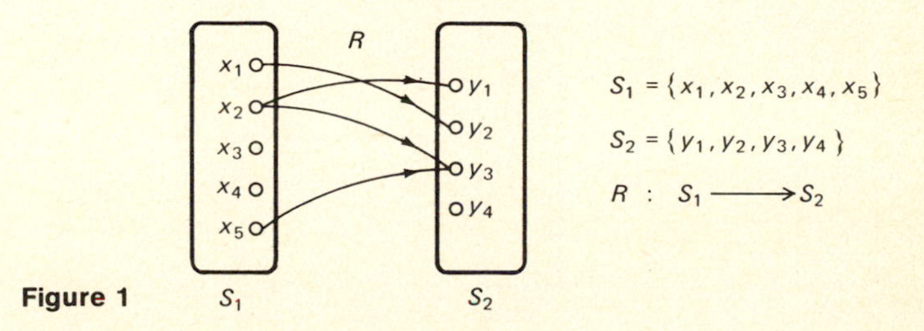

Figure 1

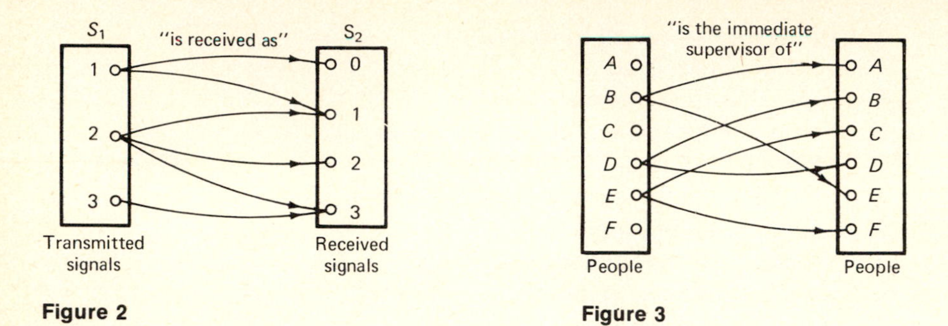

Figure 2

Figure 3

formal relations follow, in their sagittal representation. They will be used as running examples to illustrate other representations.

A communication channel is described in Figure 2. The three values of S_1 could represent three transmitted voltage levels (or possibly a space, dot, and dash respectively). The values of S_2 represent received signals. The relation "is received as" was determined by some experiments on the channel. For example, it was observed that the transmitted value of 1 is received as either 0 or 1. Such a system is often described as nondeterministic, nonideal, noisy, lousy, lossy, or crummy. Despite its slangy sound, the word *crummy* has become an accepted description of such systems.

The second running example, in Figure 3, describes the organization between people of one set, that is, $S_1 = S_2$. If x is the immediate supervisor of y, an arrow is drawn from x to y. Notice that A, C, and F are not supervisors of anyone. Who is the ultimate supervisor over everyone? This may not be too easy to see in this representation, but it will be more obvious in other representations.

In the above two running examples the sets S_1 and S_2 were rather similar, if not equal; they need not be so. For example, the relation "is skilled at" shown in Figure 4 is defined from the previous set of people to a set of tasks (such as typing, shorthand, etc.).

Representation of a relation as a set. A relation R: $S_1 \to S_2$ can be represented as a set of ordered pairs. If element x is related to element y,

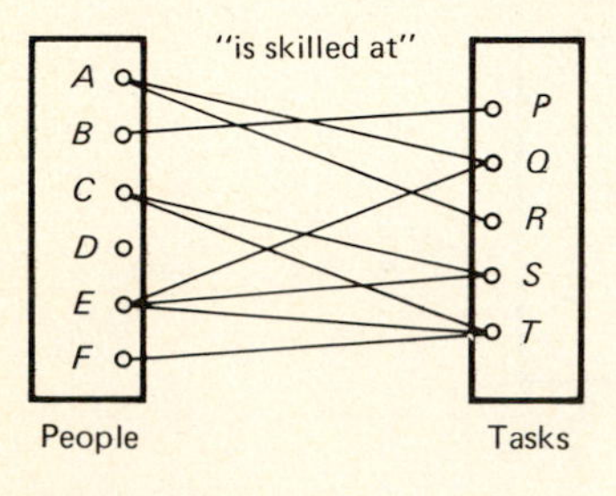

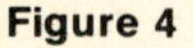

Figure 4

then the pair $\langle x, y \rangle$ is in the set. For example in the channel the symbol 1 is received as a 0, and so the ordered pair $\langle 1, 0 \rangle$, or 10, is in the set R. The entire set is

$$R = \{10, 11, 21, 22, 23, 33\}$$

This relation is illustrated graphically as the subset of Figure 5.

In general a *relation* between any sets is defined as any subset of the product of these sets.

$$R \subseteq [S_1 \times S_2]$$

It is most significant that relations can be thought of as sets since all the concepts, operations, equalities, and proofs which apply to sets can then also apply to relations. The following example illustrates how the operations of union, intersection, and complement apply to relations.

Example Three other relations L, E, W on the previous set product are indicated below and shown in Figure 6.

$x\text{L}y$ is read "x is less than y"

xEy is read "x equals y"

$x\text{W}y$ is read "x is within 1 unit of y"

Some set operations on these relations are

$$\widetilde{W} = \{13, 20, 30, 31\}$$

$$\widetilde{W} \cap L = \{12, 23\}$$

$$L \cup E = \{13, 23, 33, 11, 12, 22\}$$

We may also speak of relation E being a subset of relation W and of relations L and E being disjoint. Notice that if we use the conventional symbolism of $x < y$ for xLy and $x = y$ for xEy, the expression $L \cup E$ would become $< \cup =$, which is a proper expression but a rather confusing one.

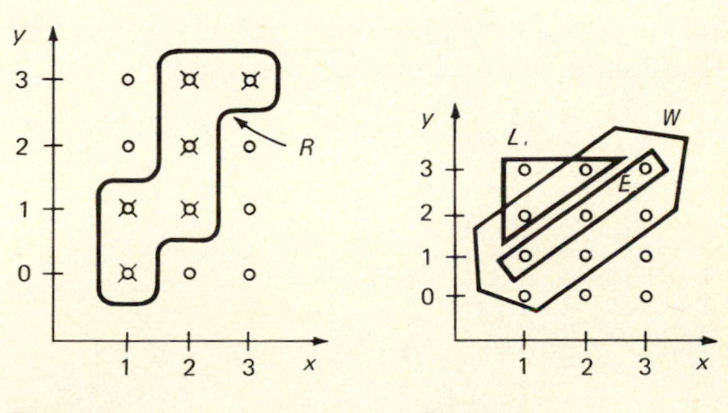

Figure 5 **Figure 6**

Representation of a relation as an operation. A relation can be interpreted as an operator R acting on an element x of set S_1 to produce an element or set of elements y from set S_2. This active operation of transforming x into y is often called a *section* and is denoted by

$$R[x] = y \qquad \text{where } x \in S_1 \text{ and } y \subseteq S_2$$

For example, the communication channel could be represented by the three equations:

$$R[1] = \{0, 1\}$$

$$R[2] = \{1, 2, 3\}$$

$$R[3] = \{3\}$$

The name section is rather appropriate since it describes the process of slicing a section of the point-set graph (Figure 9) corresponding to each x, revealing the related values of y. This type of representation is most often used for a particular type of relation, called a function, which is denoted $y = f[x]$. For functions there is a single unique y corresponding to each and every value x. Such restricted relations will be considered in a later section.

Representation of a relation as a matrix (array). A matrix is simply any array of elements (objects, symbols, numbers) usually drawn in a rectangular form with m rows and n columns. A typical element in the ith row and jth column is labeled e_{ij}. A matrix could represent a relation R by marking each element e_{ij} according to whether x_i is related to y_j. One convenient convention for this marking could be

$$e_{ij} = 1 \qquad \text{if and only if} \qquad x_i R y_j$$

The matrix representation for the relation "is received as" is given in Figure 7. It will be useful when considering interconnections of channels. Such matrices are convenient for computer computations, as well as for gaining insight by inspection. They also provide more systematic and orderly methods for complex relations.

x_i R y_j

i \ j	0	1	2	3
1	1	1	0	0
2	0	1	1	1
3	0	0	0	1

Figure 7

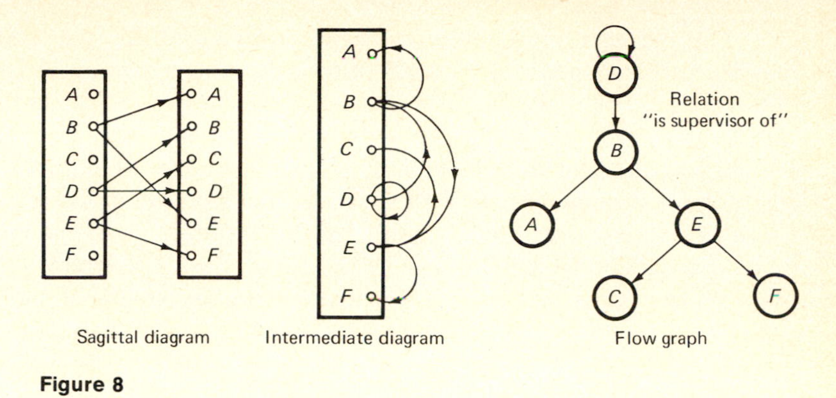

Sagittal diagram Intermediate diagram Flow graph

Figure 8

Representation of a relation as a flow graph. A type of diagrammatic representation called a *flow graph* or kinematic graph is possible when the two sets of a relation are identical, that is, $S_1 = S_2$. In such a case the sagittal representation is modified by constructing only the one set and indicating the correspondence by drawing lines joining the elements of this set. The procedure of converting a sagittal graph into a flow graph is illustrated in Figure 8 for the relation "is the supervisor of." In this case the entire treelike hierarchy of supervision is immediately evident. The flow graph representation often does provide immediate, convenient insight into the structure and properties of relations.

Even when the two sets S_1 and S_2 are not equal, as in the communication example, where $\{1,2,3\} \neq \{0,1,2,3\}$, sometimes the flow graph can be meaningfully defined on the one set $S_1 \cup S_2$. This is illustrated in Figure 9.

Summary of Representations of Relations

We have just considered many different representations of the one concept of relation. They will all be used.

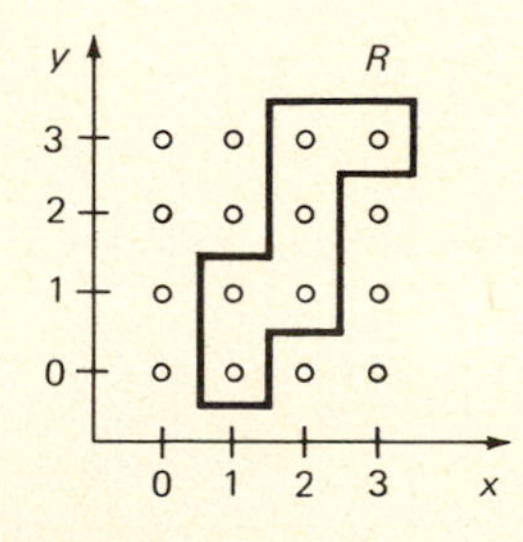

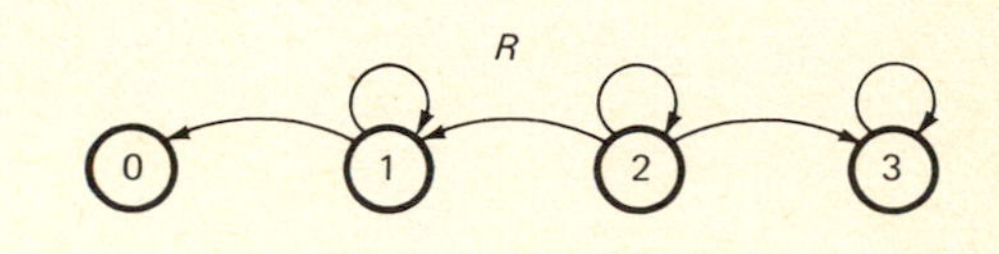

Figure 9

The *verbal* representation, xRy, is convenient for the practical reason of abstracting relations from this world, since they usually appear in this form. The representation of ordered pairs $\langle x, y \rangle$ as elements of a *set* is important for theoretical reasons, since it emphasizes that relations are sets and all that can be done with sets can be done with relations. The representation as an *operation* $R[x]$ is algebraically useful; the representation as a *matrix* is computationally convenient; the *flow graph* is intuitively convenient for the analysis of the properties of structure. These five possible representations of relations do not exhaust all methods. For example, a relation can also be represented as a *tree* with restricted branches. Try it.

In essence, the proper choice of representation can make a difficult problem into a simpler one. This choice depends both on the system and its observer.

Other Concepts: Domain, Range, Inverse

The *domain* of a relation $R\colon S_1 \to S_2$ is denoted $\text{dom}[R]$, and is defined as the set of all elements of S_1 which occur in R. Formally,

$$\text{dom}[R] = \{x\colon xRy\}$$

The *range* of a relation $R\colon S_1 \to S_2$ is denoted $\text{ran}[R]$, and is defined as the set of all elements of S_2 which occur in R. Formally,

$$\text{ran}[R] = \{y\colon xRy\}$$

The *inverse* of a relation $R\colon S_1 \to S_2$ is denoted $R^{-1}\colon S_2 \to S_1$ and defined formally as

$$R^{-1} = \{\langle y, x \rangle : xRy\}$$

Examples A relation Q is shown in the given sagittal diagram defined from the set $\{a,b,c\}$ to the set $\{0,1,2,3\}$. The domain of this relation is

$$\text{dom}[Q] = \{a,b\}$$

and the range is

$$\text{ran}[Q] = \{1,2,3\}$$

The inverse of Q can be represented as a section by

$$Q^{-1}[1] = \{a\} \qquad Q^{-1}[2] = \{b\} \qquad Q^{-1}[3] = \{a,b\}$$

or more clearly by the sagittal diagram of Figure 10. Similarly, for the relation "is the

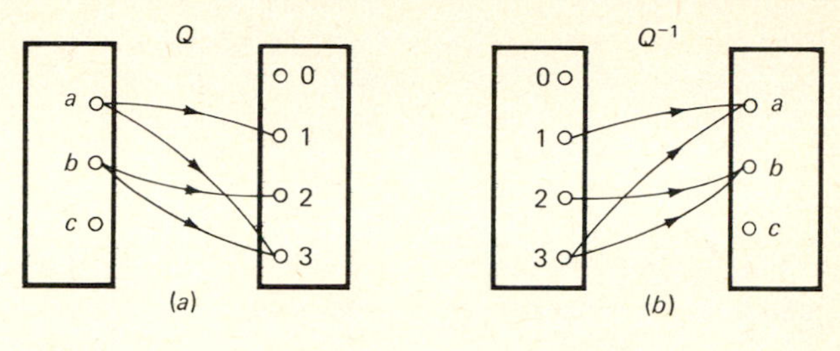

Figure 10

supervisor of" the domain is the set of all supervisors, the range is the set of all those who are supervised, and the inverse is "is supervised by."

Composition of relations. The concept of composition will be useful in interconnecting systems.

The composition of two relations R_1 and R_2 is denoted $R_1 \circ R_2$ and defined as the set of all pairs $\langle x,z \rangle$ such that for some y, x is related by R_1 to y, and y is related by R_2 to z. Formally,

$$R_1 \circ R_2 = \{ \langle x,z \rangle : \text{for some } y;\ xR_1y \text{ and } yR_2z \}$$

The composition of two previous relations Q and R is shown in Figure 11. The operation of composition is thus seen to be a way of combining two relations into one by indicating all paths from the elements of the first set to the elements of the last set. The existence of a common element y in the intermediate sets ensures that the relations are matched or compatible.

If two relations are compatible, the composition may be written as

$$z = R_2[R_1(x)]$$

which is obtained by directly substituting the operational notation

$$y = R_1[x] \qquad \text{into} \qquad z = R_2[y]$$

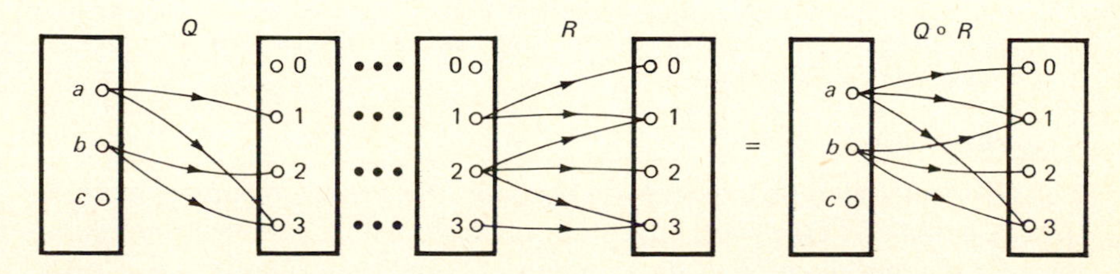

Figure 11

Example If M is the relation "is the mother of" and F is the relation "is the father of," the composition $M \circ F$ is the relation "is the paternal grandmother of."

Example If signals are relayed over two communication channels R_1 and R_2, this can be considered as one channel described by $R_1 \circ R_2$.

*Relations on One Set and Possible Properties

There is a special class of relations which are defined on one set, i.e., from one set S_1 to this same set S_1. For example the relation "is the mother of" is defined on a set of people, and the relation "is less than" is defined on a set of numbers. Such relations are often classified according to the properties of reflexivity, symmetry, and transitivity.

Definition. Relation R is *reflexive* if xRx for all x in S.

For example, relation "lives within 10 miles of" is a relation which is defined on a set of people and which is reflexive since any person x lives within 10 miles of himself. However the relation "is the mother of" is not reflexive since no one is his own mother.

Definition. Relation R is *symmetric* if xRy implies yRx.

For example, relation "is married to" is symmetric since x is married to y implies y is married to x. However, the relation "is the supervisor of" is not symmetric. Neither is the relation "is less than."

Definition. Relation R is *transitive* if xRy and yRz implies xRz.

For example, relation "is the ancestor of" is transitive since if x is the ancestor of y and y is the ancestor of z then x is the ancestor of z. However, the relation "is the mother of" is not transitive since if x is the mother of y and y is the mother of z then x is not the mother of z (but is the grandmother of z).

Example Consider the following relation defined on set $\{1,2,3,4\}$.

$$R = \{11,22,33,12,21,42,24\}$$

This relation is not reflexive since the term 44 does not appear. It is also not transitive since 12 and 24 appear but 14 does not. However, the relation is symmetric since the inverted pairs are all in the set.

*Equivalence Relations

Relations may also be classified according to whether groups of properties are satisfied. One such important relation is the equivalence relation, which is a generalization of the concept of equality or identity. It is often useful in optimizing systems.

Definition Relation R is an *equivalence relation* if it is reflexive and symmetric and transitive.

For example, the relation $M =$ "has the same mother as" is an equivalence relation which could be defined on some particular set of children. It possesses the three required properties

$$[xMx] \quad \text{and} \quad [xMy \text{ implies } yMx] \quad \text{and}$$

$$[xMy \text{ and } yMz \text{ implies } xMz]$$

Notice that an equivalence relation defined on a set partitions this set into equivalence classes in which all the elements in a partition are considered equivalent in some sense (although not identical or equal). Thus for the above relation M the set of people is partitioned into families.

It is interesting to notice that the relation "has the same mother as" has all three properties of reflexivity, symmetry, and transitivity, whereas the relation "is the mother of" has none of these three properties. Isn't that what you would expect of mothers!

Examples Some other examples of equivalence relations are:
"Receives the same grade as" defined on a set of students
"Lives in the same city as" defined on a set of people
"Is connected by a (two-way) street to" on a set of intersections

Notice that the relation "is within 2 inches of" appears to partition objects according to length; however, it is not an equivalence relation because it is not transitive.

Extending and Restricting Relations

Until now, all the relations considered have involved only two elements x and y from two sets S_1 and S_2. Such relations are sometimes known as *binary relations.*

The concept of relation may be extended to any number of sets by defining an *n-ary relation* as any subset of the product of n sets. Some examples of such nonbinary relations are:

Town x is between towns y and z.
Ship x has latitude y and longitude z.
Man x purchased object y for price z at time t.

However, great usefulness has not resulted from extending of relations but rather from restricting them in some way. One restricted type of relation which is extremely useful is the function. It is treated in the next section.

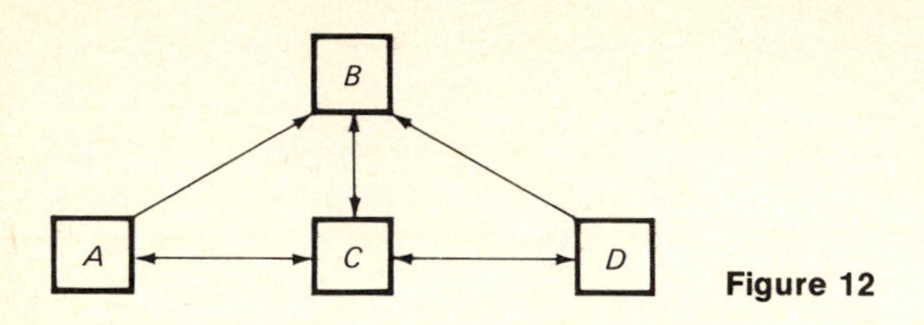

Figure 12

EXERCISE SET 1

1. The given diagram of Figure 12 is a map of a town showing some main streets (one-way streets also) and how they connect the four main parts of town symbolized A, B, C, and D:

$A =$ airport

$B =$ bus depot

$C =$ center of town

$D =$ depot (train)

Draw in all possible representations the relation "is connected directly by a one-way street to." (Note that a two-way street may be thought of as two one-way streets.)

2. How can the properties of reflexivity and symmetry be detected by observing a flow graph or a matrix?

3. For the given two relations M and N draw the sagittal diagrams for the following compositions and comment on commutativity and dualization.

(*a*) $M \circ N$ (*b*) $N \circ M$
(*c*) $(M \circ N)^{-1}$ (*d*) $N^{-1} \circ M^{-1}$
(*e*) $N \circ N^{-1}$ (*f*) $M^{-1} \circ M^{-1}$

4. Construct one channel which is equivalent to the (cascade) interconnection of two crummy channels given in Figure 14. Comment.

5. Classify the following relations according to their properties of reflexivity, symmetry, and transitivity. Which are equivalence relations?

(*a*) "Receives the same grade as" defined on the set of all students
(*b*) "Is the supervisor of" defined in Figure 3
(*c*) $\{11,22,33,12,21,42,24\}$ defined on the set $\{1,2,3,4\}$
(*d*) "Has the same age and sex as" defined on a set of people
(*e*) "Has the same age or sex as" defined on a set of people
(*f*) "Is mutually exclusive of" defined on a set of sets

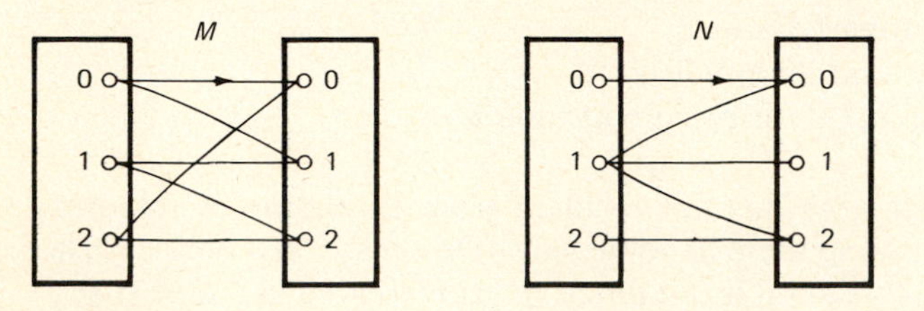

Figure 13

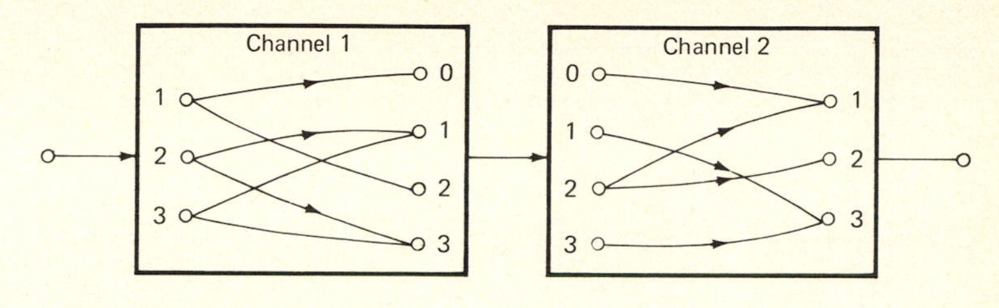

Figure 14

2. FUNCTIONS

A *function* $f: S_1 \to S_2$ is any rule, (computation, method, process, algorithm, operation, code, labeling, routine, machine, device, mapping, indication, diagram, construction), or other method that associates with every element in set S_1 one and only one element in the set S_2.

A function could also be thought of as a relation with the two properties of being everywhere defined and also single-valued. A relation is *everywhere defined* if every element in S_1 indicates an element in S_2. A relation is *single-valued* if there is a unique element in S_2 corresponding to an element from S_1.

The sagittal diagram of a function has arrows leaving from every element of S_1 (everywhere defined), but only one arrow leaves each element of S_1 (single-valued). This is illustrated in Figure 16.

In the past you may have encountered functions in a graphical form, as in Figure 15, or an algebraic form such as

$$y(x) = ax + b$$

$$f(x) = ax^2 + bx + c$$

$$g(t) = 10 \cos (2 \pi t + 1)$$

$$p(z) = \begin{cases} 0 & \text{if } x \leq 0 \\ z & \text{if } x > 0 \end{cases}$$

Such functions often make no explicit mention of the sets S_1 and S_2 since these sets are assumed understood as real numbers, complex numbers, rational numbers, or integers.

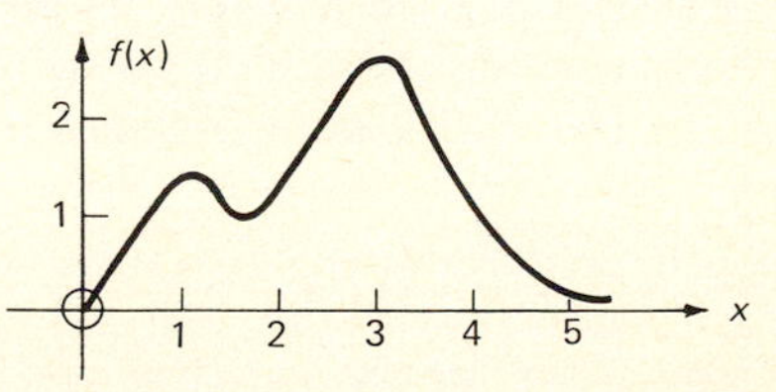

Figure 15

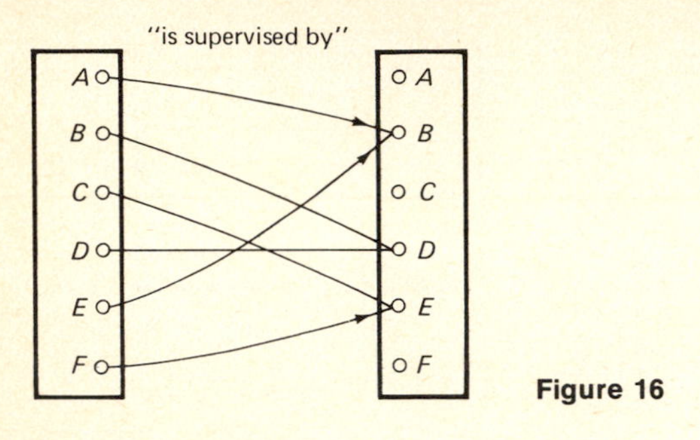

Figure 16

A function, like a relation, need not involve numbers at all. For example the function "achieves a grade of" is defined from a set of students to a set of letter grades, neither of which is numerical. It can be verified that this relation is a function by observing that every student gets a grade (everywhere defined) and that each student receives only one unique grade (single-valued). Similarly the relation "is supervised by" (Figure 16) is a function since "one can serve only one master" (single-valued) and everyone has a master (everywhere defined) even if this master is himself. The rules of a game (like "heads I win, tails you lose") also define a function.

EXERCISE SET 2

1. Which of the following relations are functions from the set $\{1,2\}$ to the set $\{2,3,4\}$?
   (*a*) $\{12\}$ (*b*) $\{13,22\}$
   (*c*) $\{31,22\}$ (*d*) $\{22,21,32\}$
2. Indicate a domain, range, and function (correspondence) which could be defined by:
   (*a*) A telephone directory
   (*b*) A calendar
   (*c*) A paint-by-numbers set
3. How many possible functions can be defined from a set of m elements to a set of n elements? *Ans.*: n^m (or is it m^n?)

3. SYSTEMS

A *system* in general is any n-tuple of sets and relations. For example, the ordinary number system consisting of the set of real numbers N and the ordinary arithmetic operations (relations) could be denoted

$$\langle N,+,\times,-,\div,<,=\rangle$$

Similarly a binary switching system could be denoted as

$$\langle \{0,\ 1\},\ \text{OR},\ \text{AND},\ \text{NOT}\rangle$$

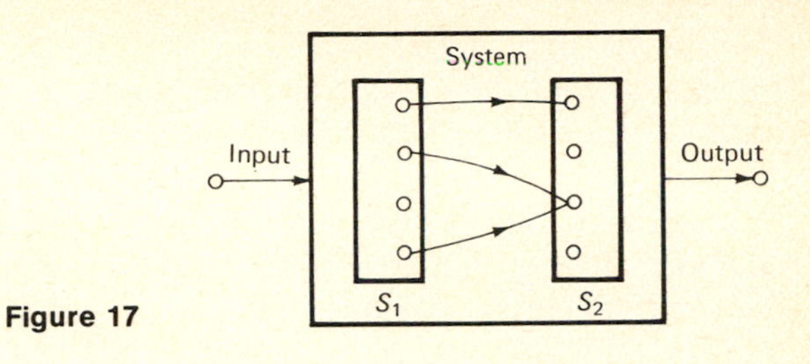

Figure 17

In systems theory we are often interested in a particular type of system known as an *abstract oriented system* which consists of:

An input set (space) S_1
An output set (space) S_2
A relation R: $S_1 \rightarrow S_2$

It is usually drawn as a black box with input and output terminals as in Figure 17. This emphasizes the fact that an actual object has been abstracted (modeled) and is available only at the terminals which represent relevant attributes or variables. The terminals which can be controlled or manipulated are called *inputs*, whereas those which can only be observed are called *outputs*. The process of determining the inputs and outputs is called *orienting*.

As an example, a relay could be abstracted, modeled, oriented, and drawn as shown in Figure 18. The input X could be the voltage applied to the coil.

$X = 0$ indicating no voltage

$X = 1$ indicating a voltage

The output could be the position of the make contact

$Y = 0$ indicating an open contact

$Y = 1$ indicating a closed contact

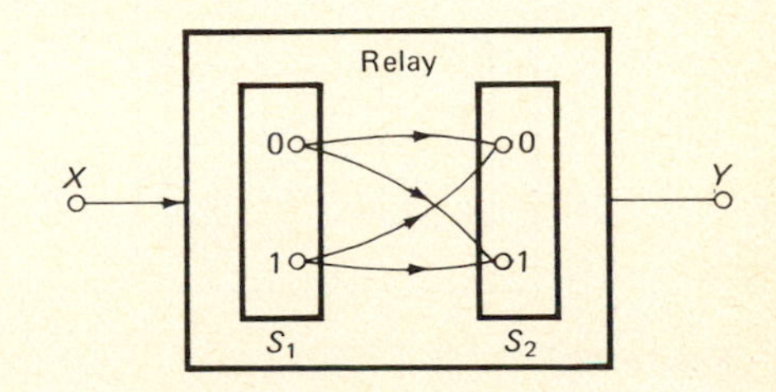

Figure 18

Notice that applying a voltage ($X = 1$) does not necessarily result in only the closed contact ($Y = 1$) since the relay may be crummy. In general, the output of a system is not necessarily a (single-valued) function of the input!

A switching system, such as a level OR component, could also be represented in such a way, by extending the input space to include all possible combinations of input values

$$S_1 = \{0,1\} \times \{0,1\} = \{00, 01, 10, 11\}$$

The diagram of Figure 19 shows both an ideal OR component and a crummy OR component. Notice that this representation is simply an extension of the table-of-combinations representation which provides for more than one output for each input (in the crummy cases). The crummy OR component could also be described as the set

$$\text{v} = \{ \langle\langle 0,0\rangle,0\rangle, \langle\langle 0,0\rangle,1\rangle, \langle\langle 0,1\rangle 1\rangle, \langle\langle 1,0\rangle,1\rangle, \langle\langle 1,1\rangle,1\rangle \}$$

or more simply without the angular brackets as

$$\text{v} = \{000, 001, 011, 101, 111\}$$

A three-variable switching system such as the MAJORITY network could be drawn similarly, as in Figure 20, by constructing the input space to be

$$S_1 = \{0,1\}^3 = \{0,1\} \times \{0,1\} \times \{0,1\}$$

A system having two input variables and two output variables, such as the binary half adder, could be represented as a relation from one cross product to another cross product:

$$+ : \{0,1\} \times \{0,1\} \rightarrow \{0,1\} \times \{0,1\}$$

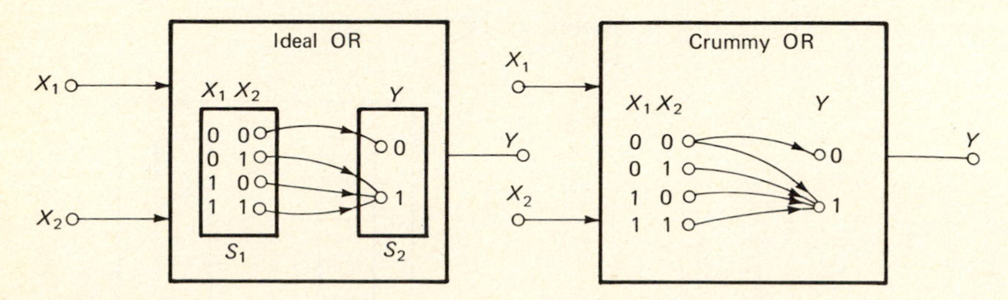

Figure 19

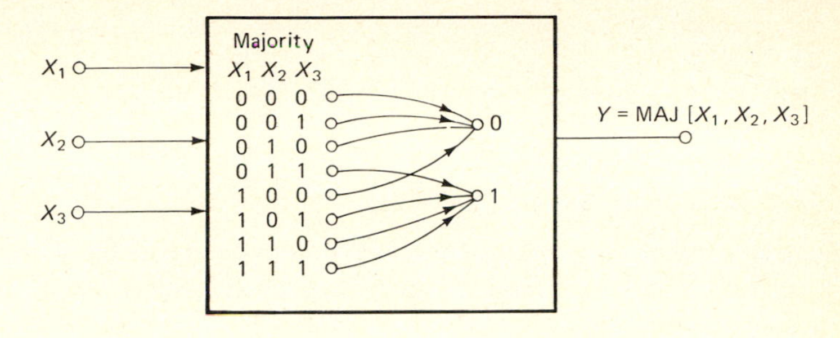

Figure 20

This half adder of Figure 21 could be written in a simpler but more abstract form as

$$+ = \{0000, 0101, 1001, 1110\}$$

where each n-tuple represents the ordered combination $X_1X_2Y_1Y_2$; for example, 1110 reads $X_1 = 1$ and $X_2 = 1$ and $Y_1 = 1$ and $Y_2 = 0$.

EXERCISE SET 3

1. Draw the abstract oriented representation for:
(*a*) The NOT component
(*b*) The EXCLUSIVE-OR
(*c*) The full adder

2. A maximum system $max[X_1, X_2]$ defined on the sets $A \times B \rightarrow (A \cup B)$ has output Y which is the numerically higher value of the two input values X_1 and X_2. Draw this system for the sets:
(*a*) $A = B = \{0, 1\}$ and comment!
(*b*) $A = B = \{0, 1, 2\}$
(*c*) $A = \{0, 1\}$; $B = \{-2, 0, 2\}$

3. How many possible systems can there be having two inputs and one output, i.e.,

$$S: A \times B \rightarrow C$$

if there are i values for A, j values for B, and k values for C? How many of these are functions? Check this for the binary case where $A = B = C = \{0, 1\}$. Check it also for the

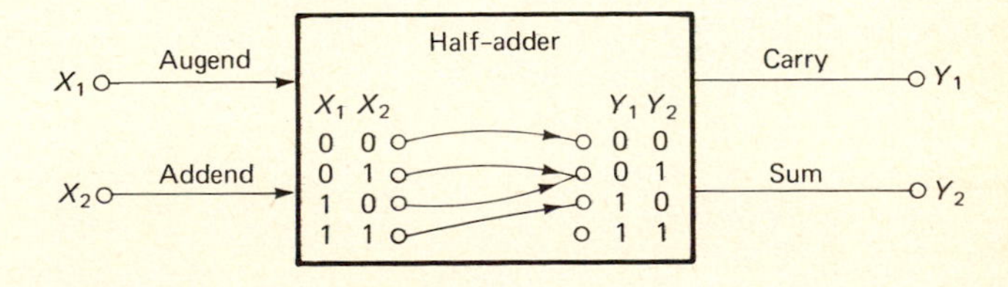

Figure 21

three-valued case to show that there are 19,683 functions (like those in Chapter 1, Figure 84*b* and *c*).

4. How many two-valued switching functions of four variables, such as $F(A, B, C, D) = A \vee \bar{B}(C \vee \bar{D})$, are possible? *Ans.*: (2^{16})

4. FUNCTION-ON-A-RELATION

A particularly interesting function is one which is defined on a relation. It assigns a unique value $z = f(x,y)$ to each *n*-tuple $\langle x,y \rangle$ of a relation. Some examples are:

1. City x is connected to city y by a road of length z.
2. Node x connects to node y through resistor z.
3. Signal x is received as signal y with probability z.
4. Behavior x follows behavior y after stimulus z.
5. Man x buys item y for cost z.
6. Condition x follows condition y after a time z.
7. Person x can perform task y for a wage z.
8. Action x when in state y results in utility z.
9. Variable x influences variable y to an extent z.
10. Machine x performs task y with efficiency z.

For a more detailed example, consider a transportation system consisting of various relevant points (such as intersections, or towns) connected together by one-way paths (such as highways, airways). One useful relation could be

$$C = \text{is directly connected to}$$

and a function defined on this relation could be the distance $d(x, y)$ between any points x and y of the relation. Such a system is represented in Figure 22 as a labeled flow graph and as a matrix.

The labeled flow graph is most like the actual map of this system and for this reason may be intuitively preferred. However, the matrix repre-

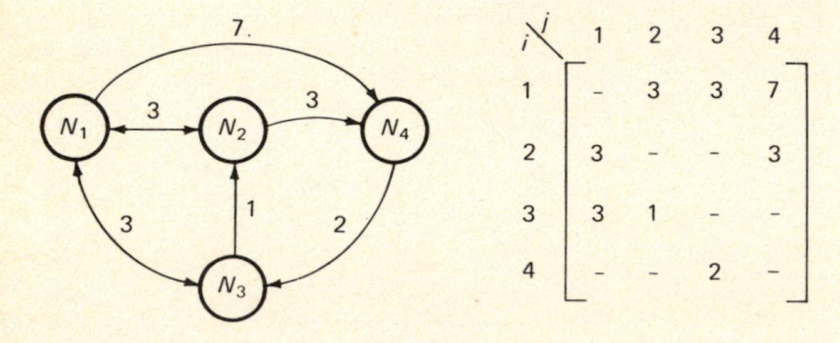

Figure 22

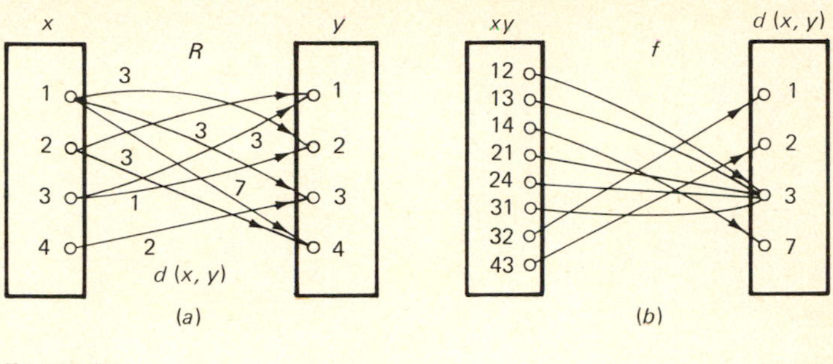

Figure 23

sentation will be more convenient for purposes of analysis and computation. This function-on-a-relation can also be represented by a sagittal relation and a sagittal function, as shown in Figure 23. All these forms will be useful.

For any relation there may be many functions of interest. For example, consider again the previous transportation system. We could define on this relation the function v_{ij}, the average velocity between points i and j. This function-on-a-relation is shown in Figure 24. Notice that the velocities are not symmetric, $v_{ij} \neq v_{ji}$.

Another function defined on this relation is the proportion P_{ij} of vehicles which leave point i to go to j. Notice in Figure 25 that these proportions are between 0 and 1 and that each row of the matrix sums to 1.

Other functions-on-relations which could be defined for this transportation system are:

d_{ij} describing the traffic density
f_{ij} describing the fraction of time congestion is encountered
n_{ij} describing the number of traffic lights
q_{ij} describing the quality of the road nonnumerically as grade A, B, C

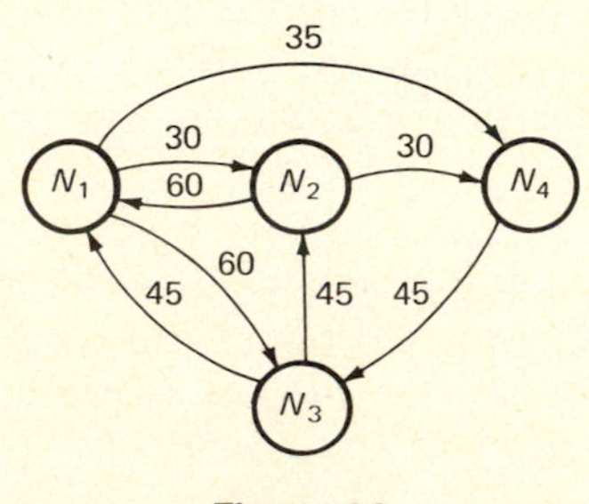

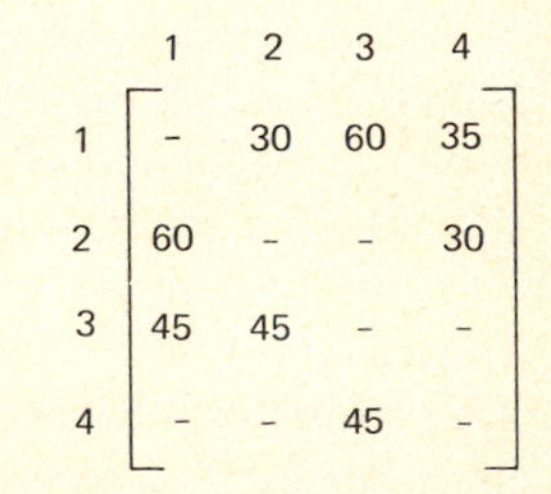

	1	2	3	4
1	-	30	60	35
2	60	-	-	30
3	45	45	-	-
4	-	-	45	-

Figure 24

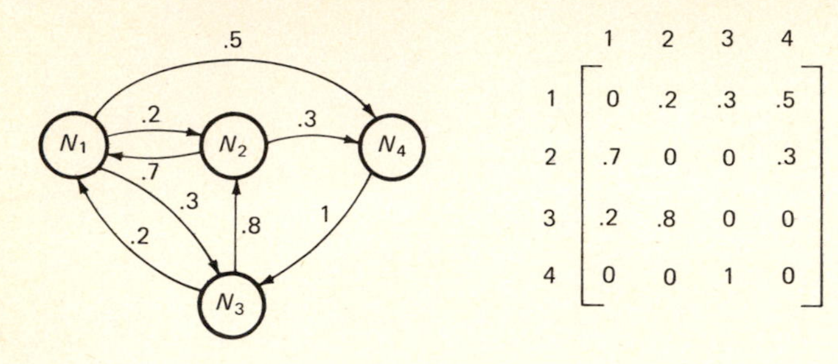

Figure 25

Operations on Arrays (Matrices)

In solving many systems problems it is often convenient to define and use some operations on arrays (or matrices or tables).

For example, if we are interested in the total cost c_{ij} of travel between any two points and we know the cost of fuel a_{ij} and the cost of the vehicle operator b_{ij}, we can define an *addition* $C = A + B$ of corresponding elements of the matrices

$$c_{ij} = a_{ij} + b_{ij}$$

This array addition is illustrated in Figure 26.

0	1	1	3
1	0	-	2
2	1	0	-
-	-	1	0

+

0	2	1	2
1	0	-	1
1	2	0	-
-	-	1	0

=

0	3	2	5
2	0	-	3
3	3	0	-
-	-	2	0

Figure 26

Similarly, if we wish to determine the average time t_{ij} to travel between any nodes, we can divide each distance d_{ij} by its corresponding velocity v_{ij} as in Figure 27. This operation, which can be called *array division*, is meaningful in this particular example, but it may not have meaning for any other systems.

In the previous example the times t_{ij} are given in hours. These times can be converted to minutes by multiplying all entries of the array by 60. This leads to one possible definition of a product, the *product of an array (or matrix) and a number (or scalar)*, which is expressed formally as $B = k * A$ and corresponds to

$$b_{ij} = k * a_{ij}$$

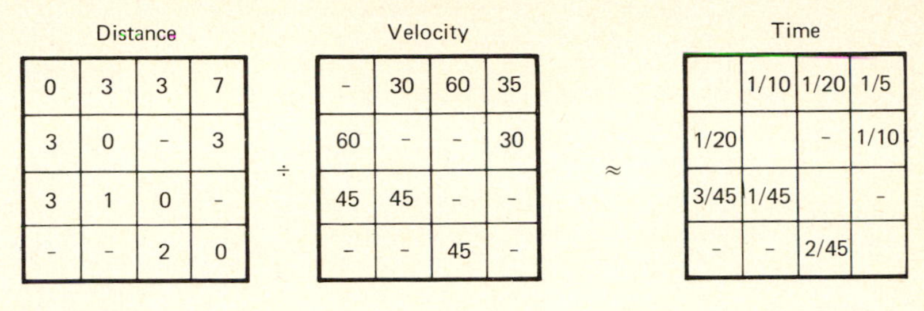

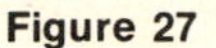

Figure 27

Product of Matrices (Arrays)

A very common matrix operation, the *matrix product*, will be useful in studying probabilistic and stochastic systems. Given a matrix A having elements a_{ij} and a matrix B having elements b_{ij}, their product $C = A \times B$ has elements c_{ij} defined as

$$c_{ij} = a_{i1} * b_{1j} + a_{i2} * b_{2j} + \cdots + a_{ir} * b_{rj}$$

$$= \sum_{k=1}^{r} a_{ik} * b_{kj}$$

This can be visualized by the diagram of Figure 28, where element c_{ij} is determined by taking the ith row of matrix A and the jth column of matrix B, multiplying the corresponding elements, and summing these products. Notice that r, the number of columns of the first matrix A, must equal the number of rows of the second matrix B.

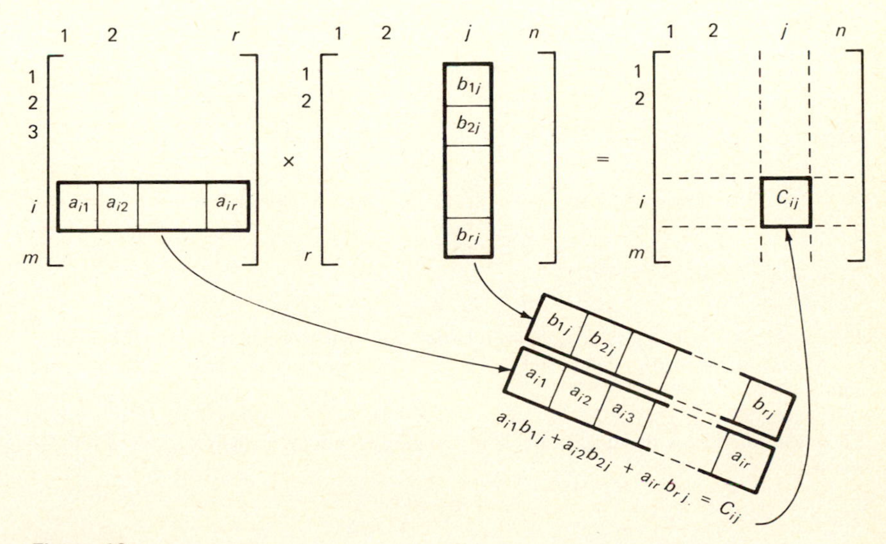

Figure 28

$$\begin{bmatrix} 0, & 1 \\ 2, & 3 \\ 4, & 5 \end{bmatrix} \times \begin{bmatrix} 6, & 7 \\ 8, & 9 \end{bmatrix} = \begin{bmatrix} 0*6+1*8, & 0*7+1*9 \\ 2*6+3*8, & 2*7+3*9 \\ 4*6+5*8, & 4*7+5*9 \end{bmatrix} = \begin{bmatrix} 8, & 9 \\ 36, & 41 \\ 64, & 73 \end{bmatrix}$$

Figure 29

Example A numerical example of matrix multiplication is given in Figure 29. Notice that the resulting product is another matrix rather than a number.

EXERCISE SET 4

1. For the transportation system of this section:

(*a*) Construct, by inspection, the matrix describing the minimum distance between any two points.

(*b*) Construct, by inspection, the matrix describing the minimum time between any two points.

(*c*) If a taxi charges $1 per mile plus $1 per minute construct a matrix of minimum profit between any two points.

2. The set diagram of Figure 31 Chapter 2 describes an opinion poll concerning some product or program. Construct the function-on-a-relation "a fraction z of people of type x have opinion y" in both sagittal and matrix form:

(*a*) Where x is from set {young, nonyoung} and y is from set {like, dislike}.

(*b*) Where x is from set {women, men, boys, girls} and y is from set {like, dislike}.

3. For the four given matrices compute (if possible) the products:

(*a*) $B \times A$	(*b*) $A \times B$
(*c*) $B \times B$	(*d*) $A \times C$
(*e*) $C \times A$	(*f*) $B \times C$
(*g*) $A \times D$	(*h*) $D \times A$
(*i*) $C \times D$	(*j*) $C \times C$
(*k*) $(B \times A) \times D$	(*l*) $B \times (A \times D)$

$$A = \begin{bmatrix} 0 & 1 \\ 2 & 3 \\ 4 & 5 \end{bmatrix} \quad B = \begin{bmatrix} 0 & 1 & 2 \\ 1 & 2 & 0 \\ 0 & 3 & 2 \end{bmatrix} \quad C = [6 \quad 7 \quad 8] \quad D = \begin{bmatrix} 1 \\ 2 \\ 3 \end{bmatrix}$$

4. From the following product of an n-tuple (vector) and a matrix, determine the value of A, B, C.

$$[A \quad B \quad C] = [1 \quad 2 \quad 3] \times \begin{bmatrix} 4 & 0 & 5 \\ 0 & 6 & 7 \\ 8 & 0 & 9 \end{bmatrix}$$

†5. Show that the composition of two relations can be performed by taking the product of the corresponding matrices and using AND and OR operations instead of multiplication and addition.

†Problems marked with a dagger are more complex or require mathematics not discussed in the text.

5. ALGORITHMS

A particularly useful type of function is the algorithm. An *algorithm* is any finite method of specifying the computation of a function. It is usually a complex, dynamic function involving repetition (iteration or recursion). The concept of an algorithm is very important because the solution to most problems will be an algorithm rather than the numbers or other symbols which result from applying the algorithm! Once you know the algorithm—the method–it is a rather straightforward process to get the numbers from a computer.

In this section, algorithms will usually be represented as flow diagrams, which are similar to flow graphs. The flow diagrams are general and not limited to any particular computers. They can readily be converted for computation on any digital computer.

The study of algorithms will be motivated by the following example. Consider making a loan (mortgage) of $3,000 for a period of 5 years (60 months) with payments of $50 a month and interest at 1 percent per month on the previous month's balance. The problem is to find the amount of "balloon payment" which must be made at the end of the 5 years to pay off all the remaining money owed. Attempt a guess before reading on.

Notice that if there were no interest charge, the $50 payments for 60 months would pay off the original $3,000 leaving no balloon payment. However with this innocent looking 1 percent interest the balloon payment is well over $1,000! Any method for computing this value would be an algorithm. One method of computation follows.

The computation starts quite directly:

Original sum	$3,000
Interest of 1%	30
Leaving amount to pay off	$3,030
Less a payment of	50
Leaving balance for second month	$2,980

This same process is repeated, with different numbers, for the second month.

Balance after first month	$2,980.00
Plus the 1% interest	29.80
Leaving	$3,009.80
Less payment	50.00
Sum (balance) for third month	$2,959.80

This can be generalized for month t.

Balance after the month t	$S(t)$
Plus the $R\%$ interest	$R * S(t) = I(t)$
Leaving	$S(t) + R * S(t)$
Less a payment of	$P(t)$
Sum (balance) for month $t + 1$	$S(t) + R * S(t) - P(t)$

This equation is of the general sequential form

$$S(t + 1) = S(t) + R * S(t) - P(t)$$

When repeated for 60 months, this process finally yields a solution of $1,366.61 as the required balloon payment.

A digital computer is ideally suited for performing such repetition. We can imitate (or simulate) the computer as follows.

First, let us realize that a computation proceeds in stages. At each stage in this example we need know only three quantities:

The month T
The sum S
The interest I

which can be thought of as three numbers each stored in a cell or register. At any stage the contents of all the registers can be described by an n-tuple called an *instantaneous description* or snapshot. In this case the 3-tuple would be

$$\langle \text{month, sum, interest} \rangle$$

or symbolically

$$\langle T, S, I \rangle$$

The contents of these registers change from stage to stage. This change could be specified by the verbal algorithm (or program) of Figure 30. Notice that each of the 10 instructions is an imperative command rather than a declarative statement or a question.

A *computation* can be defined as a sequence of instantaneous descriptions which ultimately halts. For example, the above algorithm generates the following computation.

T	S	I
<?,	????????,	?????>
<?,	3000.00,	?????>
<1,	3000.00,	?????>
<1,	3000.00,	30.00>
<1,	2980.00,	30.00>
<2,	2980.00,	30.00>
<2,	2980.00,	29.80>
<2,	2959.80,	29.80>
. . . . .	. . . . . .	. . . . .
<61,	1366.61,	14.03>

This sequence ultimately halts having printed out 1366.61.

An algorithm could also be represented diagrammatically for greater convenience or insight. For example, the flow diagram of Figure 31 describes the verbal algorithm of Figure 30. Notice the great correspondence between these two representations.

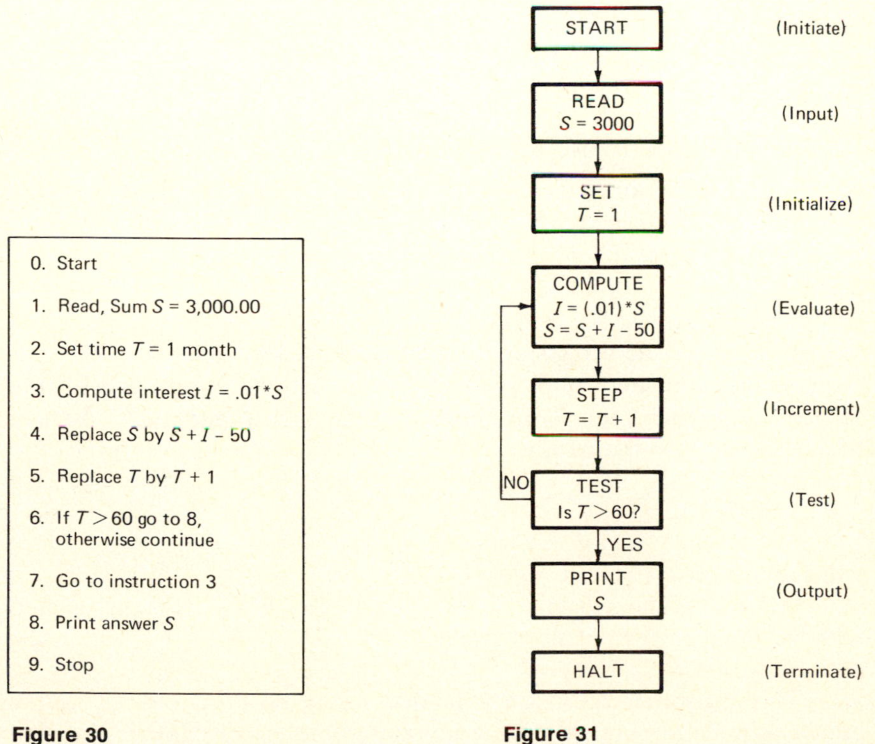

Figure 30

Figure 31

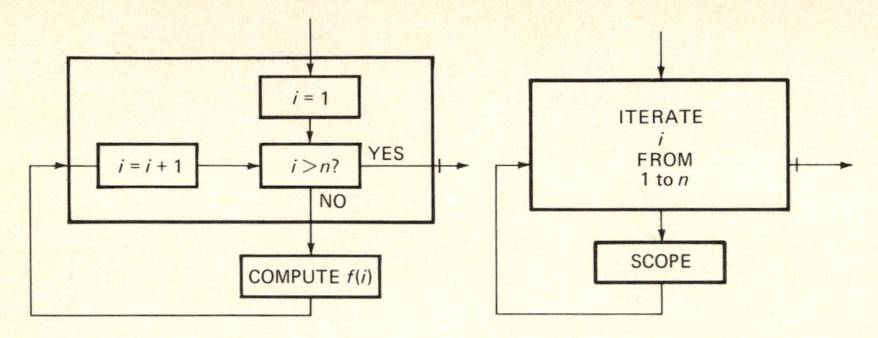

Figure 32

There are eight fundamental types of blocks or instructions: initiate, input, initialize, evaluate, increment, test, output, and terminate. However, some of these can be combined into larger blocks.

For example, Figure 32 shows an iteration instruction, which is sometimes called a DO or a FOR instruction. The *iteration* instruction indicates that a set of instructions (within the loop or scope) are to be repeated or iterated N times before continuing to any other instructions. Since there are two exits from the iteration block, the final exit arrow is marked.

Constructing Complex Algorithms—Sub-algorithms

Given a list of N positive (nonzero) numbers indexed by N integers as

$$A(1), A(2), A(3), \ldots, A(N)$$

let us find the largest number in this list.

First we set up a register B which is started at value 0. Then it is sequentially compared to all elements of the list and is replaced by any number larger than itself. The algorithm is given in Figure 33. You may wish to apply it to the numbers

5, 3, 5, 7, 2

which are stored in five A registers.

Notice that the register C simply keeps track of the index associated with the number in B. This is unnecessary now but will be useful in the next algorithm.

This algorithm could be extended easily to print out all the N numbers in decreasing (or at least nonincreasing) order of size. We simply go through the previous algorithm N times, each time printing out the

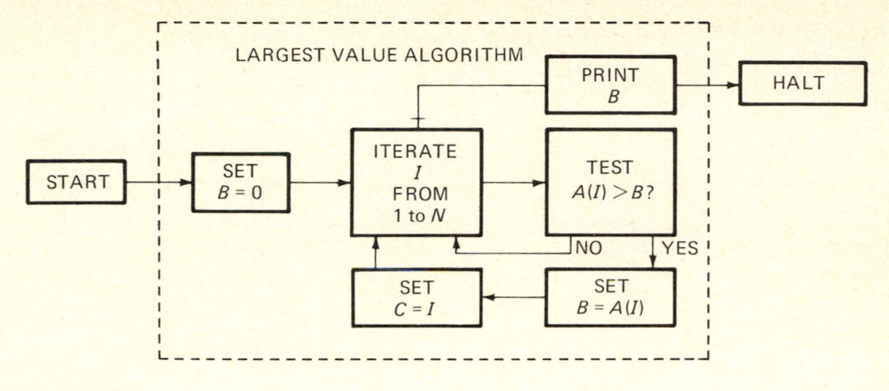

Figure 33

largest value, then setting this register to 0 and starting all over again to get the next largest value. This "*sorting*" algorithm is shown in Figure 34.

Notice that the original algorithm is a part of the larger sorting algorithm. By using such sub-algorithms (or sub-routines) we can build up algorithms of great complexity. Other algorithms for sorting are given at the end of this section.

Example An algorithm for the multiplication of two matrices is given in Figure 35*a*. The registers of Figure 35*b* indicate the values corresponding to the matrices of Figure 29. It may be instructive to apply this algorithm to the given data and check with the result given in Figure 29.

Any flow diagram can be converted into some programming language. It may be a low-level machine language (of 0s and 1s) or a high-level language (such as Fortran, Algol, Basic, or Cobol) similar to our

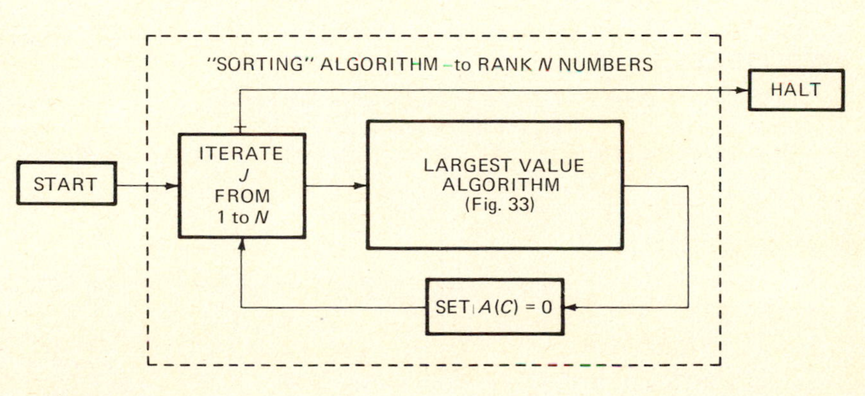

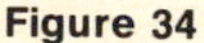

Figure 34

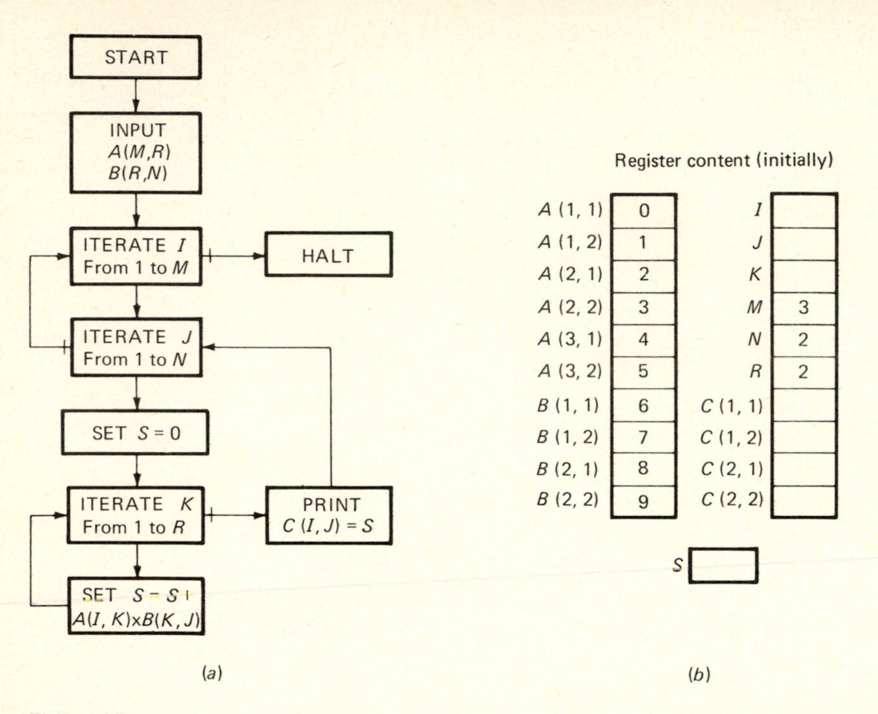

Figure 35

spoken language or mathematical formulas. There are also intermediate languages (such as Assembly Language) between these levels. The original algorithm (for computing the balloon payment) is written in the Basic language in Figure 36. Notice the great resemblance to Figure 30. Also the entire matrix product of Figure 35 can be done by the single Basic instruction

$$\text{MAT} \quad C = A * B.$$

EXERCISE SET 5

1. Modify the flow diagram of Figure 31 for the following cases:
   (*a*) Print the sum remaining at each month.
   (*b*) Compute the time to pay off the entire amount with no balloon payment.

```
10  LET S = 3000
20  LET T = 1
30  LET I = (.01)*S
40  LET S = S + I - 50
50  LET T = T + 1
60  IF   T > 60  GO TO 80
70  GO  TO  30
80  PRINT S
90  END
```

Figure 36

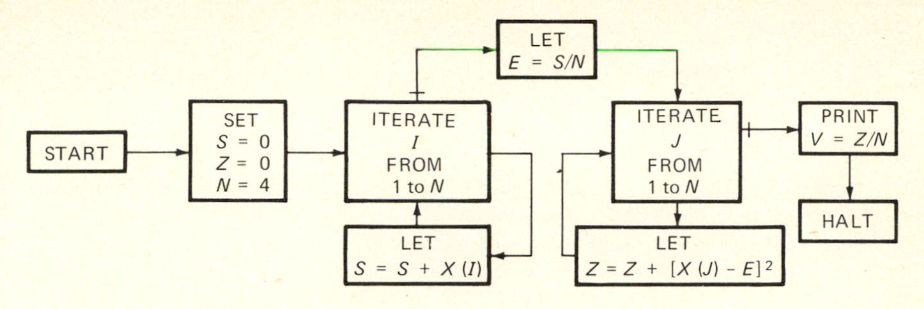

Figure 37

(*c*) Compute the accumulated interest over 5 years. Comment!

(*d*) Compute the average annual interest rate, if the rate is 1 percent for any sum greater than \$2,000 but 2 percent for any sum less than \$2,000.

(*e*) Use fewer instructions than in Figure 30.

2. Write out the complete sequence of instantaneous descriptions for the balloon-payment computation if the loan is for \$100, for 5 years, with payments of \$20 per year, and an annual interest of 12.5 percent on the unpaid balance (rounded off to the next lowest dollar).

Ans.: See Chapter 7, Figure 15

3. Find the value computed by the algorithm of Figure 37 when applied to the four numbers

$$X(1) = 1 \qquad X(2) = 2 \qquad X(3) = 2 \qquad X(4) = 3$$

What function is computed by such an algorithm when applied more generally to N numbers $X(1), X(2), \ldots, X(N)$?

4. Construct another algorithm for sorting numbers in decreasing order by successively inverting pairs of numbers which are increasing.

5. Construct yet another sorting algorithm which starts by interchanging the first number with the largest in the list, then interchanging the second number with the largest of the remaining list, and continuing in this way.

SUMMARY

The concept of relation is very general. By defining a relation as a subset of a product set, we emphasize all possibilities of the product set and realize that by dealing only with a subset of these we have imposed restrictions or constraints on the system. These constraints will be useful.

It is also important to realize that a system is associated with the more general concept of relation, rather than with the particular concept of a function.

The various concepts, representations, and properties of systems will be applied in detail to probabilistic, sequential, and stochastic systems. All the systems will be relations, but they will have some functions and algorithms associated with them.

PROBLEMS

1. Consider the relation G defined from the set $\{0, 1, 2, 3\}$ to the set $\{0, 1, 2\}$ given by the following set of ordered pairs

$$G = \{00, 11, 20, 31\}$$

(*a*) Find the domain and range of G.
(*b*) Find the inverse and complement of G.
(*c*) Check the properties of reflexivity, symmetry, transitivity.
(*d*) Check the properties of single-valuedness and everywhere defined.
(*e*) Show that the composition $G \circ G^{-1}$ is an equivalence relation (by checking the properties or by giving the equivalence classes).

2. How many relations can be defined from set $\{0, 1\}$ to set $\{a, b\}$? Enumerate (list) all the relations in a systematic way. (*Hint*: Use a table-of-combinations.)

3. How many relations can be defined from one set having m elements to another set having n elements? *Ans*.: 2^{mn}

4. A *modulo-n-sum* system has two inputs, say A and B, each of which can have n values $0, 1, 2, \ldots, n\text{-}1$. The output, denoted $A \oplus B$, is the remainder when n divides the arithmetic sum of the input values. Construct this system in sagittal form and matrix form:
(*a*) For $n = 2$ (and comment)
(*b*) For $n = 3$

5. A "*weak*" *order relation* on a set S is defined as having all three properties of reflexivity, antisymmetry, and transitivity. A "*strict*" *order relation* on a set S is defined as having all three properties of nonreflexivity, nonsymmetry, and transitivity, where

R is *nonreflexive* if xRx for no x in set S.
R is *antisymmetric* if xRy and yRx implies $x = y$.
R is *nonsymmetric* if xRy implies $y\bar{R}x$.
Classify the following order relations as weak or strict:
(*a*) "x is less than y" defined on a set of real numbers
(*b*) "x is a subset of y" defined on all subsets of a set
(*c*) "x is the ancestor of" defined on a set of human beings
(*d*) "x divides y" defined on a set of integers *Ans*.: weak

6. Construct the relation "x can safely donate blood to y" defined on all eight blood types of Figure 30, Chapter 2. Represent it as a sagittal diagram, a flow graph, and a matrix.

7. Show that to get the eighth power of a matrix you need only perform three matrix products. Find the sixth and eighth power of the matrix of Figure P-7, and comment.

$$\begin{bmatrix} .5 & .5 & 0 \\ .6 & 0 & .4 \\ .3 & .3 & .4 \end{bmatrix}$$

Figure P-7

8. Draw the following equivalence relation in flow-graph form and comment on its structure.

$$\{11, 22, 33, 44, 55, 66, 14, 15, 26, 41, 45, 51, 54, 62\}$$

9. Construct an algorithm to check the matrix representation of a relation for the properties of

(*a*) Reflexivity (*b*) Symmetry

(*c*) Transitivity (*d*) Equivalence
(*e*) Single-valuedness

10. Construct an algorithm to do the following computations:
(*a*) Convert a number in base 10 to base 2.
(*b*) Print out all prime numbers between 0 and a given N.
(*c*) Prove, by complete induction, a switching equality such as the generalized complementarity property (for fixed n)

$$\overline{(X_1 \vee X_2 \vee X_3 \vee \cdots \vee X_n)} = \bar{X}_1\bar{X}_2\bar{X}_3 \cdots \bar{X}_n$$

11. The variance of a set of numbers $X(1), X(2), \ldots, X(n)$ occurring with probabilities $p(1), p(2), \ldots, p(n)$ is defined by the "static" formula

$$V[X] = \sum_{i=1}^{n} p(i) * \left[X(i) - \sum_{i=1}^{n} p(i) * X(i)\right]^2$$

Convert this into the more "dynamic" flow-diagram form.

Ans.: See Fig. 88*a* in Chap. 5

12. Construct an algorithm to generate a table-of-combinations for four variables, then use it to generate:
(*a*) All binary numbers from 0 to 15.
(*b*) All subsets of the set $\{a, b, c, d\}$.
(*c*) All relations from set $\{a, b\}$ to set $\{c, d\}$.
(*d*) All binary operators.

13. For the given algorithm (operating on the given matrix A) show the sequence of 2-tuples $\langle I, J\rangle$ in the computation. What is the purpose of F (flag)? What does this algorithm do?

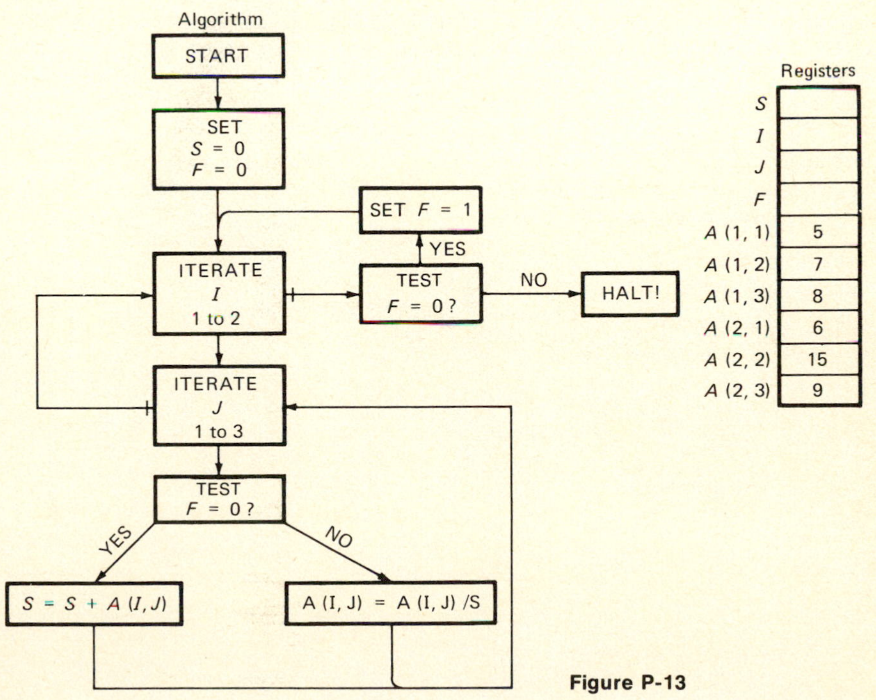

Figure P-13

14. Construct an algorithm in flow-diagram form indicating:

(*a*) How to operate some device, such as a camera, camp stove, etc.

(*b*) How to go through a process, such as developing a film, etc.

15. The population movement between three parts of a city labeled A, B, C (representing a swank neighborhood, a suburb, and a slum portion) from one year to the next is shown in the given sagittal diagram. Initially the population is distributed with 10 percent in A, 40 percent in B, and 50 percent in C.

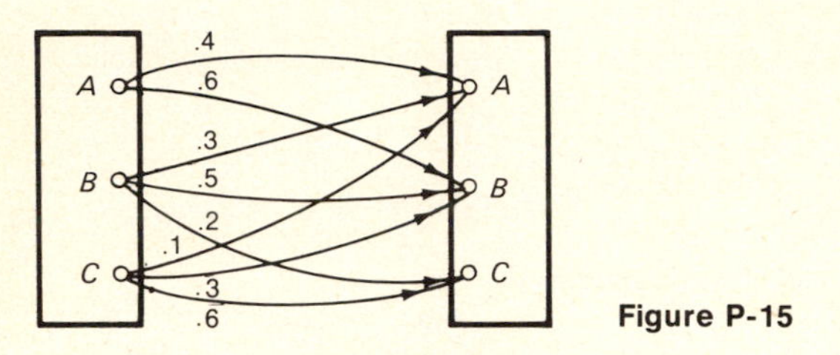

Figure P-15

(*a*) Represent this system as a flow graph.

(*b*) Find the population distribution after 1 year. *Ans.*: 21%, 41%, 38%

(*c*) Find the distribution after 2, 3, and 4 years.

(*d*) Express the population distribution between year t and $t + 1$ by a matrix product.

(*e*) Express the population distribution after n years as a matrix product.

(*f*) Find the distribution after 8 years (by performing only three matrix products) and comment on the ultimate population distribution.

16. For the relation shown in Figure 4, select the smallest number of people to make certain that someone can cover any task. How many solutions are there to this problem? [*Hint*: Write out symbolically that you need to select B (for task P) and A or E (for task Q) and C or E or F (for task T). Use the switching algebra to rewrite this into the form where the solution is evident.]

17. *Project*. Find in your experience some meaningful example of a relation. Represent it in various ways, analyze it, then attempt to use it. (For example, you may consider the relation "course x is a prerequisite for course y" or "course x is taught at time y" and attempt some optimal plan or schedule.)

CHAPTER 4
Probabilistic Systems

Instantaneous Nondeterministic Systems

A goal of systems theory is to discover some law, order, or regularity in physical systems. This regularity usually takes the form of an output expressed as a function of an input. In some systems, however, the output may not be uniquely determined by the input. For example, a relay may momentarily malfunction, for when a voltage is applied to the coil, the contact may occasionally be open (called *sluggishness*) or when no voltage is applied, its contact may be closed (called *sticking*). It appears that such systems are more appropriately modeled by a relation than by a function. One possible model of such *nonunique* systems is the *probabilistic model*, which is studied in this chapter. Others, such as sequential and stochastic models, will be considered in other chapters.

Before attempting to model such a system we could ask why the output is not a single-valued function of the input. Is it because physical systems are inherently not functions but are relations? Or is it because systems are functions but all the inputs are not known (or hidden)? Or is it perhaps that even if all the inputs and functions were known, the output is still too difficult or time-consuming to evaluate? These questions ask in essence: What is the nature of reality? It may be an interesting question, but in complex systems we accept the fact that we cannot know reality, we can only attempt to model this reality to accomplish some goal. This goal may be to create systems, to optimize, to control, or to make decisions despite any such constraints as this nonfunctionality. Fortunately, however, the mathematical and engineering results obtained do not depend on the answer to such philosophical questions.

There is diversity of thought concerning other matters also. Some of this diversity can be illustrated by asking the simple question: What is the probability that upon flipping a coin it lands with the head facing upward? Why? Some answers in no particular order follow:

1. The probability is 0.5; it's obvious.
2. The probability is $\frac{1}{2}$, since one side will come up as often as the other side.
3. If it's a two-headed coin, the probability of a head is 1.
4. After the coin is thrown the probability is 1 or 0.
5. Someone told me the probability is 0.5. Don't ask me why.
6. After flipping a coin 100 times it came up heads 53 times, so the probability is 0.53.
7. After flipping a coin many times, say T, and plotting the portion of times a head comes up, and taking the limit as T approaches infinity, the probability tends to this limit of 0.50.
8. The probability must be slightly less than $\frac{1}{2}$, since the coin may land on its edge.
9. If I bet x dollars on heads with "rational" men, they insist on betting x dollars also; thus indicating 1 : 1 odds or a probability of $\frac{1}{2}$.
10. If I'm betting on heads, the probability of heads is small; this is known as the *law of perversity of inanimate objects.*
11. The probability will deviate slightly from the value of $\frac{1}{2}$ if the coin is bent or "bulgier" on one side.
12. It depends on how you flip the coin; you may be able to control the probability.
13. If you knew the initial velocity and rotational momentum of the coin, its aerodynamic properties, and the hardness properties of the surface it falls on, you could predict exactly how it would fall; no probabilities or uncertainty are involved.
14. It depends on the stars and the month you were born and Lady Luck.
15. I don't care what numerical value the probability has, just call it p and ask more interesting questions such as: What is the probability of more than r heads in n throws?
16. From my experience it seems that heads occur a little more than half the time.
17. If there were many heads before this flip, the probability of another head is less than 0.5, because the probabilities should even up after a while. (The coin has memory?)
18. If there were many heads before this flip, the probability of another head is larger than 0.5, because the coin is "loaded."
19. From theoretical geometric considerations such as assuming a coin as a perfect cylinder with parallel ends a distance d apart, the probability should be somehow shown to be a function of d.
20. I don't know.

Some of these views are inadmissible; others are associated with such names as classical, geometric, a priori, equiprobable, subjective, Laplace, von Mises, relative frequency, and axiomatic. We will consider some of these in greater detail later. Notice that many of these views are in agreement on a value of probability near 0.5, even though it was arrived at in many different ways. Thus again there is unity despite diversity.

1. MODELING AND REPRESENTING PROBABILISTIC SYSTEMS

Modeling a Crummy Relay

Consider a system consisting of a simple relay operated by some environment, such as an elevator, computer, or control system as in Figure 1.

Orientation This system has a natural input-output assignment or orientation. The input X is anything which can be controlled, which in this case is the voltage applied to the relay coil. Let

Input value $X = 1$ represent voltage applied to coil

Input value $X = 0$ represent no voltage applied to coil

The output Y is anything that can be observed, which in this case is the relay contact position. Let

Output value $Y = 1$ represent a closed contact

Output value $Y = 0$ represent an open contact

Without further knowledge about this system we might expect that it could be modeled by a function as in Figure 2. Such a system was considered in Chapter 1.

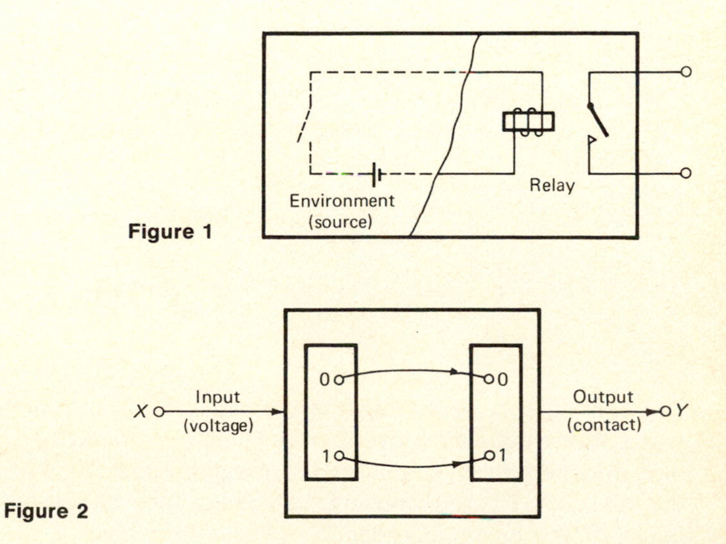

Figure 1

Figure 2

Observation More accurate modeling of such a physical system must start with observation, which may be accompanied by experiment. An observation consists of a sequence of inputs and corresponding outputs, each input-output pair sometimes called a *trial.* The inputs may come from the undisturbed original environment, or they may be imposed by an experimenter who could flip a coin and apply a voltage whenever a head comes up. We will assume that the interval between trials is sufficiently long to give the relay ample time to respond to the input, thus avoiding any unnecessary dependence on time (for now). A representative observation or input-output sequence could be

Trial	12345678910	. . . 20	. . . 30	. . . 40
Input	0101111111	0011111111	0111100011	0111111101
Output	0100110101	1010110111	0010000101	0010111001

If a coin was used to determine the input, it was a rather badly bent one since there are many more ones than zeros applied.

Basic Events Notice that there are only four possible outcomes to this experiment. These outcomes, which will be called basic events, are:

Outcome	*Condition*	*Description*
$X = 0$ and $Y = 0$	No voltage applied and contact open	Proper operation
$X = 0$ and $Y = 1$	No voltage applied and contact closed	Sticking
$X = 1$ and $Y = 0$	Voltage applied and contact open	Sluggishness
$X = 1$ and $Y = 1$	Voltage applied and contact closed	Proper operation

For convenience in writing, these events are often represented in many ways. For example the event, $X = 1$ and $Y = 0$, may be written as X_1Y_0, or it may be written simply as 10, or again it may be written as e_2 where the subscript 2 is any convenient labeling such as the decimal representation of the binary number 10. These four basic events are not only

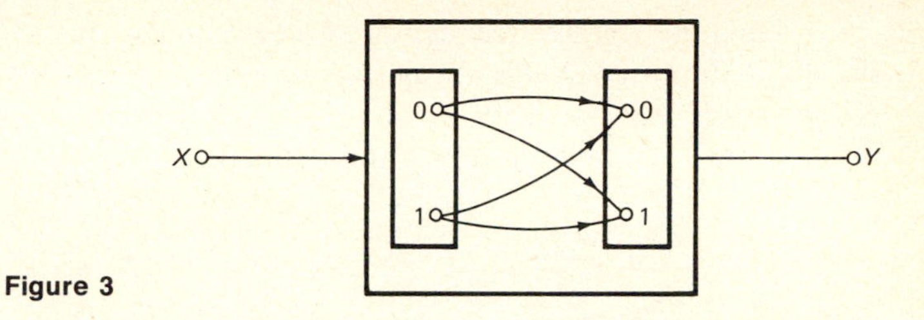

Figure 3

exhaustive (include all possibilities) but are also mutually exclusive; i.e., (no two of these events can occur simultaneously).

Notice also that for any input, say $X = 1$, there may correspond either of two outputs, $Y = 0$ or $Y = 1$, thus suggesting a model which is a relation, as in Figure 3.

Relative Frequency This relation, however, does not reflect one aspect of the behavior of this system, which is that some events occur more frequently than other events. For example, the event $X_1 Y_1$ occurs almost ten times as frequently as event $X_0 Y_1$. This suggests that there may be a regularity in the frequency of occurrence of events. The fraction of times that an event occurs is called its relative frequency. The *relative frequency* of event E is symbolized $f[E]$ and defined more formally as

$$f[E] = \frac{N_T[E]}{T}$$

where $N_T[E]$ is the number of occurrences of event E in T trials. For example, after 10 trials the event $X_1 Y_1$ occurs 5 times and thus has a relative frequency of $5/10 = 0.5$. This relative frequency can be computed as the number of trials T increases, and often it tends to a stable value. Such a regularity is demonstrated graphically for this event $e_3 = X_1 Y_1$ in Figure 4.

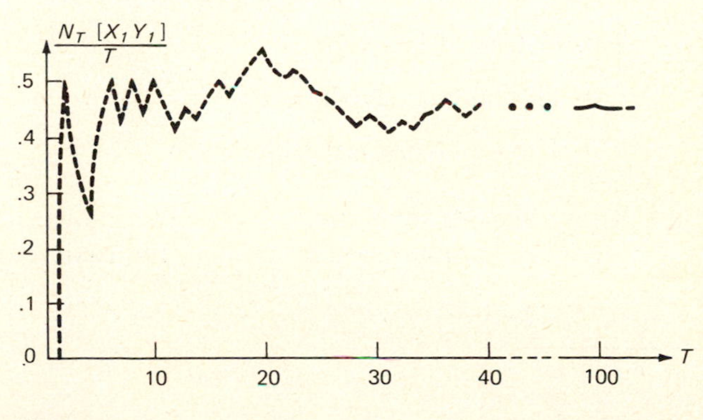

Figure 4

When the relative frequency of an event has this sort of regularity, the stable limiting value is a way (one of many) of defining the probability of the event. Formally, this relative-frequency definition of the probability of event E is denoted $p[E]$ and defined

$$p[E] = \lim \frac{N_T[E]}{T}$$

Of course the larger the value of T, the more accurate the probability, but there is no need to increase T indefinitely since the accuracy is determined by a (practical) goal. We need not pretend that this value can be known with absolute precision.

The probability of the above-considered event $e_3 = X_1 Y_1$ can be estimated from Figure 4 as 0.45. The probabilities of the remaining basic events can be determined similarly. These basic events and their probabilities are summarized as the function of Figure 5 in a table-of-combinations form.

The model of a crummy relay has now been converted from a relation to a function. However, this function does not uniquely determine an output for each input but determines the probability of an input-output combination. The relay is still crummy, but it is crummy in a regular sort of way!

Composite Events Until now we have considered only basic events. The probabilities of all the basic events are positive, and their sum equals 1. There are, however, other events which may be of significance in this system. For example, we may wish to know the probabilities of the following events (which are arbitrarily labeled):

$A =$ The system malfunctions (makes an error)

$B =$ The input has value 1

$C =$ The output has value 1

$D =$ The input or output has value 1

These and all other such events are known as *composite events* since they can be expressed in terms of the basic events.

	X	Y	p[XY]
$e_0 =$	0	0	$.20 = p_0$
$e_1 =$	0	1	$.05 = p_1$
$e_2 =$	1	0	$.30 = p_2$
$e_1 =$	1	1	$.45 = p_3$

Figure 5

For example, event A (The system malfunctions) simply consists of the basic events e_1 (sticking) or e_2 (sluggishness). Formally this is

$$A = X_0 Y_1 \vee X_1 Y_0 = e_1 \vee e_2$$

The probability of a composite event can be found by simply adding the probabilities of its individual basic events, since they are mutually exclusive. For example,

$$\begin{aligned}
p[A] &= p[e_1 \vee e_2] && \text{since } A = e_1 \vee e_2 \\
&= p[e_1] + p[e_2] && \text{since basic events are mutually exclusive} \\
&= p_1 + p_2 && \text{a convenient (but unnecessary) abbreviation} \\
&= 0.05 + 0.30 && \text{substituting} \\
&= 0.35 && \text{adding}
\end{aligned}$$

The probability that the system malfunctions is 0.35, which might be interpreted as "over a third of the time the system is in error." This probability, which was obtained from the model of a relay, can be verified if necessary by computing the relative frequency directly from the original observations, i.e.,

$$\frac{N_{40}\ [(X = 0 \text{ and } Y = 1) \text{ or } (X = 1 \text{ and } Y = 0)]}{40} = \frac{14}{40} = 0.35$$

Of course, once a model is developed, it essentially summarizes all the data, and so it is much simpler to use it than to return to the original data or observations. This model, containing four basic events and their probabilities, is sufficient to determine the probabilities of all possible composite events. For this system there are $2^4 = 16$ composite events. They are shown with their probabilities on the table-of-combinations of Figure 6. The value 1 indicates which of the four basic events are in the composite event.

X	Y	E_0	E_1	E_2	E_3	E_4	E_5	E_6	E_7	E_8	E_9	E_{10}	E_{11}	E_{12}	E_{13}	E_{14}	E_{15}
0	0	0	0	0	0	0	0	0	0	1	1	1	1	1	1	1	1
0	1	0	0	0	0	1	1	1	1	0	0	0	0	1	1	1	1
1	0	0	0	1	1	0	0	1	1	0	0	1	1	0	0	1	1
1	1	0	1	0	1	0	1	0	1	0	1	0	1	0	1	0	1
$p[E_i]$	=	0	.45	.30	.75	.05	.50	.35	.80	.20	.65	.50	.95	.25	.70	.55	1.0

Figure 6

The probabilities of the events B, C, and D can be computed from the model by first expressing these events in terms of basic events and then adding the corresponding probabilities. For example,

$$B = \text{The input has value 1}$$
$$p[B] = p[e_2 \vee e_3] = p_2 + p_3 = 0.30 + 0.45 = 0.75$$

$$C = \text{The output has value 1}$$
$$p[C] = p[e_1 \vee e_3] = p_1 + p_3 = 0.05 + 0.45 = 0.50$$

$$D = \text{The input or output has value 1}$$
$$p[D] = p[e_1 \vee e_2 \vee e_3] = 0.05 + 0.30 + 0.45 = 0.80$$

Notice that the event D is related to events B and C by

$$D = B \vee C$$

however the probabilities do not simply add, i.e.,

$$P[B \vee C] \neq p[B] + p[C]$$

This nonadditivity results because the events B and C are not mutually exclusive; they both have the event e_3 in common. This example should illustrate the importance of the restriction on additivity; probabilities are additive only when events are mutually exclusive. Formally this could be written as

$$p[E_i \vee E_j] = p[E_i] + p[E_j] \quad \leftrightarrow \quad E_i M E_j$$

This is read "the probability of the OR of two events equals the sum of the probabilities of the individual events if and only if the events are mutually exclusive." The concept of mutual exclusiveness can be viewed in two ways:

$$E_i M E_j \quad \leftrightarrow \quad E_i \cap E_j = \phi$$

or alternatively

$$E_i M E_j \quad \leftrightarrow \quad p[E_i \cap E_j] = 0$$

Before giving a general model of a probabilistic system let us consider one more example to abstract from later.

Modeling a Two-dice System

A die is a cube with one to six dots arranged on each side so that the opposite sides sum to 7. On the throw of one die the outcome must be a number from the set $\{1,2,3,4,5,6\}$. If two dice are thrown, there are 36 possible outcomes, corresponding to the elements of the set product

$$\{1,2,3,4,5,6\} \times \{1,2,3,4,5,6\}$$

From geometric, or a priori, or equiprobable considerations we could assume all 36 of the basic events equally likely and therefore having the probability $\frac{1}{36}$. However, if the dice are shortened in the 1-6 direction (called *ace-six flats*), experimental and relative frequency considerations would yield unequal probabilities.

The ideal and a nonideal situation are shown in Figure 7 to illustrate two common representations, the point graph and matrix representation. Notice that a table-of-combinations representation having 36 rows is possible but not convenient.

Some events from the family of events are illustrated below with their corresponding sets on the point graph of Figure 8.

$P =$ The first die X shows three dots

$Q =$ The second die Y shows more than four dots

$R =$ Both dice show the same number of dots

$S =$ The sum of the dots is 11

$T =$ The sum of the dots is less than 6.

The probability of any such event is again determined by summing

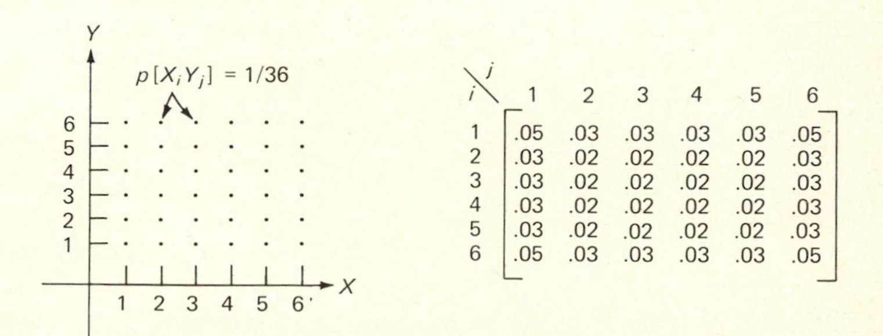

Figure 7

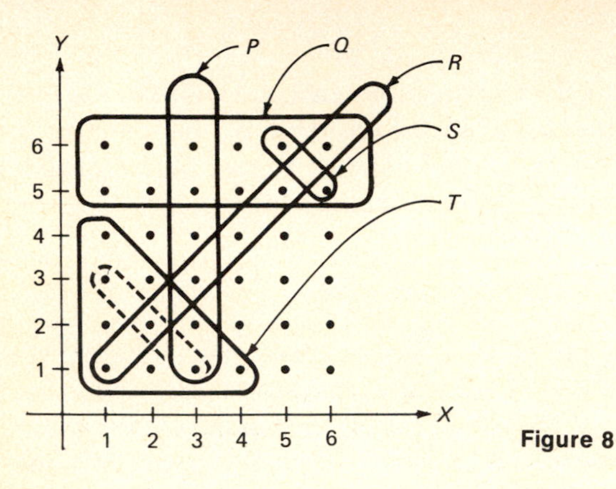

Figure 8

the probabilities of the corresponding basic events e_{ij}. For example, the probability of the event

M = The sum of the dots is 4

can be computed as

$$\begin{aligned} p[M] &= p[e_{13} \vee e_{22} \vee e_{31}] \\ &= p_{13} + p_{22} + p_{31} \\ &= \tfrac{1}{36} + \tfrac{1}{36} + \tfrac{1}{36} \\ &= \tfrac{3}{36} \\ &= 0.083 \end{aligned}$$

General Model of a Probabilistic System

The coin, relay, and dice are examples of probabilistic systems. These three systems, along with many others involving traffic, manufacturing, communicating, and gambling, have much in common. The following model is a generalization abstracted from such probabilistic systems. These concepts (like all mathematical concepts) can be related to the concepts of set theory.

A finite *probabilistic system* consists of three entities: an event space, a family of events, and a probability function.

An *event* represents any outcome, happening, occurrence, or result. It is often identified as being a set or subset. However, we will consider an event as a description, proposition, or statement *associated* with a set. Some examples of events are "the coin lands heads up," or "the relay malfunctions," or "the sum of the dots on the faces of the two dice is 7." Events, as used here, are only indirectly related to sets.

An *event space* Ω (omega) is essentially a universe; even its symbol resembles an inverted U. It consists of basic events e_i which are exhaustive (include all possible outcomes) and which are mutually exclusive (have no events occurring jointly). The event space makes precise what is to be considered as an outcome. Whether the event "the coin lands on edge" is a basic event depends on your goals and how you set up the system. You could make the model simple by not including such a basic event, and should it occur, you need only ignore it and flip again.

A *family of events*, denoted F, consists of all combinations of the basic events. Each such combination, called a *composite event*, could take the form of events connected by OR, AND, and NOT. Thus if there are n basic events in an event space, there are 2^n composite events in this family of events. For example, in the crummy-relay case there are 4 basic events and 16 composite events, but in the two-dice system there are 36 basic events and therefore 2^{36} (over 60 billion) composite events. Since there are often many events in a family, and since we are usually interested in very few of these events, the family of events is seldom written out completely.

A *probability function*, denoted p, simply assigns to each event E (in the family F) a number $p[E]$ between 0 and 1, with the restriction that the following properties hold:

1. *Positivity* The probability of any event is positive or zero.

$$P[E] \geqslant 0$$

2. *Normality* The probabilities of all basic events sum to 1.

$$\sum_{i=1}^{n} p[e_i] = 1$$

3. *Additivity* (restricted) The probability of the event "E_i or E_j" is the sum of the probabilities of the individual events if the events are mutually exclusive.

$$p[E_i \vee E_j] = p[E_i] + p[E_j] \qquad \text{if } E_i M E_j$$

Notice that there is no indication of how the probabilities are assigned by this function. For the ideal dice we used an equiprobable assignment, for the crummy relay we used a relative frequency assignment, and for the coin we found a great many assignments.

It is important to realize that the probability of any event E can be determined by constructing E from the basic events and then adding the corresponding probabilities of the basic events since they are mutually exclusive. Thus basic events are the building blocks of any event in the family.

Alternative Notation Since the notation and nomenclature of probability are not standardized, a list of equivalent expressions used in other texts may be helpful. For example, an event space is often referred to as a sample space, event set, or partition. A basic event is called an outcome, a simple event, a fundamental event, a basic result, or a sample point. The family of events is often referred to as a field or a Borel field.

A probabilistic system is often given in a more formidable looking symbolic form as follows:

A *finite probabilistic system* is a 3-tuple $\langle \Omega, F, p \rangle$ where

$\Omega = \{e_1, e_2, \ldots, e_n\}$ is an event space of basic events e_i which are exhaustive: $e_1 \vee e_2 \vee \cdots \vee e_n = \Omega$ and mutually exclusive: $e_i \cap e_j = \phi$ for $i \neq j$

$F = \{E: E \subseteq \Omega\}$ is a family of events

$p: F \rightarrow \{x: 0 \leq x \leq 1\}$ is a probability function satisfying axioms

Positivity	$p[E] \geq 0$
Normality	$p[\Omega] = 1$
Additivity	$p[E_i \vee E_j] = p[E_i] + p[E_j]$ if $E_i M E_j$

Now that we have a model we should be able to analyze it and obtain some useful properties, results, or theorems. Before doing this, however, let us consider more examples and representations and some other concepts.

Representation of a Probabilistic System In the examples considered so far we have introduced many representations. The sagittal function, the table-of-combinations, and the set diagram were convenient for the crummy relay system; whereas the point graph and matrix notation were convenient for the dice system. A tree could also be used. The crummy relay system is shown in Figure 9 represented in all these ways to summarize and illustrate the variety of representations.

As if these representations were not sufficient, there is an additional, most important representation (sagittal function-on-a-relation) which will be introduced later in Section 2 (Analysis).

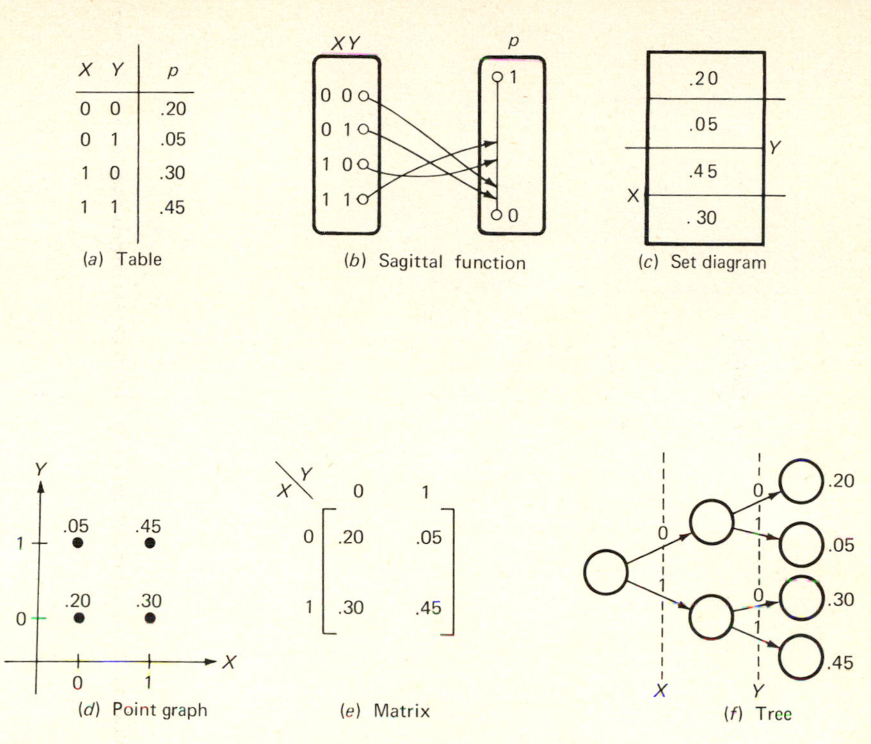

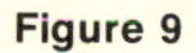

Figure 9

EXERCISE SET 1

1. It is surprisingly instructive (even though tedious) to flip a coin, and to plot the relative frequency as you proceed. Continue until you are satisfied that equilibrium has been reached. Comment!

2. Determine the following probabilities from the model of the crummy relay system and comment.

(*a*) The input has value 1, or the output has value 0.
(*b*) The input has value 1, and the system malfunctions. *Ans.*: 0.45
(*c*) The input has value 0, and the system malfunctions.
(*d*) The input and output have the same value. *Ans.*: 0.65
(*e*) Neither input nor output has value 0.
(*f*) $e_0 \vee e_1 \vee e_2$ (*g*) $e_0(e_1 \vee e_2)$
(*h*) $\bar{B} \vee C$ (*i*) $A(B \vee \bar{C})$
(*j*) $X_0 \vee X_1 Y_1$ (*k*) $X_0 \vee Y_1$

3. In many games involving dice the sum of the dots on the two faces is most important. Determine the probabilities of all 11 possible sums for the ideal case. Sketch these events on a point graph.

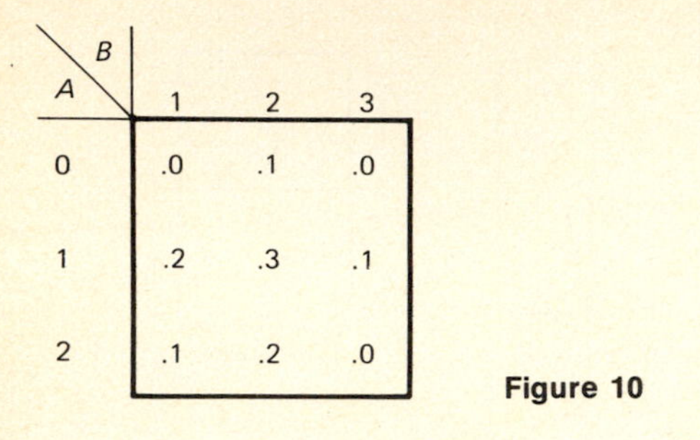

Figure 10

4. The matrix given in Figure 10 describes a system consisting of two production lines in some factory, the first line producing A items a day and the second line producing B items per day. Find the probability of the following events:

(*a*) Both production lines make the same number of items.
(*b*) Production of line B exceeds that of line A.
(*c*) Production of A and B differs by at most one item.
(*d*) The total number of items produced is exactly 3. *Ans.*: 0.4
(*e*) $A < 2$ and $B \geq 2$.
(*f*) $A < 2$ or $B > 2$.
(*g*) $A = 3$.

Other Examples of Probabilistic Systems

In this section we will consider some examples which illustrate some difficulties in modeling and indicate additional methods of assigning and computing probabilities.

A Survey. Consider the following results of a survey which is intended to relate age and sex to the opinion about a product or a program. The results of Figure 11 are given in the form of a set diagram with the numbers indicating numbers of people. Let us now identify the elements of the probabilistic system $\langle \Omega, F, p \rangle$, the event space, the family of events, and the probability function. The first choice for the event space is likely

(*a*)

		L	
	20	30	
	10	15	M
Y	5	15	M
Y	5	0	

(*b*)

		L	
	$p_0 = .20$	$p_1 = .30$	
	$p_2 = .10$	$p_3 = .15$	M
Y	$p_6 = .05$	$p_7 = .15$	M
Y	$p_4 = .05$	$p_5 = .00$	

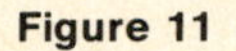

Figure 11

to be Y, M, and L, which is incorrect since some of these categories can occur jointly. The basic events are instead the 8 (2^3) combinations of these events.

$$\begin{aligned}\Omega &= \{\bar{Y}\bar{M}\bar{L}, \bar{Y}\bar{M}L, \bar{Y}M\bar{L}, \bar{Y}ML, Y\bar{M}\bar{L}, Y\bar{M}L, YM\bar{L}, YML\} \\ &= \{Y, \bar{Y}\} \times \{M, \bar{M}\} \times \{L, \bar{L}\}\end{aligned}$$

The family of events consists of all possible combinations of these eight basic events. There are thus 2^8 or 256 different composite events in this family.

The probability function which is natural in this case is the ratio

$$p[E] = \frac{n(E)}{n(\Omega)}$$

where $n(E)$ is the number of people corresponding to event E and $N(\Omega)$ is the total number of people surveyed. This method of assigning probabilities is sometimes called the *classical* assignment, although if each of the eight ratios tends to a constant value as the number of people surveyed increases, the assignment becomes a *multiple* relative-frequency assignment.

The set diagram in Figure 11*b* indicates the resulting event space and probability function. The subscripts on the probabilities denote the binary designation of the basic subsets (events). It is easily seen that the probability function has the three necessary properties of positivity, normality, and (restricted) additivity. From such a set diagram the probabilities of any events can be determined directly by summing the probabilities corresponding to any set. For example,

$$\begin{aligned}p[L] &= p[e_0 \vee e_2 \vee e_4 \vee e_6] \\ &= p[e_0] + p[e_2] + p[e_4] + p[e_6] \\ &= p_0 + p_2 + p_4 + p_6 \\ &= 0.2 + 0.1 + 0.05 + 0.05 \\ &= 0.40\end{aligned}$$

From this result it is a temptation to say: The probability that any person dislikes this program is 0.4. However, the universe is not the set of all people, but only those interviewed in a survey. The statement must be modified to: The probability that any surveyed person dislikes the program is 0.4. These two statements are equivalent and interchangeable if the set of sampled people is large, if the set is a representative cross section of all people, and if the process of taking the survey does not interfere with the system. These and other such factors are outside the realm of the theory but are nevertheless very important in practice.

Probabilities in Noncrummy Relays Probabilities have been used to describe malfunctions of crummy relays, but probabilities may also apply to ideal or deterministic relays to describe frequencies of operation. Consider, for example, the system of Figure 12. It may have been the input to the crummy relay of Figure 1. The frequency of operation $f(e_i)$ of the three relays is summarized on the given table-of-combinations. From these frequencies the probabilities $p(e_i)$ are computed by the relative frequency assignment. The probabilities are also shown on a set diagram for convenience. The event "current flows in network N" can be written as "current flows when B or, A and not C, are closed." This corresponds to the algebraic expression

$$N = B \vee A \cdot \bar{C}$$

which corresponds to the basic events

$$e_2 \vee e_3 \vee e_4 \vee e_6 \vee e_7$$

having the probability

$$p_2 + p_3 + p_4 + p_6 + p_7 = 0.75$$

Notice this value of 0.75 coincides with the previous probability of flow of input current to the crummy relay.

Notice again that the concept of probability has been applied in two different ways to the system consisting of an environment connected to a

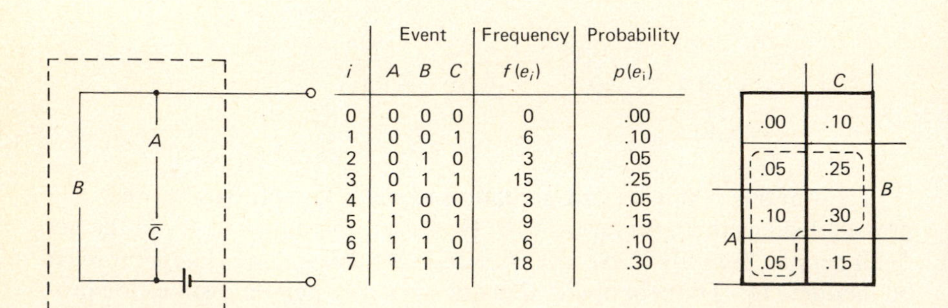

	Event			Frequency	Probability
i	A	B	C	$f(e_i)$	$p(e_i)$
0	0	0	0	0	.00
1	0	0	1	6	.10
2	0	1	0	3	.05
3	0	1	1	15	.25
4	1	0	0	3	.05
5	1	0	1	9	.15
6	1	1	0	6	.10
7	1	1	1	18	.30

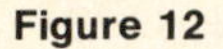
Figure 12

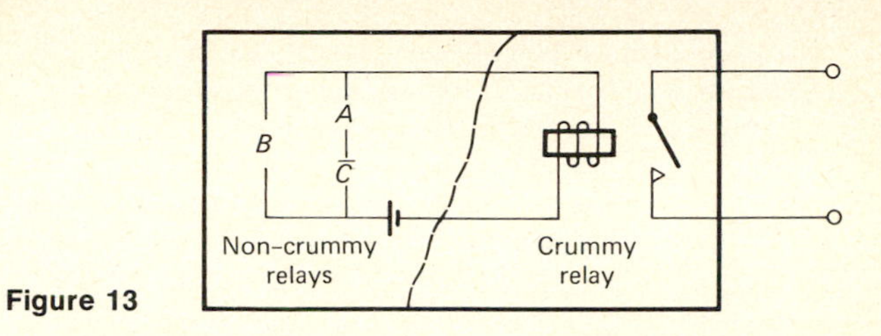

Figure 13

crummy relay (in Figure 13). Probabilities described the frequency of operation of the ideal environment (input), and they also described the error or malfunction of the crummy relay. Probabilities were applied to both crummy and noncrummy relays in the same system. Later, probabilities will also be used to describe the catastrophic failure (or death?) of relays and other systems.

EXERCISE SET 2

1. The following networks are constructed from relays of Figure 12. Find the probabilities that each of the following networks is closed.

(*a*) ABC (*b*) $A \vee B \vee C$
(*c*) $A(B \vee C)$ (*d*) $A \vee BC$
(*e*) $\bar{A} \vee B$ (*f*) $\bar{A} \vee AB$
(*g*) $\overline{B \vee C}$ (*h*) $(\overline{B \vee C}) \vee \bar{B}\bar{C}$

2. A system (man or weapon) has at most four trials to succeed and stops after the first success. Construct the corresponding event space and represent it as a tree. How many events are in the family of events? *Ans.*: 32

3. A certain manufactured item is classified for defects in three categories: materials, workmanship, and color. In a typical lot of 100 items, it is known that:

10 have a defect in material only.
10 have a defect in workmanship only.
10 have a defect in color only.
10 have a defect in color but not material.
30 have a defect in material but not color.
20 have exactly two types of defects.
20 have a defect in color.

Construct a probabilistic system (table-of-combination form and also set-diagram form) corresponding to these data. Find the probability of each of the following events, indicating which representation is most convenient for determining each probability:

(*a*) An item has no defects. *Ans.*: 0.4
(*b*) An item has three types of defects.
(*c*) An item has at least one type of defect.
(*d*) An item has exactly one type of defect. *Ans.*: 0.3

(*e*) An item has at most two types of defects.
(*f*) An item has defects in material or workmanship.
(*g*) An item has defects in material and not color.
(*h*) An item has defects in neither color nor material. *Ans.*: 0.5
(*i*) An item has defects in material and color, or workmanship.
(*j*) An item has defects in material and, color or workmanship.

2. ANALYSIS OF PROBABILISTIC SYSTEMS

In this section we will consider many types of probabilities (conditional, joint, marginal, forward, and reverse probabilities) and study some properties of, and relations between, probabilities. Then, in the following section, these concepts will be applied to synthesis and optimization.

Conditional Probability

Often, because of experience, observation, or the passage of time, some additional information becomes known. This information may indicate that some event has occurred or that some restrictions have been imposed on the system. These changing conditions may cause a change in probabilities. The concept of conditional probability provides a way for revising probabilities on the basis of any additional knowledge.

The *conditional probability* of an event E occurring, given that an event F has occurred, is denoted by $p[E \mid F]$ and defined as

$$p[E \mid F] = \frac{p[E \cdot F]}{p[F]}$$

This definition is easily seen from the set diagram of Figure 14, which illustrates that the given event reduces the original universe to F. That part of E which is in the new universe F is now $E \cdot F$. The ratio of these probabilities then leads to the definition

$$p[E \mid F] = \frac{p[E \cdot F]}{p[F]}$$

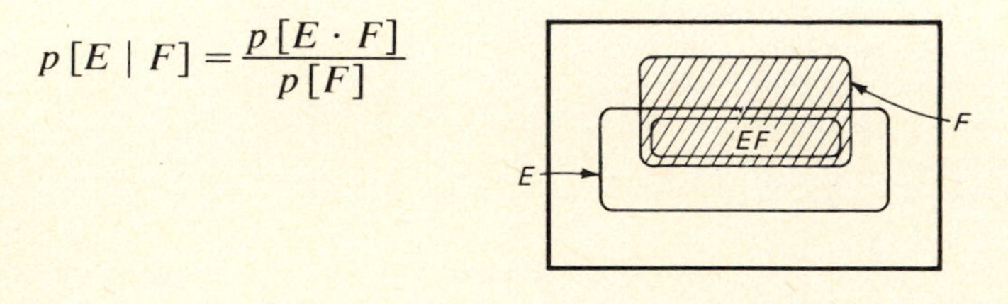

Figure 14

This definition is often rewritten as a product rather than a quotient and in this form called the *probability product property*

$$p[EF] = p[E \mid F] * p[F]$$

Notice especially that this product property indicates that individual probabilities are not multiplied, in general, to determine joint probabilities, i.e.,

$$p[EF] \neq p[E] * p[F]$$

Example The original crummy relay repeated in Figure 15 has differing probabilities of error depending on how it is used.

If the input has value 0, the error is

$$p[\text{Error} \mid X_0] = \frac{5}{25} = 0.20$$

However, if the input has value 1, the probability of error is

$$p[\text{Error} \mid X_1] = \frac{30}{75} = 0.40$$

Notice that this system is twice as likely to err when the input value is 1. Later this observation will be used to make systems more reliable.

Conclusions from Conditional Probabilities. Suppose the survey of Figure 16 yields the following information (which is actually fictitious) relating smoking S and lung cancer C. The probability of a person's having cancer is

$$p[C] = 0.03 + 0.12 = 0.15$$

However, given that a person smokes, the probability of having cancer becomes

$$p[C \mid S] = \frac{p[C \cdot S]}{p[S]} = \frac{0.12}{0.40} = 0.30$$

If the person does not smoke, the probability of cancer is

$$p[C \mid \bar{S}] = \frac{p[C\bar{S}]}{p[\bar{S}]} = \frac{.03}{0.60} = 0.05$$

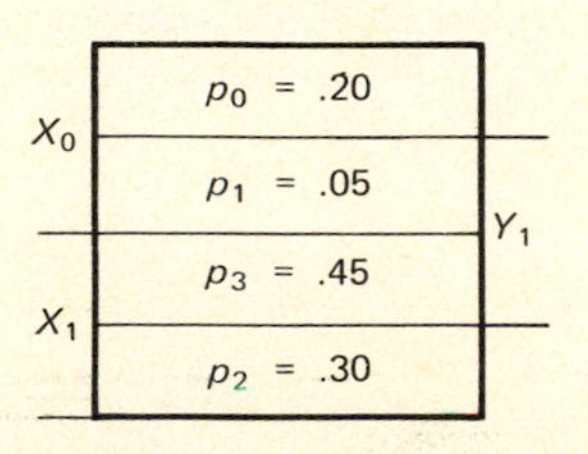

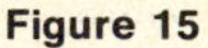
Figure 15

.57
.03
C
.12
S
.28

Figure 16

Notice that if a person smokes, his probability of cancer increases by a factor of 2 and if he does not smoke it decreases by a factor of 3.

These probabilities seem to indicate that smoking causes cancer. However the following probabilities seem to indicate that cancer causes smoking!

1. The probability of a person smoking is $p[S] = 0.40$.
2. If a person has cancer his probability of smoking doubles to

$$p[S \mid C] = \tfrac{12}{15} = 0.8$$

3. If a person does not have cancer, his probability decreases to

$$p[S \mid \bar{C}] = \tfrac{28}{85} = 0.33$$

This example illustrates the point that the probabilities *alone* do not orient the relationship according to cause and effect. In fact there may be some other factors causing smoking and cancer to occur together.

Similarly we should be careful not to conclude that medicines cause illness because both are found together!

EXERCISE SET 3

1. In the television survey of Figure 11, find the probability that a person likes the program if it is known that:

(*a*) The person is a male.
(*b*) The person is a young female.
(*c*) The person is young or male.
(*d*) The person is neither female nor young.

Find also the following probabilities:

(*e*) $p[(M \vee Y) \mid \bar{L}]$
(*f*) $p[M \mid (L \vee Y)]$ *Ans.*: 0.5
(*g*) $p[(M \vee Y) \mid (L \vee \bar{Y})]$

2. The table of Figure 17 indicates the probabilities of congestion of three freeways A, B, C during a typical rush period. The value of 1 indicates occurrence of congestion.

(*a*) Which freeway is least congested?
(*b*) Find the probability that B is congested given that:

1. A is congested.
2. A is not congested.
3. C is congested.
4. A and C are congested.
5. A or C are congested. *Ans.*: 0.5
6. Neither A nor C is congested.

A	B	C	$p[ABC]$
0	0	0	.4
	0	1	.0
0	1	0	.0
0	1	1	.1
1	0	0	.0
1	0	1	.3
1	1	0	.2
1	1	1	.0

Figure 17

		D		
.25	.00	.00	.05	
.15	.10	.05	.00	S
.05	.10	.05	.00	
.20	.00	.00	.00	
	M			

(G labels the left side of the map.)

Figure 18

*(c) Show how knowing the congestion of A and C *completely determines* the congestion of B. Draw a sagittal diagram of this function.

3. Suppose the given system of Figure 18 describes how good grades G are related to smoking S, drinking D, and marital status M (married) of students at some school.

(a) Find the probability that a student gets good grades G:

1. Given that he drinks
2. Given that he smokes
3. Given that he is married
4. Given no additional information

(b) Which of the three variables D, M, S are the grades least dependent on?

Ans.: Smoking

(c) What other conclusions can you draw from these data?

Properties of Probabilistic Systems

Before considering other aspects and applications of probabilistic systems it would be well to establish some general theorems or properties which hold for all probabilistic systems. The proof of such theorems starts from any true statement and proceeds step by step, finally arriving at the required conclusion.

An unusual aspect of proofs in probabilistic systems is that they involve two algebras, the Boolean algebra, which applies to events, and the ordinary algebra, which applies to numbers. For this reason it is important to be consistent in notation, realizing that expressions such as $P[A+B]$ and $p[A] \vee p[B]$ are meaningless and confusing. The justification of each step of a proof may involve properties of these two algebras, as well as the various probabilistic properties of positivity, normality, and restricted additivity.

Some examples of proofs follow. The resulting theorems will be used and applied later in the chapter. The first theorem shows a formal way of arriving at a result which we have previously determined in a less formal way.

Theorem 1

$$p[\bar{E}] = 1 - p[E]$$

Proof

$\bar{E} \vee E = \Omega$	property of complementarity (events)
$p[\bar{E} \vee E] = p[\Omega]$	property of substitution (events)
$p[\bar{E} \vee E] = 1$	property of normality (probability)
$p[\bar{E}] + p[E] = 1$	property of restricted additivity (probability)
$p[\bar{E}] = 1 - p[E]$	property of transposition (ordinary algebra)

Theorem 2

$$p[E \vee F] = 1 - p[\bar{E} \cdot \bar{F}]$$

Proof

$E \vee F = (\overline{\bar{E} \cdot \bar{F}})$	an equality of events (dualization)
$p[E \vee F] = p[\overline{\bar{E} \cdot \bar{F}}]$	substitution of equivalent events
$= 1 - p[\bar{E} \cdot \bar{F}]$	Theorem 1

Notice that there may be alternative ways of specifying the probability of an event such as $E \vee F$. For example,

Theorem 3

$$p[E \vee F] = p[E] + p[F] - p[EF]$$

These theorems can be extended to three events as follows.

Theorem 4

$$p[E \vee F \vee G] = 1 - p[\bar{E} \cdot \bar{F} \cdot \bar{G}]$$

Other expressions are not always so simple to extend to more events. For example,

$$P[E \vee F \vee G] \neq p[E] + p[F] + p[G] - p[EFG]$$

Instead we can prove another theorem.

Theorem 5

$$p[E \vee F \vee G] = p[E] + p[F] + p[G] - p[EF] - p[EG] - p[FG] + p[EFG]$$

Proof

$$\begin{aligned} p[E \vee F \vee G] &= p[(E \vee F) \vee G] \\ &= p[(E \vee F)] + p[G] - p[(E \vee F)G] \\ &= p[E] + p[F] - p[EF] + p[G] - p[EG \vee FG] \\ &= p[E] + p[F] + p[G] - p[EF] - p[EG] \\ &\quad - p[FG] + p[EGFG] \end{aligned}$$

Some other interesting theorems which could be proved follow.

Theorem 6

$$p[E \underline{\vee} F] = p[E] + p[F] - 2p[EF]$$

Theorem 7

$$p[E \downarrow F] = 1 - p[E] - p[F] + p[EF]$$

Theorems involving conditional probabilities may be proved in similar ways.

Theorem 8 The Extended Product Principle

$$p[E_1E_2E_3] = p[E_1 \mid E_2E_3] * p[E_2 \mid E_3] * p[E_3]$$

Proof

$$\begin{aligned} p[E_1E_2E_3] &= p[E_1(E_2E_3)] \\ &= p[E_1 \mid E_2E_3] * p[E_2E_3] \quad \text{from definition of conditional probability} \\ &= p[E_1 \mid E_2E_3] * p[E_2 \mid E_3] * p[E_3] \end{aligned}$$

By commuting the events we can obtain another, more commonly used result.

$$p[E_1E_2E_3] = p[E_3E_2E_1] = p[E_1] * p[E_2 \mid E_1] * p[E_3 \mid E_1E_2]$$

The general form of the extended product principle is

$$p[E_1E_2E_3 \cdots E_n] = p[E_1] * p[E_2 \mid E_1] * p[E_3 \mid E_2E_1] * p[E_4 \mid E_3E_2E_1] \cdots * p[E_n \mid E_{n-1} \cdots E_2E_1]$$

Restricted Results

Many of the convenient mathematical properties, such as additivity, do not hold in general for probabilistic systems. However, they often do hold under restricted conditions. For example,

$$p[E \vee F] \neq p[E] + p[F] \qquad \text{in general}$$

$$p[E \vee F] = p[E] + p[F] \qquad \text{if } EMF$$

There are other such results which hold for restricted conditions. For example we know that conditional probabilities cannot be expressed as a simple quotient;

$$p[E \mid F] \neq \frac{p[E]}{p[F]} \qquad \text{in general.}$$

However, we can determine the particular condition for which this simple quotient holds, by rewriting it as

$$\frac{p[EF]}{p[F]} = \frac{p[E]}{p[F]}$$

$$p[EF] = p[E]$$

$$EF = E$$

$$E \subseteq F$$

In other words

$$p[E \mid F] = \frac{p[E]}{p[F]} \qquad \text{if } E \subseteq F$$

EXERCISE SET 4

1. Prove Theorems 6 and 7.
2. Extend Theorem 6 to three events, $p[E_1 \veebar E_2 \veebar E_3]$
3. Why is it meaningless to try to expand $p[E_1 \downarrow E_2 \downarrow E_3]$?

4. Prove all the following alternative ways of extending the product principle:

(*a*) $p[ABC] = p[A]p[B \mid A]p[C \mid AB]$
(*b*) $p[ABC] = p[AB \mid C]p[C]$
(*c*) $p[ABC] = p[B \mid AC]p[C \mid A]p[A]$

5. Indicate what special conditions (if any) are necessary for each of the following results to be true. Justify briefly.

(*a*) $p[E \vee F] = p[E]$ *Ans.*: $F \subseteq E$
(*b*) $p[E \vee F] = p[EF]$
(*c*) $p[E \vee F] = p[E \downarrow F]$
(*d*) $p[E \vee F] = p[E \underline{\vee} F]$
(*e*) $p[E \vee F] = p[F \vee E]$ *Ans.*: always

6. When are the following results (concerning conditional probability) valid?

(*a*) $p[A \mid B] = p[B \mid A]$
(*b*) $p[(A \vee B) \mid C] = p[A \mid C] + p[B \mid C]$
(*c*) $p[\bar{A} \mid \bar{B}] = 1 - p[\bar{A}/B]$
(*d*) $p[\bar{A} \mid \bar{B}] = 1 - p[A \mid \bar{B}]$

A Systems View of Probabilistic Systems

In this section the previously modeled crummy relay system will be used to illustrate some system-theoretic concepts, alternative representations, mathematical tools, philosophical principles, computational results, and practical applications. The crummy relay system and its model are repeated in Figure 19 for convenience of reference. This system is represented as a matrix called the *joint-probability matrix*, which indicates the probability of the joint occurrence of the input and output.

One convenience of the joint-probability matrix representation is that the probability of any input value can be determined very simply by summing the row corresponding to this value. For example, the probability that the input has value 1 is determined as

$$p[X = 1] = p[X_1] = p[X_1 Y_0 \vee X_1 Y_1]$$
$$= \sum_j p[X_1 Y_j] = 0.3 + 0.45 = 0.75$$

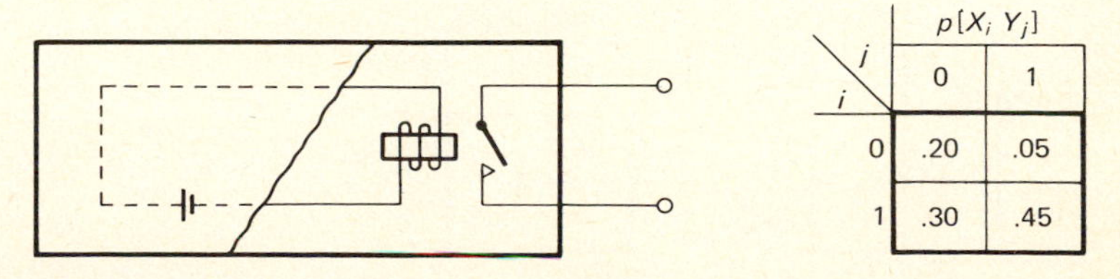

Figure 19

In general

$$p[X_i] = \sum_j p[X_i X_j]$$

Notice that summing over an index such as j has the effect of removing this index from the resulting expression $p[X_i]$ at the left. Similarly the probability of an output's having value j can be determined by summing all probabilities in the jth column, i.e.,

$$p[Y_j] = \sum_i p[X_i Y_j]$$

The probabilities such as $p[X_i]$ or $p[Y_j]$, in which a value has been summed out, are known as *marginal probabilities* and are often written in the margins of the joint probability matrix, as shown in Figure 20*a*. An algorithm describing this computation is given in Figure 23.

Forward Conditional Probabilities In addition to joint and marginal probabilities, we will also use conditional probabilities. A forward conditional probability indicates the probability of an output value j, when the input value i is given or known. This forward probability has the form $p[Y_j \mid X_i]$, which can be calculated directly from the joint and marginal probabilities as

$$p[Y_j \mid X_i] = \frac{p[X_i Y_j]}{p[X_i]} = \frac{p[X_i Y_j]}{\sum_j p[X_i Y_j]}$$

For example,

$$p[Y_1 \mid X_0] = \frac{p[X_0 Y_1]}{p[X_0]} = \frac{0.05}{0.25} = 0.2$$

Similarly all four possible forward probabilities could be calculated and represented on a forward probability matrix as in Figure 20*b*. Comparing the two matrices of Figure 20 shows that an element of the forward matrix

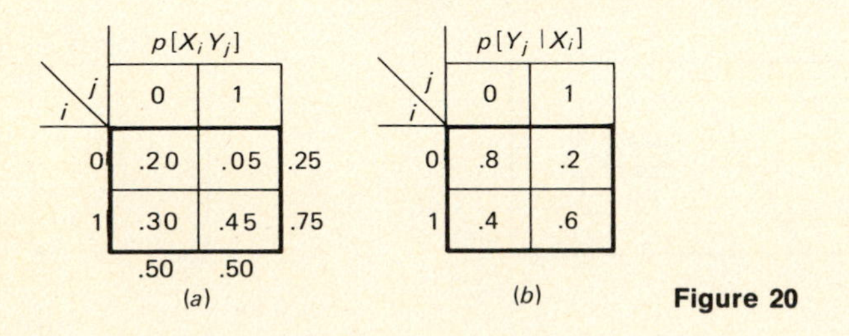

Figure 20

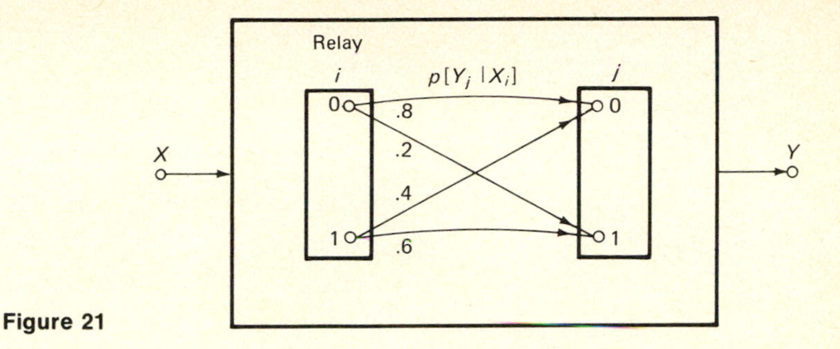

Figure 21

is determined by taking the corresponding element of the joint matrix and dividing it by the marginal probability at the right of this matrix.

Sagittal Representation A final and important representation of a system is the sagittal representation illustrated in Figure 21. Notice that the forward conditional probabilities are simply functions-on-relations. This representation indicates all the significant probabilities in a graphic way. These probabilities are important since most systems, such as relays and communication channels, are specified most meaningfully in such a way. The reason for this is that the forward probabilities separate the system from its environment. This will be illustrated next.

It may not have been apparent that the original crummy relay system consisted of both the relay and its environment. It also may not have been apparent that the probability of error, then computed as 0.35, did not describe the relay alone but depended on the environment (which had input value 1 with probability 0.75). To illustrate this important point, let us study this same crummy relay (which is now detached from any source by conditional forward probabilities) when it is connected to another source which has value 1 with a probability of 0.1 as in Figure 22. Notice that the environment can also be described by a sagittal function-on-a-relation. It has only one input value which cannot be controlled or changed; i.e., the environment behaves as a source or generator of probabilities.

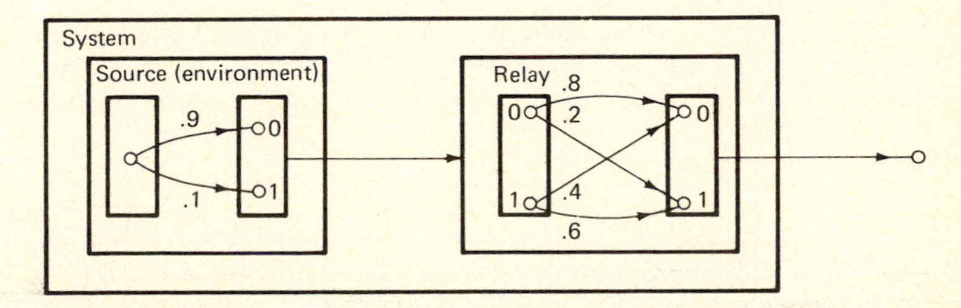

Figure 22

The probability of error (or malfunction) is still in general

$p\,[\text{Error}] = p\,[\text{Sticking or sluggishness}]$	
$= p\,[X_0 Y_1 \vee X_1 Y_0]$	by decomposing events
$= p\,[X_0 Y_1] + p\,[X_1 Y_0]$	mutual exclusiveness
$= p\,[Y_1 \mid X_0]\; p\,[X_0] + p\,[Y_0 \mid X_1]\; p\,[X_1]$	conditional-probability
$= (0.2) * (0.9) + (0.4) * (0.1)$	substituting in this case
$= 0.22$	numerical solution

Comparing this probability of error of 0.22 with the previous value of 0.35 leads to the conclusion that the probability of error is indeed a property not of the relay alone but also of the source, or environment. The probabilities of the source entered in computing the joint probabilities from the forward conditional probabilities.

This example illustrates a philosophical and practical problem of concern to any observer or experimenter. It shows that some properties of systems (such as the probability of error) can be influenced by the inputs the experimenter uses to determine his information. In short, the experimenter is a part of any system that he studies.

In some cases, such as the above, the problem is not terribly difficult since it is still possible to separate the experimenter from the system by using conditional probabilities. In other fields such as physiology, biology, sociology, psychology, and modern physics the experimenter may not be able to separate his effect completely from the system. Even a blind man measuring the size of a soap bubble encounters this same philosophical (and of course practical) difficulty.

EXERCISE SET 5

1. *System identification.* It is desirable to relate the weather on one day to the reading of a barometer on the previous day, to develop a probabilistic predicting system. The barometer B has values R, S, F corresponding to rising, still, and falling. The weather W has values G or B (good or bad) according to some well-defined criteria. A "typical" three-week observation follows, indicating the day, the barometer reading on this day, and the weather on this day.

Day	1	2	3	4	5	6	7		8	9		· · ·			14				· · ·				21
Barometer	S	F	F	F	R	F	F		S	F	S	F	S	S	S		R	S	S	F	S	S	R
Weather	B	G	B	B	G	G	G		G	B	B	G	G	G	G		G	G	B	G	B	B	B

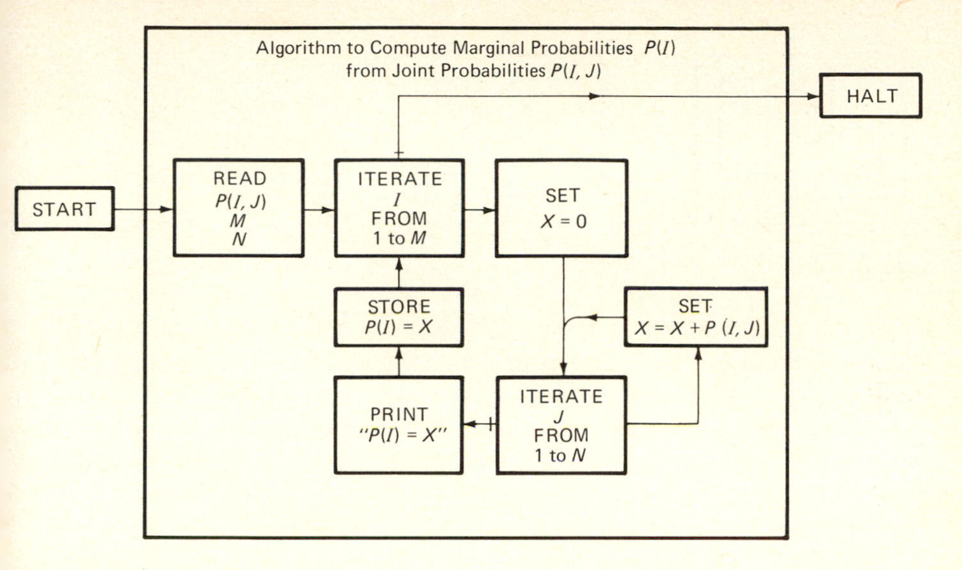

Figure 23

From this sequence construct a probabilistic prediction system by indicating the forward conditional probabilities $p[W_j \mid B_i]$ in the form of a sagittal function-on-a-relation. A similar barometer will be used later (in Figure 96) as an information instrument for decision making. (*Hint*: $p[W_G \mid B_R] = 1$.)

2. The algorithm of Figure 23 is to compute the marginal probabilities $p[X_i]$ from the joint probabilities $p[X_i Y_j]$. Extend this algorithm to compute the forward probabilities $p[Y_j \mid X_i]$.

Reverse Probabilities So far we have considered only forward probabilities $p[Y_j \mid X_i]$, which are important in the specification and decomposition of systems. We could also investigate conditional probabilities of the form $p[X_i \mid Y_j]$, which are known as reverse conditional probabilities or simply as reverse probabilities. Such probabilities are significant in systems where the output is observed and it is desired to know the probability that a particular input caused this output. It is an attempt at relating a cause to a given effect.

For example, in the original crummy system (redrawn in Figure 24), if the output (contact position) is observed to have value 1, what would you guess is the probability that the input (voltage) value is also 1? Similarly if the observed output has value 0, how likely is it that the input causing this has value 0 also? Of course in an ideal noncrummy case these above probabilities would be 1. However, in this nonideal case it would be extremely beneficial to attempt an intuitive guess at the probabilities. Before computing the values would you guess the probabilities

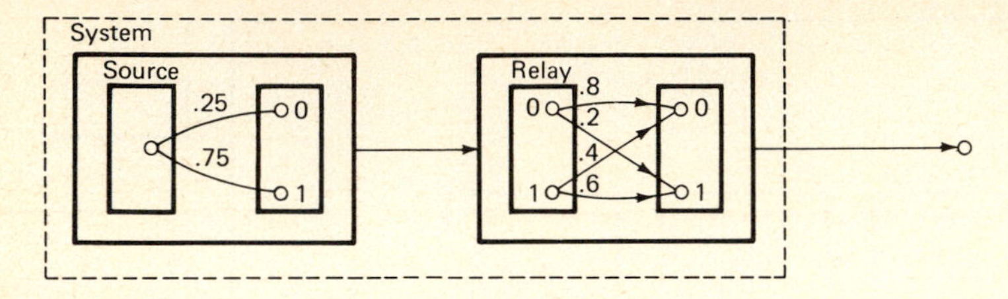

Figure 24

are in the vicinity of 0.9, or 0.8, or perhaps even 0.6? You should hazard a guess at these two probabilities to see just how wrong unaided intuition can be in even such simple cases.

The exact evaluation of these reverse probabilities is straightforward:

$$p[X_1 \mid Y_1] = \frac{p[X_1 \cdot Y_1]}{p[Y_1]}$$

$$= \frac{p[Y_1 \mid X_1]\, p[X_1]}{p[Y_1 \mid X_1]\, p[X_1] + p[Y_1 \mid X_0]\, p[X_0]}$$

$$= \frac{(0.6)\,(0.75)}{(0.6)(0.75) + (0.2)(0.25)} = 0.9$$

This can be interpreted intuitively as: "if the output is observed to have value 1, the input causing it also has value 1 about 90 percent of the time."

The computation of reverse probabilities can be generalized from the above particular example to the following result, which is known as *Baye's theorem*:

$$p[X_i \mid Y_j] = \frac{p[Y_j \mid X_i]\, p[X_i]}{\sum_i p[Y_j \mid X_i]\, p[X_i]}$$

Later we will find a more convenient and systematic method of computation using matrices.

The second reverse probability, that the output value 0 was caused by an input value 0, can be similarly computed and yields a more startling result

$$p[X_0 \mid Y_0] = \frac{(0.8)(0.25)}{(0.8)(0.25) + (0.4)(0.75)} = 0.4$$

This result is startling because it indicates that an output value of 0 is unlikely to be caused by an input value of 0. When the output has value 0, we are extremely uncertain about the input value.

This unexpected result is indeed a correct one and can be obtained in another way which may lead to more insight. Consider the physical, analogous system of Figure 25 consisting of balls marked 0 and 1 falling through various paths. There are two contributions to this reverse probability $p[X_0 \mid Y_0]$: First, a large fraction 0.8 of the small portion 0.25 of the input 0s become output 0s. Second, a small fraction 0.4 of a large portion 0.75 of the input 1s become output 0s. Of the outputs that have value 0, a fraction $(0.25)(0.8) = 0.20$ is caused by an input value of 0, whereas a fraction $(0.4)(0.75) = 0.30$ is caused by an input of 1. Of the total portion $(0.2 + 0.3)$ of the outputs having value 0, only 0.2 is actually caused by a zero, yielding the probability

$$p[X_0 \mid Y_0] = \frac{0.2}{0.2 + 0.3} = 0.40$$

which agrees with the original conclusion. These reverse conditional probabilities could be improved by improving the relay (naturally!), but they can also be improved by decreasing the probability of the source's having value 1.

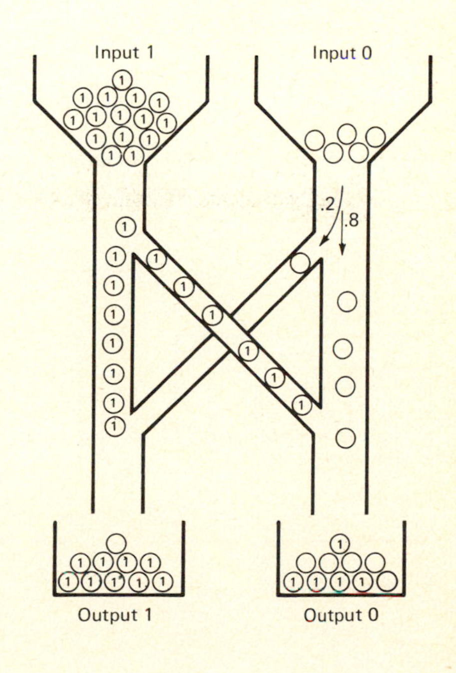

Figure 25

A Communication System This example illustrates an application of probability theory to communications, as well as showing a systematic method of computing all joint and reverse probabilities.

Consider the communication system of Figure 26, consisting of a source generating three symbols and a channel transmitting these three symbols. The three symbols could correspond to a negative voltage (−1), a zero voltage (0), and a positive voltage (+1). The symbols could also correspond to a dot, dash, and space. Let us compute the probability of error of this system, and especially compute the reverse conditional probabilities $p[X_i \mid Y_i]$ for all three values of *i*. First, the forward probabilities from the given channel are entered on a matrix (Figure 27*a*). The input probabilities $p[X_i]$ may be associated with each row (at right) in preparation for the next step.

The elements $p[Y_j \mid X_i]$ of every row in the forward matrix are all multiplied by the input probability $p[X_i]$ associated with this row, yielding the joint probabilities

$$p[X_i \cdot Y_j] = p[Y_j \mid X_i] * p[X_i]$$

The columns (and rows) of this joint matrix (Figure 27*b*) can now be summed, yielding the marginal probabilities of the output (and input).

Each element $p[X_i Y_j]$ of the joint matrix is now divided by the marginal probability of output $p[Y_j]$ below each column, yielding the reverse probabilities

$$p[X_i \mid Y_j] = p[X_i \cdot Y_j] / p[Y_j]$$

These are shown in the reverse matrix of Figure 27*c*.

The elements on the diagonals of the matrices are usually most

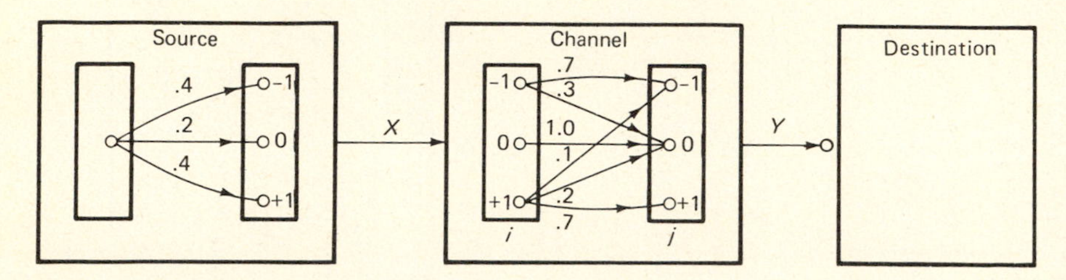

Figure 26

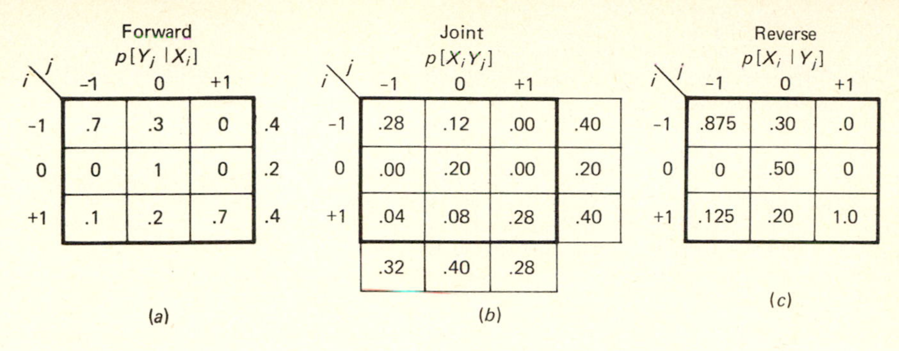

Forward $p[Y_j \mid X_i]$

i \ j	-1	0	+1	
-1	.7	.3	0	.4
0	0	1	0	.2
+1	.1	.2	.7	.4

(*a*)

Joint $p[X_i Y_j]$

i \ j	-1	0	+1	
-1	.28	.12	.00	.40
0	.00	.20	.00	.20
+1	.04	.08	.28	.40
	.32	.40	.28	

(*b*)

Reverse $p[X_i \mid Y_j]$

i \ j	-1	0	+1
-1	.875	.30	.0
0	0	.50	0
+1	.125	.20	1.0

(*c*)

Figure 27

useful. For example, the probability of correct operation can be obtained by summing the diagonal elements on the joint matrix, i.e.,

$$p[\text{Correct operation}] = p[X_{-1}Y_{-1}] + p[X_0Y_0] + p[X_1Y_1]$$

$$= \sum_i p[X_iY_i]$$

$$= 0.28 + 0.20 + 0.28 = 0.76$$

The probability of error is the complement of this event:

$$p[\text{Error}] = 1 - p[\text{Correct operation}]$$

$$= 1 - \sum_i p[X_iY_i]$$

$$= 1 - 0.76 = 0.24$$

The probability of error is not the only measure of performance of a probabilistic system; reverse probabilities provide an informative measure. For example, the diagonal entries on the reverse matrix indicate that if the symbol received is $+1$, the symbol sent was certainly a $+1$ and if the symbol received is -1, the receiver is slightly less certain that a -1 value was sent. However, if the symbol received is 0, the receiver is *most* uncertain as to which symbol was transmitted. When he receives a value 0, he may as well flip a coin to determine whether input value zero caused this output zero. By both these methods his probability of being correct is 0.5.

EXERCISE SET 6

1. There are three machines in a factory producing the same item. Machine M_1 produces 10% of the total output, M_2 produces 60%, and M_3 produces 30%. Machine M_1 has 10% of its output defective, whereas M_2 has 30% defective, and M_3 has 20% defective.

(*a*) Represent this system as a sagittal function-on-a-relation.
(*b*) If an item is selected at random from the output of this factory and is found to be defective, what is the probability that it came from machine M_2? If it is good, what is its probability of coming from machine M_2?
(*c*) What is the probability of obtaining a defective item from this factory?
Ans.: 0.25

2. The diagram of Figure 28 shows a manufacturing system which produces items with a probability of defect of 0.6. These items are then examined by an inspector who observes an item having quality i (good or defective) and classifies it with possible error as quality j.
(*a*) Determine the probability of error of this system. *Ans.*: 0.24
(*b*) If the item is actually good, what is the probability of its being classified as good?
(*c*) If the item is classified good, what is the probability of its being actually good?
Ans.: 0.7

3. A rather common exercise in probability concerns two bags (or urns or boxes) each containing various colored balls (or beads or marbles). Suppose bag A has a red and a white ball and bag B has three red and one blue ball. One bag is chosen at random (equiprobably), and a ball is drawn from it. What is the probability that the ball comes from bag B:
(*a*) If the ball is red? *Ans.*: 0.6
(*b*) If the ball is blue?
(*c*) If the ball is red and the bag B has twice the probability of being chosen because of its larger size? *Ans.*: 0.75

4. Construct an algorithm (in flow-diagram form) to determine the joint and reverse probabilities from the forward probabilities and the source probabilities.

3. SYNTHESIS AND OPTIMIZATION OF PROBABILISTIC SYSTEMS

Consider the previous relay as a binary communication channel (Figure 29) which has a large conditional probability of error (0.4) if the input has value 1 and a smaller conditional probability of error (0.2) if the input has

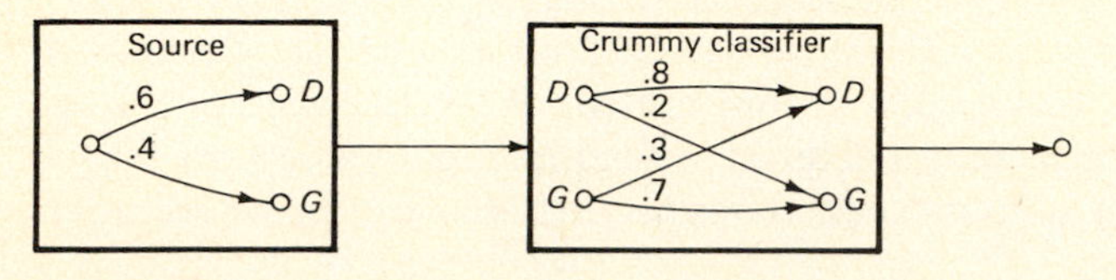

Figure 28

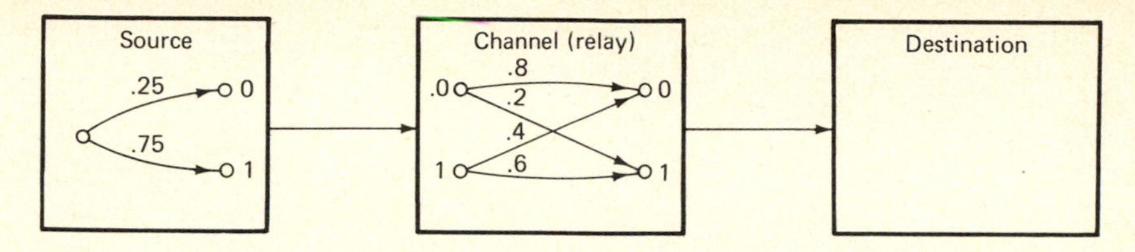

Figure 29

value 0. Connected to the indicated source, the system has a probability of error of

$$\begin{aligned} p[\text{error}] &= p[X_1 Y_0] + p[X_0 Y_1] \\ &= p[Y_0 \mid X_1]\, p[X_1] + p[Y_1 \mid X_0]\, p[X_0] \\ &= (0.4)(0.75) + (0.2)(0.25) \\ &= 0.35 \end{aligned}$$

This system probability of error is high because the large input probability (0.75) is associated with the large conditional probability of error (0.4). If this large input probability could be associated with the small conditional probability of error, the probability of system error could be reduced. This can in fact be done by NOTting the source, thus producing the equivalence source given in Figure 30. With this new source, the probability of error becomes

$$\begin{aligned} p[\text{Error}] &= p[Y_0 \mid X_1]\, p[X_1] + p[Y_1 \mid X_0]\, p[X_0] \\ &= (0.2)(0.75) + (0.4)(0.25) \\ &= 0.25 \end{aligned}$$

The probability of system error has been reduced from 0.35 to 0.25 by this process, which is called *encoding* of the source. However, the encoding has also changed the transmitted signal, but it can be regained by

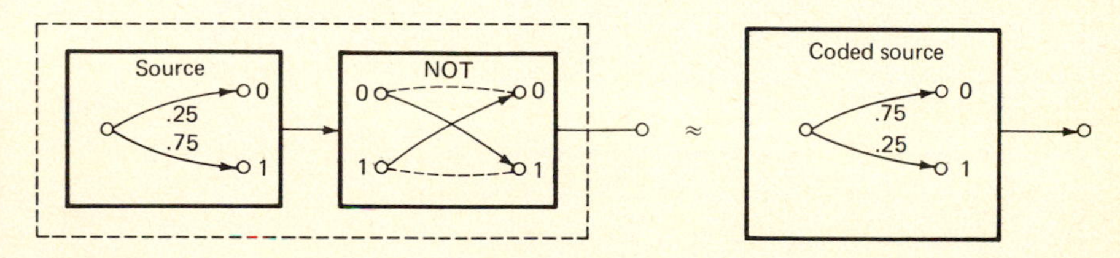

Figure 30

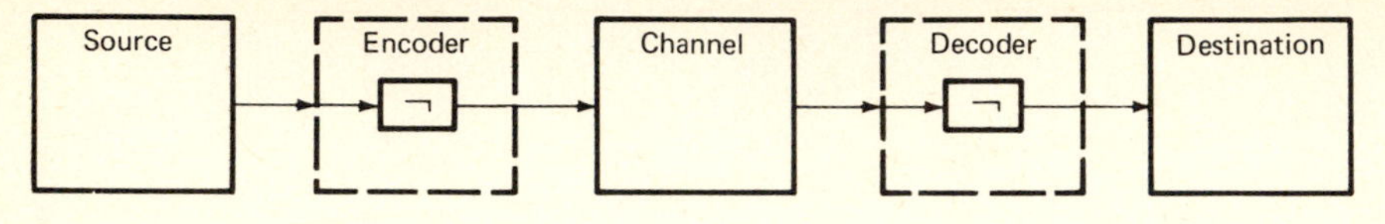

Figure 31

again NOTting the output from the channel (a process called *decoding*). The new, coded, more reliable system is shown in Figure 31. This is the general form of communication systems as studied in information theory. There are also other methods of encoding and decoding for many other goals, such as speed, cost, and secrecy.

EXERCISE SET 7

1. Determine the reverse conditional probabilities for both the original and the coded systems, and compare.

2. For the given *Z channel* of Figure 32, how must the source probabilities be related for this coding method to decrease the probability of error?

3. For the given *binary erasure channel* (BEC) of Figure 33, determine the probability of error and all reverse probabilities P_{ii} (do some by inspection). Construct an encoder and decoder, and find the probability of error of this new coded system. *Ans.*: 0.26, 0.14

4. For the given three-valued communication system of Figure 34 find the probability of error. How many possible encoders could you consider? Find the optimal encoder, decoder, and the resulting probability of error without comparing all possible cases. *Ans.*: 0.31, 6, 0.19

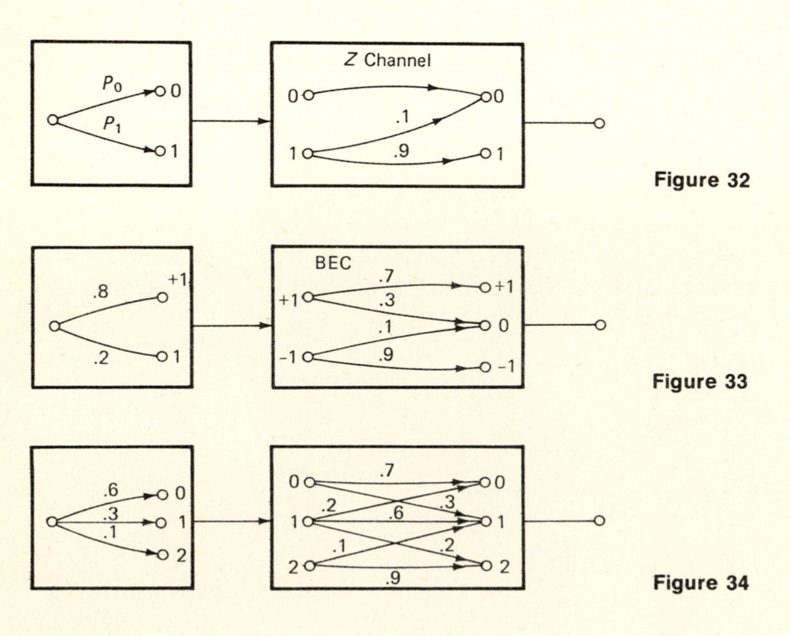

Figure 32

Figure 33

Figure 34

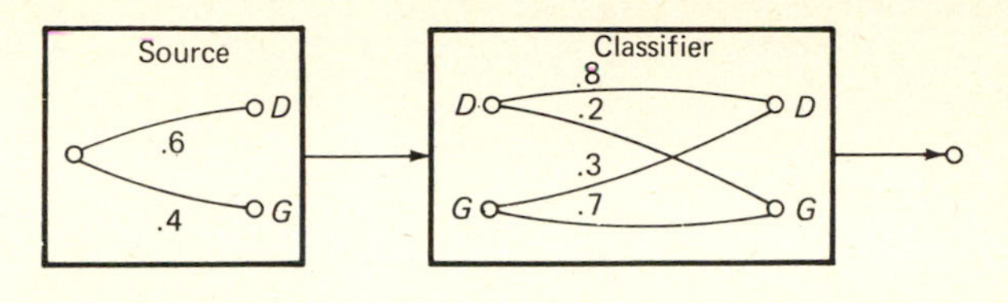

Figure 35

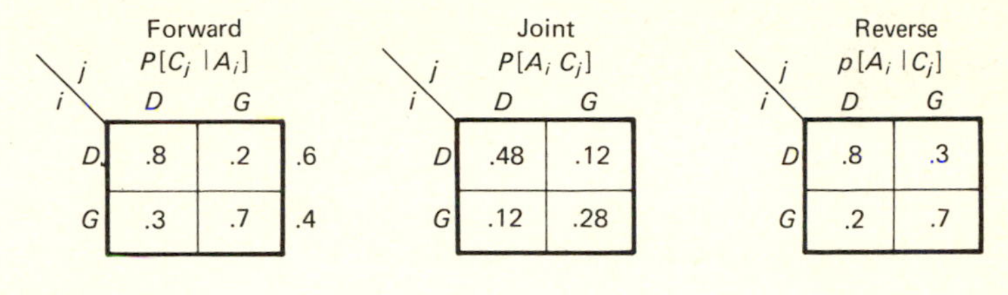

Forward $P[C_j \mid A_i]$

$i \backslash j$	D	G	
D	.8	.2	.6
G	.3	.7	.4

Joint $P[A_i\, C_j]$

$i \backslash j$	D	G
D	.48	.12
G	.12	.28

Reverse $p[A_i \mid C_j]$

$i \backslash j$	D	G
D	.8	.3
G	.2	.7

Figure 36

Optimizing Crummy Classification

The diagram of Figure 35 shows a very crummy manufacturing system which produces items with a probability of defect of 0.6. These items are then examined by an inspector (man, machine, or animal) which observes an item whose actual quality is A_i (where i is good or defective) and classifies it with possible error as quality C_j. The classifier is described by its forward conditional probabilities $p[C_j \mid A_i]$. All the forward, joint, and reverse probabilities are shown in the matrices of Figure 36. If an item has been classified as good, the probability of its actually being defective is $p[A_D \mid C_G] = 0.3$. This is a considerable improvement over the previous probability of defect of 0.6.

This system could be improved by taking only these (classified) good items and feeding them into a second classifier as in Figure 37. The corresponding matrices are given in Figure 38. Notice that the input probabilities to the second classifier are modified by the condition that they have been classified as good. Now if this system has classified an object as good, the probability that it is actually defective is 0.11, another improvement.

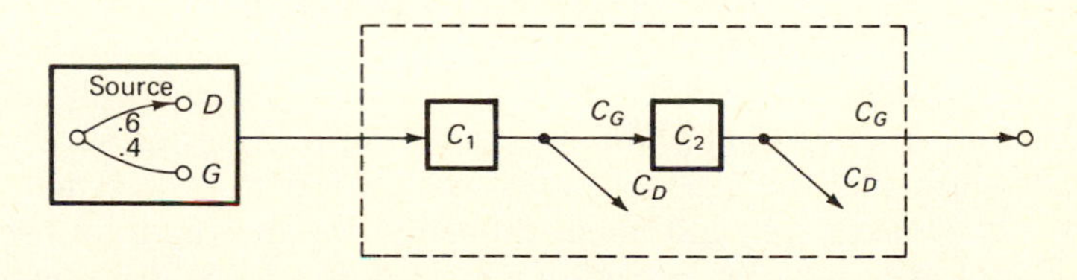

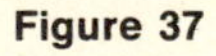
Figure 37

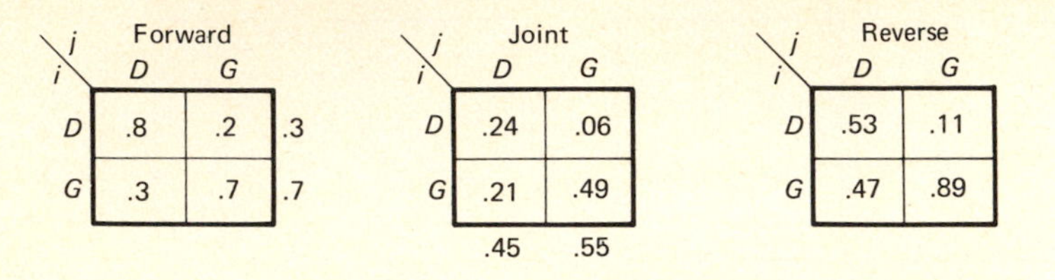

Figure 38

To decrease further the probability of defect we could follow with further classifiers. However, there is a consequence to this action: the portion of items in the preferred class decreases. In other words, there is a lower yield.

Such a classifying system was actually tested by a pharmaceutical manufacturer who trained pigeons to peck at improperly formed pills. Of course the pigeons were isolated from the pills by glass walls. In spite of its great efficiency this system was abandoned, presumably because of people's prejudice against pigeons.

EXERCISE SET 8

1. Show that in the case of two classifiers of Figure 37, only 22 percent of the original items are in the final preferred class.

2. Find the probability of defect in the preferred class if a third classifier is used. What portion of the items are in the preferred class? Compare the probability of error of the three classifiers. *Ans.*: 0.03, 0.14

3. Construct an algorithm (in flow-diagram form) to determine the probability of defect of items which have gone through N stages of classification. Use some previously constructed algorithms as sub-algorithms.

Interconnection and Equivalence of Probabilistic Systems

Given two systems (described by their forward probabilities) connected in cascade, it is possible to find a single system which is equivalent to this interconnection. For example, the two channels of Figure 39 could be replaced by the one indicated equivalent channel. This may be done in many ways, two of which are illustrated below.

For any input-output combination $X_i Y_j$ of the equivalent system the probability $p[Y_j \mid X_i]$ can be determined by following all paths leading from input X_i of the first system to output Y_j of the second system, as in Figure 40. The probability of each such path is determined by the product property of conditional probabilities, and the total probability is the sum of these path probabilities since the paths are mutually exclusive.

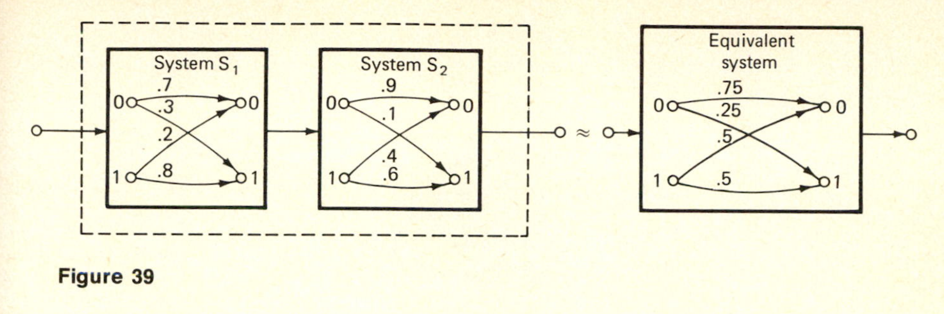

Figure 39

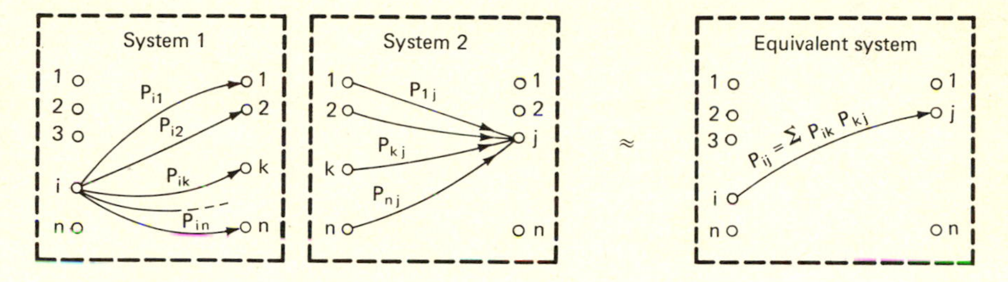

Figure 40

Notice that this is essentially the concept of composition of functions-on-relations. We can formalize this method as follows. Let E_{ij} be the event of going from input X_i to output Y_j:

$$E_{ij} = E_{i1}E_{1j} \vee E_{i2}E_{2j} \vee E_{i3}E_{3j} \vee \cdots \vee E_{in}E_{nj}$$

The corresponding probability is

$$P_{ij} = P_{i1}P_{1j} + P_{i2}P_{2j} + P_{i3}P_{3j} + \cdots + P_{in}P_{nj}$$

where p_{ik} = forward probability of first system

P_{kj} = forward probability of second system

P_{ij} = forward probability of equivalent system

This probability can be rewritten as

$$p_{ij} = \sum_{k=1}^{n} p_{ik} * p_{kj}$$

Notice that this expression is simply the definition of matrix multiplication. This indicates that the forward matrix of the equivalent system can be determined by multiplying the forward matrices of the component systems.

For example, the system which is equivalent to the cascade combination of the previous two channels can be determined by matrix multiplication as

$$S_1 \times S_2 = \begin{bmatrix} 0.7 & 0.3 \\ 0.2 & 0.8 \end{bmatrix} \times \begin{bmatrix} 0.9 & 0.1 \\ 0.4 & 0.6 \end{bmatrix}$$

$$= \begin{bmatrix} (0.7)(0.9) + (0.3)(0.4) & (0.7)(0.1) + (0.3)(0.6) \\ (0.2)(0.9) + (0.8)(0.4) & (0.2)(0.1) + (0.8)(0.6) \end{bmatrix}$$

$$= \begin{bmatrix} 0.75 & 0.25 \\ 0.50 & 0.50 \end{bmatrix}$$

This systematic method of multiplying matrices is often more convenient than following through all possible paths directly.

EXERCISE SET 9

1. For the above cascaded system determine the probability of error if it is connected to a source having input probability:

(*a*) $p[X_0] = 0.4$

(*b*) $p[X_0] = 0.5$

(*c*) $p[X_0] = 0.7$

2. If the two channels of Figure 39 were commuted (interchanged so that channel 1 follows channel 2), determine the equivalent channel and find the probability of error when connected to the same sources as in Problem 1 above. Compare the results of these problems. Try to generalize.

3. (*a*) Determine the probability of error of the given binary communication system of Figure 41 by following through all paths.

(*b*) Determine the probability of error of this system if channels A and B are commuted; use matrices.

†(*c*) Under what conditions does the commuting not change the probability of error?

Ans.: $P_0 = P_1$ or $ad = bc$

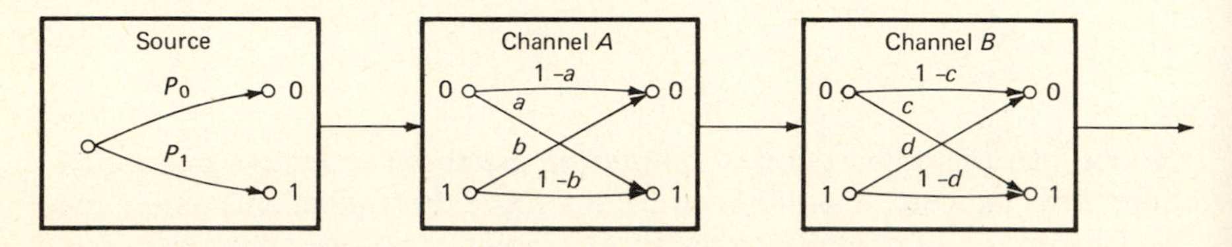

Figure 41

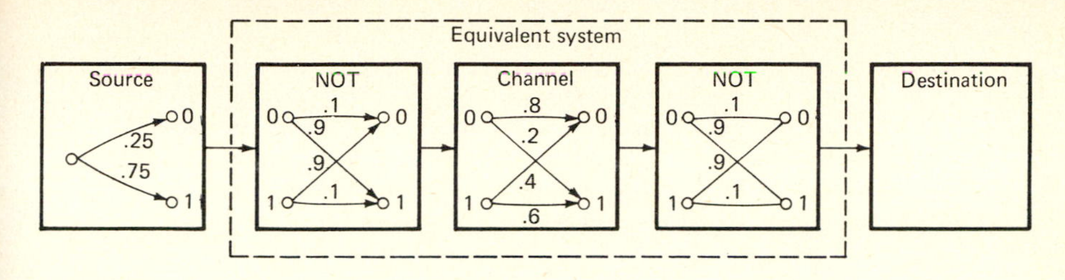

Figure 42

Extended Cascading. The previous methods may be extended to any number of cascaded systems by successively taking two systems at a time. For example, consider the coded communication system of Figure 31 but with crummy NOT elements as shown in Figure 42. The one channel which is equivalent to these three channels can be determined by the following succession of matrix products:

$$\begin{bmatrix} 0.1 & 0.9 \\ 0.9 & 0.1 \end{bmatrix} \times \begin{bmatrix} 0.8 & 0.2 \\ 0.4 & 0.6 \end{bmatrix} \times \begin{bmatrix} 0.1 & 0.9 \\ 0.9 & 0.1 \end{bmatrix} =$$

$$\begin{bmatrix} 0.44 & 0.56 \\ 0.76 & 0.24 \end{bmatrix} \times \begin{bmatrix} 0.1 & 0.9 \\ 0.9 & 0.1 \end{bmatrix} = \begin{bmatrix} 0.548 & 0.452 \\ 0.292 & 0.708 \end{bmatrix}$$

The probability of error of this new system is now

$$\begin{aligned} p[\text{Error}] &= p[Y_1 \mid X_0]\, p[X_0] + p[Y_0 \mid X_1]\, p[X_1] \\ &= (0.452)(0.25) + (0.292)(0.75) \\ &= 0.332 \end{aligned}$$

Notice that even with these additional crummy NOT components the probability of error of the coded system (0.332) is still less than the probability of error of the original uncoded system (0.35).

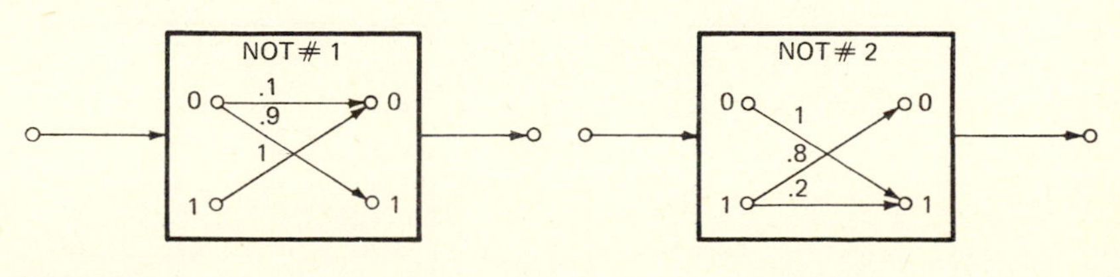

Figure 43

EXERCISE SET 10

1. For the above coded communication system, only the NOT components of Figure 43 are available. Which would you use as encoder and which as decoder to minimize the probability of error? If two NOTs of each type were available, which would you choose?

2. Consider the communication system of Figure 42 but with identical crummy NOT components of Figure 44. What is the largest probability of error p of the NOT elements to make the coded system more reliable than the uncoded system? *Ans.*: $p = 0.125$ and 1.0

Symmetric Systems. One particularly common type of system has the property of *equiprobable symmetry*; for all inputs the forward conditional probabilities of error are the same. A binary symmetric channel (BSC) and a ternary symmetric channel (TSC) are shown in Figure 45.

A consequence of this type of symmetry is that the total probability of error is not dependent on the source. For example, the probability of error of a binary symmetric system is

$$\begin{aligned} p[\text{Error}] &= pP_0 + pP_1 && \text{where } P_i \text{ are source probabilities} \\ &= p(P_0 + P_1) && \text{by distributivity} \\ &= p && \text{by normality} \end{aligned}$$

Similarly the probability of error of a ternary symmetric system is

$$\begin{aligned} p[\text{Error}] &= P_0(p + p) + P_1(p + p) + P_2(p + p) \\ &= (P_0 + P_1 + P_2)(p + p) \\ &= 2p \end{aligned}$$

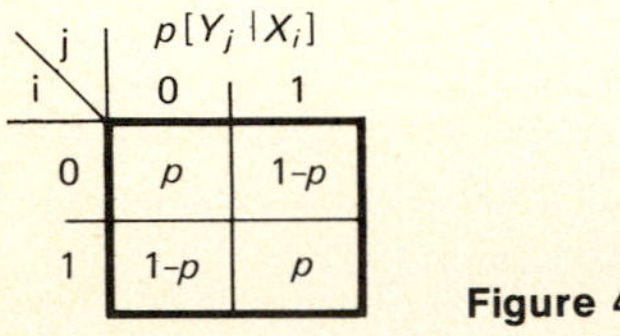

$i \backslash j$	0	1
0	p	$1-p$
1	$1-p$	p

Figure 44

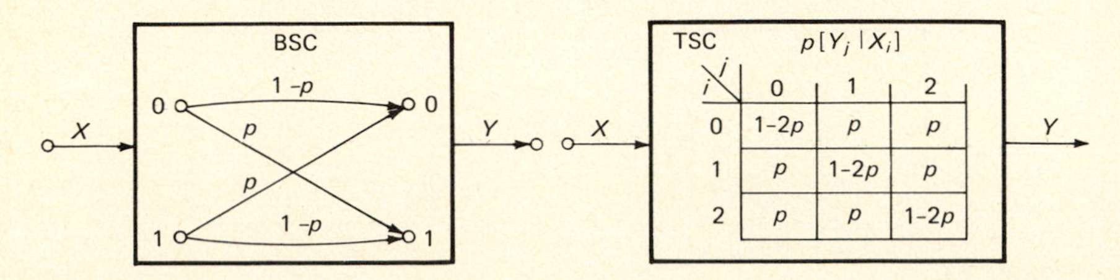

Figure 45

In general the probability of error of an n-ary symmetric system is

$$p[\text{Error}] = (n - 1)p$$

Thus a symmetric system is completely specified by one number, the probability of error, which does not depend on the input (source) probabilities. This often results in considerable convenience.

Example Two identical binary symmetric channels are cascaded to form one larger system. Find the individual probabilities of error p if the probability of error of the large equivalent system must be less than 0.375.

Solution Either from the matrix product or by following through all paths the probability of error can be determined as

$$p[\text{Error}] = (1 - p)(p) + (p)(1 - p)$$
$$= 2p - 2p^2$$

In this case

$$p[\text{Error}] = 2p - 2p^2 = 0.375 = \tfrac{3}{8}$$
$$16p^2 - 16p + 3 = 0$$
$$(p - 0.25)(p - 0.75) = 0$$
$$p = 0.25 \text{ or } 0.75$$

The solution is any value less than 0.25 or greater than 0.75. The second unexpected solution when taken to its extreme value of $p = 1$ indicates that two NOT elements make a perfect channel.

EXERCISE SET 11

1. Determine the probability of error of the systems which are equivalent to the binary symmetric channel interconnections of Figure 46. Values p and q are the conditional probabilities of error in each case. *Ans.*: b; $4p^3 - 6p^2 + 3p$

2. Prove that the probability of error of three binary symmetric channels in cascade is

$$p[\text{Error}] = a + b + c - 2(ab + bc + ac) + 4abc$$

where a, b, c are the probabilities of error of the individual systems. Comment on the effect of commuting symmetric channels.

3. Find the probability of error of a system consisting of two identical symmetric three-valued channels when connected in cascade if the forward conditional probability of error of each individual channel is p. Evaluate this for $p = 0.1$.

4. Repeat problem 3 for two four-valued symmetric channels. *Ans.*: $3(2p - 4p^2)$

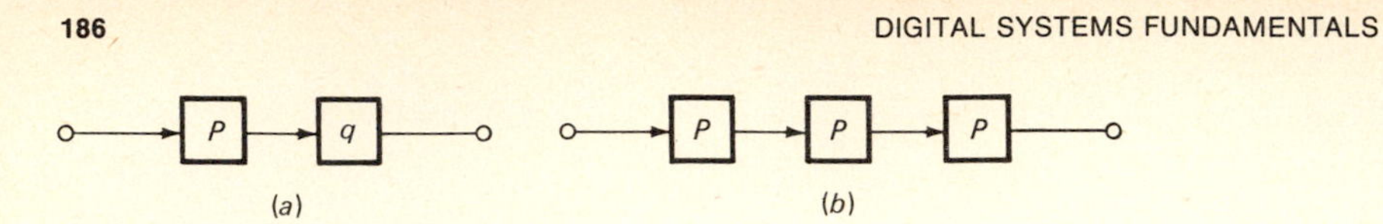

Figure 46

Multi-input systems. So far we have considered only systems with a single input. However, all these results can be extended to multi-input systems by considering each combination of input values as one input. Consider, for example, the two-input system of Figure 47, which is believed to be an OR component. The joint probabilities could be computed from the sequence using the relative frequency definition. These probabilities are listed in tabular manner in Figure 48*a*, in which a combination of input values is considered as one input, i.e.,

$$X = \langle A, B \rangle$$

The sagittal function-on-a-relation representation is readily obtained from the table of Figure 48*b*. Notice that the errors are of two types: the output has value 1 when it should be 0, and the output has value 0 when it should be 1. The probability of error is computed by summing the joint probabilities corresponding to the two types of errors:

$$\begin{aligned} p[\text{Error}] &= p[001] + p[010 \vee 100 \vee 110] \\ &= 0.1 + 0.2 + 0.1 + 0.0 \\ &= 0.4 \end{aligned}$$

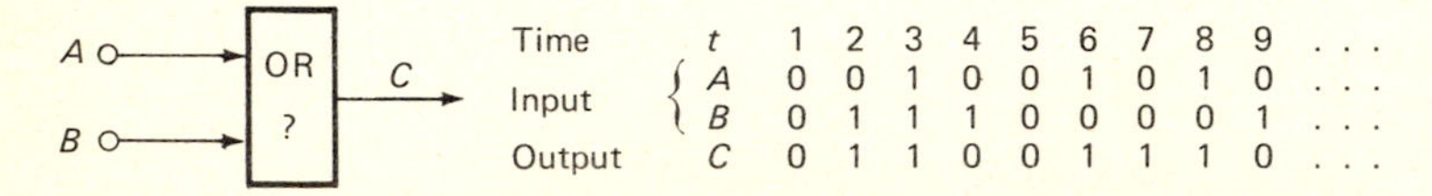

		1	2	3	4	5	6	7	8	9	...
Time	t	1	2	3	4	5	6	7	8	9	...
Input	A	0	0	1	0	0	1	0	1	0	...
	B	0	1	1	1	0	0	0	0	1	...
Output	C	0	1	1	0	0	1	1	1	0	...

Figure 47

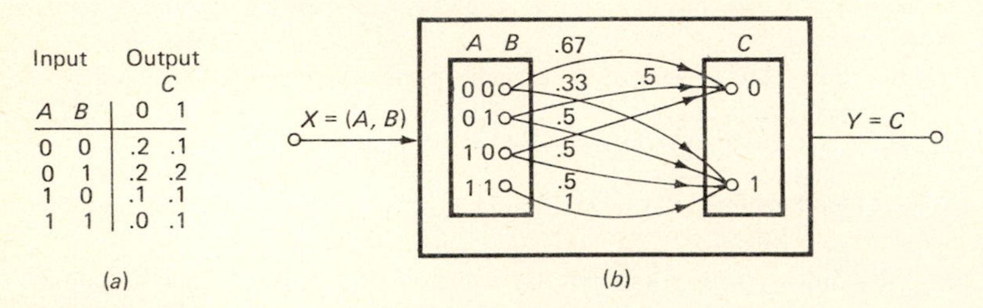

Input A	Input B	Output C = 0	Output C = 1
0	0	.2	.1
0	1	.2	.2
1	0	.1	.1
1	1	.0	.1

(a)

(b)

Figure 48

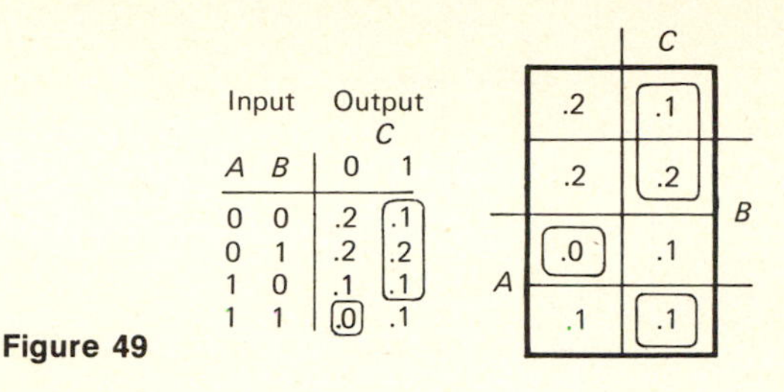

Input A	Input B	Output C = 0	Output C = 1
0	0	.2	.1
0	1	.2	.2
1	0	.1	.1
1	1	.0	.1

Figure 49

If this component (which was believed to be an OR) was actually an AND component, its probability of error could be computed from either of the diagrams of Figure 49 as

$$p[\text{Error}] = 0.1 + 0.2 + 0.1 + 0.0 = 0.4$$

Notice that the probability of error is the same value (0.4) regardless of whether the system is believed to be an AND or an OR. This illustrates the unusual fact that a crummy system could behave as either an AND or an OR equally well (or equally badly). In other words, an AND and an OR could be equivalent with respect to their probability of error.

EXERCISE SET 12

1. Because of the buckling of rods and seating of valves the hydraulic switching system of Figure 50*a* behaves probabilistically as indicated by the joint probabilities of Figure 50*b*. Find the probability of error of this system.

2. The diagrams of Figure 51 show the structure of a binary half-adder as well as the (joint) probabilistic behavior of it. Find the probability of error of this system. *Ans*: 0.30

3. Find the probability of error of the MAJORITY element described by the set diagram of Figure 52. *Ans.*: 0.28

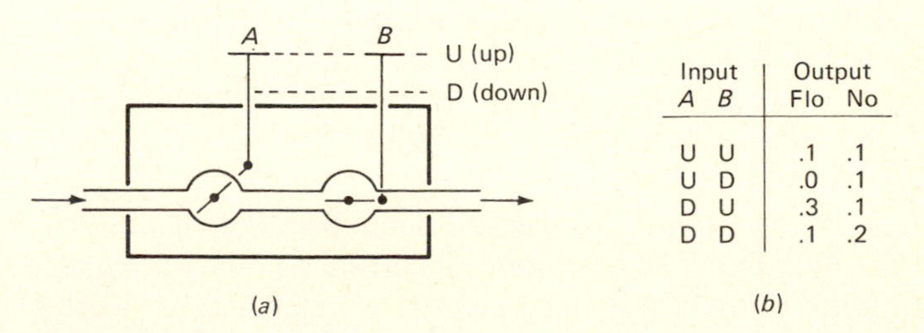

Input A	Input B	Output Flo	Output No
U	U	.1	.1
U	D	.0	.1
D	U	.3	.1
D	D	.1	.2

(*b*)

Figure 50

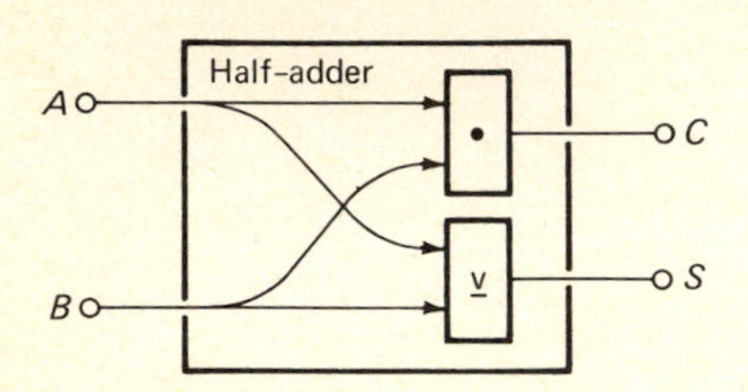

A B \ CS	00	01	10	11
	$P[A, B, C, S]$			
0 0	.10	.05	.05	.00
0 1	.03	.20	.00	.02
1 0	.02	.10	.00	.03
1 1	.00	.05	.30	.05

Figure 51

.10	.00	.05	.05
.06	.04	.07	.03
.02	.08	.15	.05
.09	.01	.12	.08

(X_3 above, X_2 right, X_1 left, Y below)

Figure 52

4. INDEPENDENT SYSTEMS—ANALYSIS AND SYNTHESIS

So far the probabilistic systems we have considered have been quite general. However, in some limited cases, there may be certain restrictions or constraints which may result in more convenient analysis. The concept of independence is one such property which holds for some systems but certainly not all. It is important, therefore, to realize that all the concepts of this section are limited; but when they do apply, they are extremely useful.

Probabilistic Independence

It is often of interest to know whether events are related or relevant to one another or, alternatively, whether the events are independent of each other. In this section we will consider the concept of probabilistic independence; in a later section this will be used to develop a measure of dependence.

Two events E and F are defined as *independent*, denoted $E \perp F$, if the probability of their joint occurrence equals the product of the probabilities of their individual occurrence. Formally this definition is written

$$E \perp F \quad \leftrightarrow \quad p[E \cdot F] = p[E] * p[F]$$

and read "event E is independent of event F if and only if the probability of the joint event E and F equals the product of the individual probabilities." This multiplicative property of independent events is mathematically very convenient. In fact it is tempting to assume independence when it actually does not hold.

In the television survey, repeated in Figure 53, the enjoyment of the program (like) is independent of the age (youth), i.e.,

$$L \perp Y \qquad \text{or equivalently} \qquad Y \perp L$$

This independence can be justified from the definition

$$p[L \cdot Y] = p[L] * p[Y]$$
$$0.15 = (0.6) * (0.25)$$

However, the enjoyment of the program is not independent of the sex of the viewer since

$$p[L \cdot M] \neq p[L] * p[M]$$

Similarly we can show that in this case sex is not independent of age, which is not what we would usually expect.

There are other ways of viewing independence intuitively. For example, the probability of one event should not be changed upon knowing that an independent event has occurred, i.e.,

$$E \perp F \leftrightarrow p[E \mid F] = p[E]$$

In the above survey the independence between opinion and age can be shown in two ways

$$p[L \mid Y] = p[L] = 0.6$$

and also

$$p[Y \mid L] = p[Y] = 0.25$$

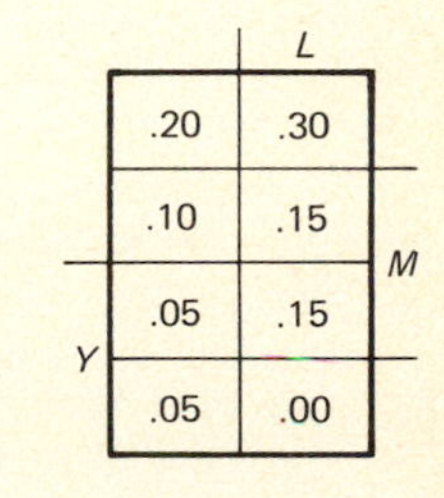

Figure 53

There is still another equivalent way of viewing independence; the probability of one event given that a second independent event has occurred should equal the probability of this event given that the second event has not occurred:

$$E \perp F \leftrightarrow p[E \mid F] = p[E \mid \bar{F}]$$

In the above case

$$p[L \mid Y] = P[L \mid \bar{Y}] = 0.6$$

and

$$p[Y \mid L] = p[Y \mid \bar{L}] = 0.25$$

All these alternative views of independence are equivalent to the original definition. This original definition involving multiplication of probabilities is easy to remember and most convenient to apply.

Example If the probability of meeting a red traffic signal is 0.6, and the probability of meeting a second red signal is 0.8, and the probability of meeting both signals when red is 0.5, then these signals are almost independent since the product of the individual probabilities (0.8) (0.6) is almost equal to the joint probability of 0.5.

Example The solution to a problem involves 70 independent computations, each of which has a probability 0.99 of being successful. The probability of performing all 70 computations successfully is

$$(0.99)(0.99)(0.99) \cdots (0.99) = (0.99)^{70} = 0.5$$

The probability of solving this problem successfully is only 0.5, a discouraging overall probability despite the encouraging individual probabilities of 0.99.

EXERCISE SET 13

1. To distinguish between mutual exclusiveness and independence construct a set diagram of two events having probabilities $p(A) = 0.3$ and $p(B) = 0.6$:

(*a*) If the events are mutually exclusive.

(*b*) If the events are independent.

2. In a study performed in a retirement home, 400 people were given a special diet, and 600 people were given a normal diet. After 10 years 300 people died, 120 of which were on the special diet. Was the death rate independent of the diet? *Ans.*: yes

3. If the output of a probabilistic system is independent of the input, how can this be determined by inspecting the sagittal diagram (of forward probabilities)?

4. Two classifiers each have probability p of accepting a defective item although they reached their decisions independently.

(*a*) Find the probability of agreement between them.

(*b*) Find the probability that both do not accept this item. *Ans.*: $2p^2 - 2p + 1$

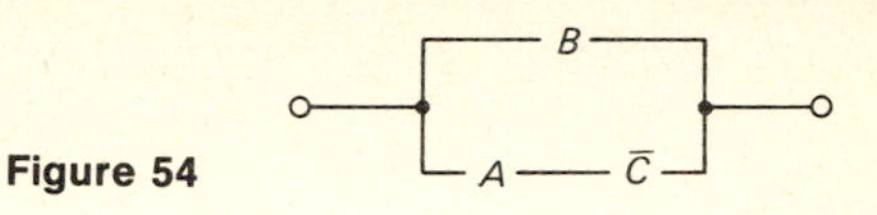

Figure 54

Independent Relay Networks

Consider a system of three ideal relays A, B, C having probabilities of operating of p_a, p_b, and p_c. Since the relays are operated from three widely separated, unrelated positions, they have been assumed independent. The contacts of these relays are connected to form the network $N = B \vee A\bar{C}$ of Figure 54.
The probability of this network's being closed is

$$\begin{aligned} p[B \vee A\bar{C}] &= p[B] + p[A\bar{C}] - p[AB\bar{C}] \\ &= p[B] + p[A]\, p[\bar{C}] - p[A]\, p[B]\, p[\bar{C}] \\ &= p_b + p_a(1 - p_c) - p_a p_b (1 - p_c) \end{aligned}$$

For the particular case $p_a = 0.6$, $p_b = 0.7$, and $p_c = 0.8$ this is

$$p[B \vee A\bar{C}] = (0.7) + (0.6)(0.2) - (0.6)(0.7)(0.2) = 0.736$$

For another particular case $p_a = p_b = p_c = p$ this value is

$$\begin{aligned} p[B \vee A\bar{C}] &= p + p(1 - p) - p^2(1 - p) \\ &= p^3 - 2p^2 + 2p \end{aligned}$$

Complex Independent Networks

The network shown in Figure 55 is called a *hammock* net. It consists of $L * W$ independent contacts C, each contact closed with the same probability p. Determine the probability of the network's being closed. Notice that the network consists of L stages each having W contacts in parallel.

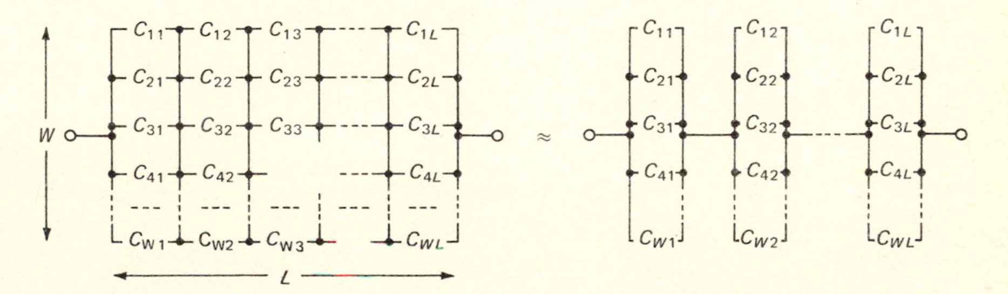

Figure 55

The probability of each contact's being closed is p.
The probability of each contact's being open is $1 - p$.
The probability that all W contacts of a stage are open is $(1 - p)^W$.
The probability of each stage's being closed is $1 - (1 - p)^W$.
The probability of all L stages being closed is $[1 - (1 - p)^W]^L$

The probability of the network's being closed is thus

$$p\,[\text{Net closed}] = [1 - (1 - p)^W]^L$$

This same result could have been obtained algebraically as

$$\begin{aligned} p\,[\text{net closed}] &= p\,[(C_{11} \vee C_{21} \vee \cdots\ C_{w1}) \\ &\qquad\qquad (C_{12} \vee C_{22} \vee \cdots\ C_{w2})(\cdots)] \\ &= [p(C_1 \vee C_2 \vee \cdots \vee C_w)]^L \\ &= [p(\overline{\bar{C}_1 \cdot \bar{C}_2 \cdot \bar{C}_3 \cdots \bar{C}_w})]^L \\ &= [1 - p(\bar{C}_1 \cdot \bar{C}_2 \cdots \bar{C}_w)]^L \\ &= [1 - p(\bar{C}_1)\,p(\bar{C}_2) \cdots p(\bar{C}_W)]^L \\ &= [1 - (1 - p)(1 - p) \cdots (1 - p)]^L \\ &= [1 - (1 - p)^W]^L \end{aligned}$$

EXERCISE SET 14

1. Consider a system of three independent relays A, B, C, operating with probabilities $p_a = 0.6$, $p_b = 0.7$, $p_c = 0.8$.

(*a*) Construct the event space as a set diagram and table-of-combinations.
(*b*) Compare this independent system to the system of Figure 12. Comment.
(*c*) Find the probability that the following networks are closed:

1. $A \vee B \vee C$ 2. $B(A \vee C)$
3. $A \vee BC$ 4. $\overline{A \vee B}$
5. $A \vee \bar{A}$ 6. $A\bar{B}C \vee \bar{A}B\bar{C}$

2. If each of the given relays of Figure 56 is operated with probability p, find the probability that the given networks are closed. *Ans.: a.* $2p^2 - p^3$; *b.* $p^6 - 2p^5 + 2p^2$

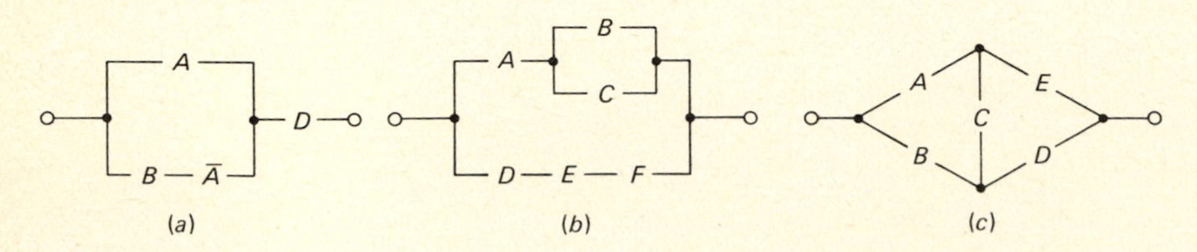

Figure 56

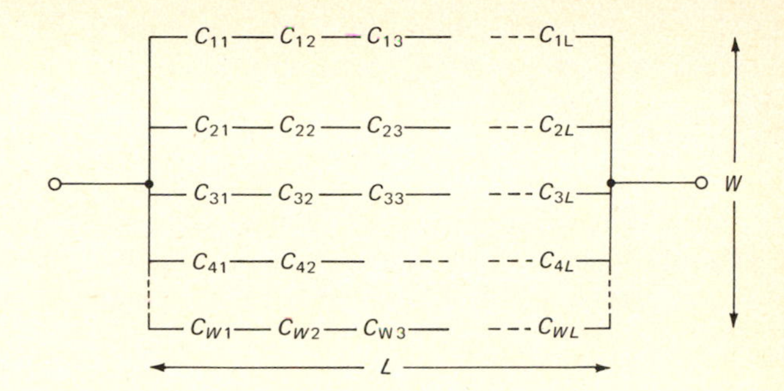

Figure 57

3. Determine the probability that the "*clothesline*" type of network of Figure 57 is closed if all contacts are independent and equal in probability.

More Reliable Systems from Less Reliable Flow Components

A commonly encountered problem is to interconnect some probabilistic systems to obtain a better system. This is particularly important when the best available components are insufficient to accomplish some goal.

For example, consider a relay in a missile (or robot) which could destroy the system on command from earth if it endangers the lives of others. The goal is not only to increase the probability of destroying itself when commanded to destroy but also to decrease the probability of destroying itself when no command is given. The real goal of obtaining a reliable system is thus a combination of these two goals. Improving one goal may worsen the other.

Suppose that because of vibration, temperature, etc., the relay on this system is as bad as the one we have been studying; i.e., when the coil is energized, the probability of error is 0.4, and when not energized, the error is only 0.2. This relay and its model are shown in Figure 58.

In an effort to increase the reliability, we may consider various interconnections of such relays. Intuitively a parallel interconnection of contacts (as shown in Figure 59) seems to satisfy the goal. Let us see why it doesn't.

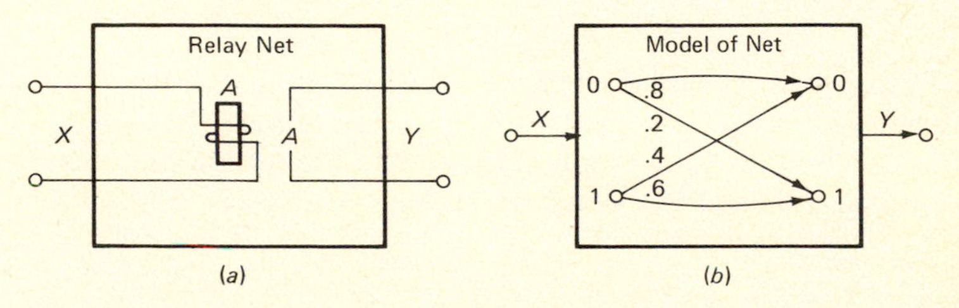

Figure 58

Since the coils are connected in parallel, the probability of being energized is the same for both relays, i.e.,

$$P[A=1]=p[A]=p[B]=p$$

The failures in the contacts are also assumed to be independent since the failure of one contact should not influence the failure of any other contact.

The probability that this parallel network is closed is

$$\begin{aligned} p[Y] &= p[Y=1] \\ &= p[A \vee B] \\ &= p[A]+p[B]-p[AB] \\ &= p[A]+p[B]-p[A]\,p[B] \qquad \text{since } A \perp B \\ &= p+p-pp \\ &= 2p-p^2 \end{aligned}$$

where p is the probability that a contact is closed.

When the coil is energized, the probability that each contact is closed is

$$p=1-0.4=0.6$$

and the probability that the network is closed is

$$\begin{aligned} p[Y\,|\,X] &= p[(Y=1)\,|\,(X=1)] \\ &= 2p-p^2 \\ &= 2(0.6)-(0.6)(0.6) \\ &= 0.84 \\ p[\bar{Y}\,|\,X] &= 1-p[Y\,|\,X] \\ &= 1-0.84 \\ &= 0.16 \end{aligned}$$

Thus, when the coil is energized, the probability of error is 0.16, which is an improvement over the previous value of 0.4.

However, we are not finished with the evaluation; the goal involves another probability. When the coil is not energized, the probability that each contact is closed is

$$p=0.2$$

and the probability that the network is closed is

$$\begin{aligned}p[Y \mid \bar{X}] &= p[(Y=1) \mid (X=0)] \\ &= 2p - p^2 \\ &= 2(0.2) - (0.2)(0.2) \\ &= 0.36\end{aligned}$$

Thus when the coil is unenergized the probability of error is 0.36, an increase over the previous value of 0.2.

The equivalent system is shown in Figure 59*b*. Comparison with Figure 58 indicates that the parallel interconnection does not satisfy the goal. The interconnection increases the probability of destroying the system on command, but it also unfortunately increases the probability of destroying the system when no such command has been given!

We can similarly construct the equivalent system for two relay contacts in series (coils still in parallel), but this series interconnection also will not satisfy the goal. Try it.

Ultimately it may occur to us that some sort of series-parallel network may satisfy the goal. For example, consider the network of Figure 60.

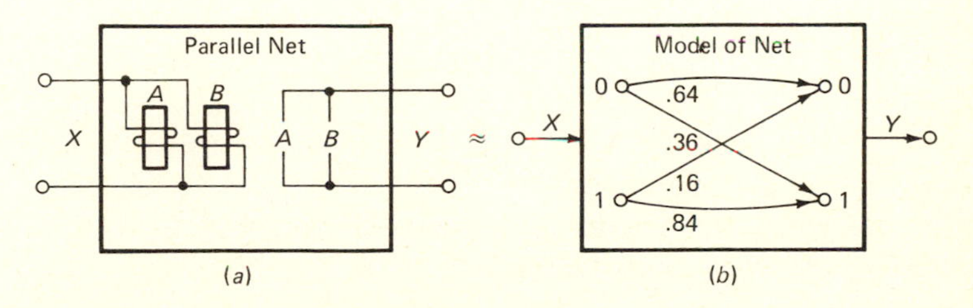

Figure 59

$$\begin{aligned}p[Y] &= p[Y=1] \\ &= p[(A \vee B)(C \vee D)] \\ &= p[A \vee B]\, p[C \vee D] \\ &= (2p - p^2)(2p - p^2) \\ &= (2p - p^2)^2\end{aligned}$$

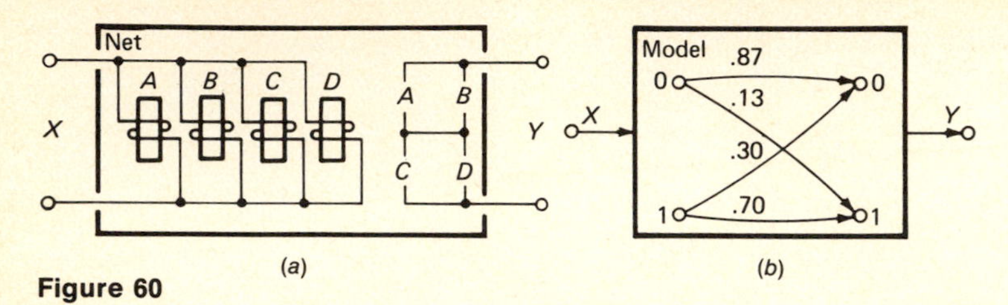

Figure 60

If the relay is energized,

$$p = 1 - 0.4 = 0.6$$

$$p[Y \mid X] = p[(Y = 1) \mid (X = 1)]$$

$$= [(2)(0.6) - (0.6)(0.6)]^2 = 0.70$$

$$p[\bar{Y} \mid X] = 1 - 0.70 = 0.30$$

The probability of error is 0.30 (when energized). If the relay is unenergized,

$$p = 0.2$$

$$p[Y \mid \bar{X}] = p[(Y = 1) \mid (X = 0)]$$

$$= (0.36)^2 = 0.13$$

The probability of error is 0.13 (when unenergized). Comparing this network to the original network indicates that the goal has been satisfied. We created one reliable relay from four less reliable ones!

Notice that the network of Figure 60 is a special case of the hammock network, with length $L = 2$ and width $W = 2$. This suggests a generalization to all hammock networks. Recall that the probability of closure of a hammock network is

$$H(p) = [1 - (1 - p^w)]^L$$

Suppose the conditional probabilities of error of relays are given as

$$p[Y_0 \mid X_1] = p_{10} = a \qquad \text{(sluggishness)}$$

and

$$p[Y_1 \mid X_0] = p_{01} = b \qquad \text{(sticking)}$$

When the relays are energized, the probability of contacts closing is

$$p = 1 - p_{10} = 1 - a$$

and the probability of a hammock network closing is

$$p[Y|X] = H(1-a)$$

The probability of error in this case is

$$\begin{aligned} p[\bar{Y}|X] &= 1 - H(1-a) \\ &= 1 - [1 - [1 - (1-a)]^{w}]^{L} \\ &= 1 - (1 - a^{w})^{L} \end{aligned}$$

Now if the relay is unenergized, the probability of contacts closing is

$$p = p_{10} = b$$

and the corresponding probability of error is

$$p[Y|\bar{X}] = H(b)$$

$$= [1 - (1-b)^{w}]^{L}$$

The overall probability of error is

$$p[\text{Error}] = p[X] * p[\bar{Y}|X] + p[\bar{X}] * p[Y|\bar{X}]$$

If we are given N crummy relays with their conditional probabilities of error a and b, we can select various values of length L and width W of the hammock network (so that $L * W \leq N$) to determine which combination yields the lowest probability of error. A computer would be most convenient for such an evaluation.

EXERCISE SET 15

1. If contacts are connected so that $N = AC \vee BD$, show that the goal (of a more reliable system) is not satisfied.

2. Given 12 identical relays, as in Figure 58, indicate intuitively a more reliable hammock network which would also make the resulting equivalent system more symmetrical.

3. If the relay contacts are arranged like the clothesline network of Figure 57, determine the overall probability of error in terms of the source probabilities and the forward conditional probabilities of error a and b.

†**4.** Plot $H(p)$ versus p for the networks of Figure 58 to 60 and comment on the intersection.

More Reliable Systems from Less Reliable Level Components

One method of improving the reliability of systems is to make use of extra or redundant components, as in Figure 61. For example, if only crummy binary channels are available, a message may be sent over three such channels. At the destination the majority value is assumed correct. Similarly, any odd number of classifiers could inspect some item, with the majority vote taken as the conclusion. Also, multi-input components, such as ORs, may be triplicated, as shown in Figure 61*b*. Intuitively, such interconnections of systems appear to decrease the probability of error. Also, it appears that for best results the failure of one component should not influence the failure of other components; the components should be independent. A more quantitative view of such systems follows.

Consider a system of three crummy binary systems connected as in Figure 61 to an ideal majority element. This equivalent system fails only when two or three of the individual systems fail. This can be enumerated as

$$p[\text{Error}] = p[A_F B_F C_F \vee A_F B_F C_S \vee A_F B_S C_F \vee A_S B_F C_F]$$

Since these four events are mutually exclusive, the expression can be rewritten

$$p[\text{Error}] = p[A_F B_F C_F] + p[A_F B_F C_F] + p[A_F B_F C_F] + p[A_F B_F C_F]$$

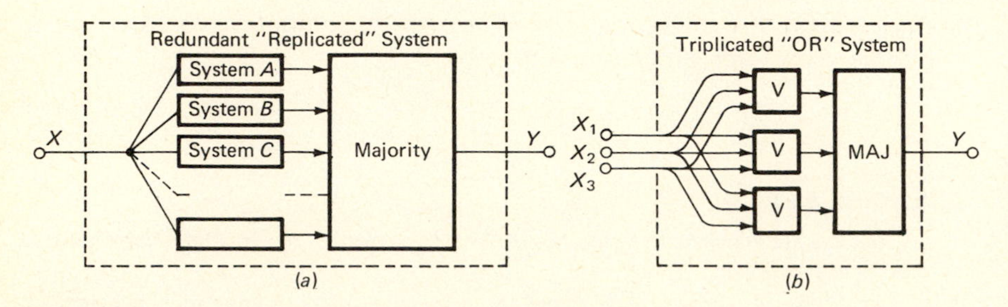

Figure 61

If all the systems are identical and independent, failing with probability p and succeeding with probability $q = 1 - p$, the probability of error of the equivalent system becomes

$$\begin{aligned} p[\text{Error}] &= ppp + ppq + pqp + qpp \\ &= p^3 + 3p^2 q \\ &= p^3 + 3p^2(1-p) \\ &= 3p^2 - 2p^3 \end{aligned}$$

For the particular case when $p = 0.1$, the probability of error of the equivalent system becomes

$$p[\text{Error}] = 3p^2 - 2p^3 = 3(0.1)^2 - 2(0.1)^3 = 0.028$$

The probability of error has been reduced from 0.1 for an individual system to 0.028 for a system of three such systems. The probability of error has been reduced to less than one-third of the original error. It can be reduced farther by using more individual systems.

EXERCISE SET 16

1. Comment on the effectiveness of the equivalent system of Figure 61 if the probability of error of each component system is 0.01. If it is 0.5. If it is 0.7.

2. Plot the probability of error of the equivalent system versus the probability of error of the component systems, noticing the S-shaped nature of such relations. If the probability of error of the equivalent system must be less than 0.1, what is the maximum probability of error of each component?

3. If the three OR components of Figure 61*b* have differing but still symmetrical probabilities of error, such as $p_1 = 0.1$, $p_2 = 0.2$, and $p_3 = 0.3$, find the probability of error of the equivalent triplicated system.

4. Determine the probability of error of a system as in Figure 61 using five identical, independent, symmetrical components, having probability of error p.
Ans.: $6p^5 - 15p^4 + 10p^3$

†5. Consider the parity error-detecting system of Chapter 2, Figure 39, where each channel has a probability of error of $p = 0.1$. Determine the probability of error of this entire system if the errors occur independently. *Ans*: $7p^4 - 12p^3 + 6p^2 \approx .06$

More Optimization

Two nonsymmetric "conservative" classifiers independently classify an item, which is then accepted if both classify the item as good. Construct one classifier which is equivalent to this interconnection of Figure 62 and analyze this system.

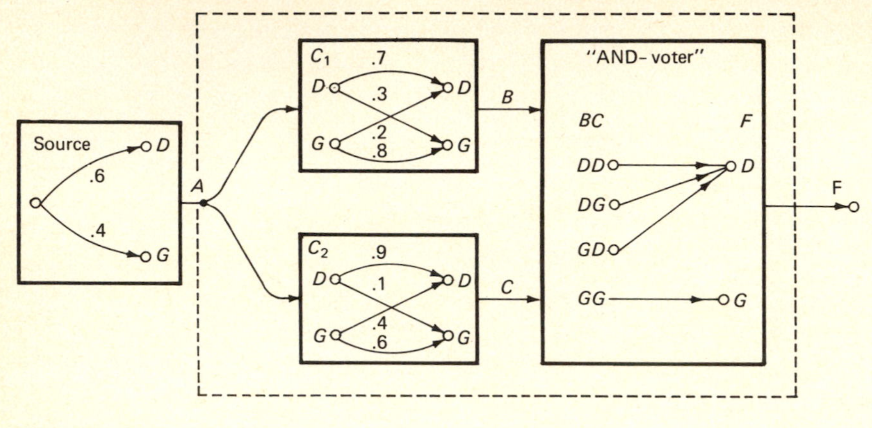

Figure 62

The equivalent system can be determined by following through all paths from each input to each output. For example, if the input is a defective item, the paths are as shown in Figure 63. Note that the actions (outputs) of the classifiers are assumed independent. Evaluating a similar set of paths when the input is a good item yields the equivalent system of Figure 64.

The probability of error of this system is rather high:

$$p[\text{Error}] = (0.6)(0.03) + (0.4)(0.52) = 0.226$$

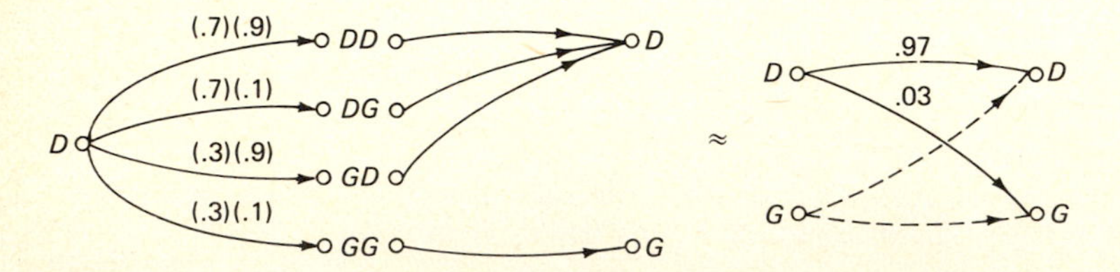

Figure 63

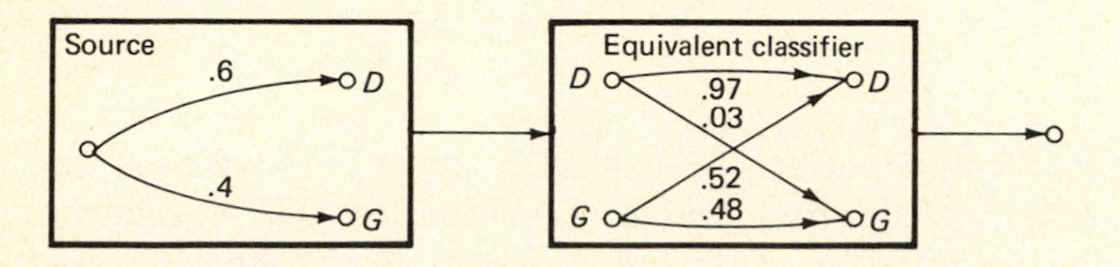

Figure 64

However, the reverse probability of an actually defective item given that it is classified as good is rather low:

$$p[A_D \mid C_G] = \frac{(0.6)(0.03)}{(0.6)(0.03) + (0.4)(0.48)} = 0.09$$

Notice that the equivalent system is indeed cautious about classifying the defective items as good.

Example Draw the sagittal function-on-a-relation corresponding to the equivalent crummy AND element which results from the interconnection of independent NOT and OR elements of Figure 65.

Solution This problem is rather complex, even though the method involves simply following all possible paths for each input combination. The *decomposition* of Figure 66 is helpful. For example, when $A = 1$ and $B = 0$,

$$p[CD = 10] = (1) * (0.9) = 0.9 \qquad \text{and} \qquad p[CD = 11] = 0.1$$

$$p[F = 1] = (0.9) * (0.37) + (0.1) * (1) = 0.433$$

EXERCISE SET 17

1. Suppose the system of Figure 62 is modified by changing the AND element to an element having three output values: agreement indicating defective D, agreement indicating good G, or no agreement at all N.
   (*a*) Determine the probability of error. *Ans.*: 0.05

   (*b*) Determine the probability of disagreement. *Ans.*: 0.38

   (*c*) If an element is classified as good, what is the probability that it is actually defective?
   (*d*) Do this problem if the classifiers are symmetric.
2. Using the NOT and OR components defined in Figure 65, determine the network equivalent to $\bar{A} \vee B$ and also $A \vee \bar{B}$.

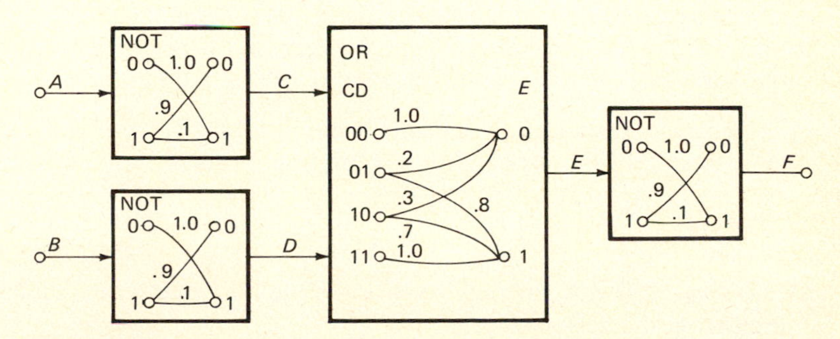

Figure 65

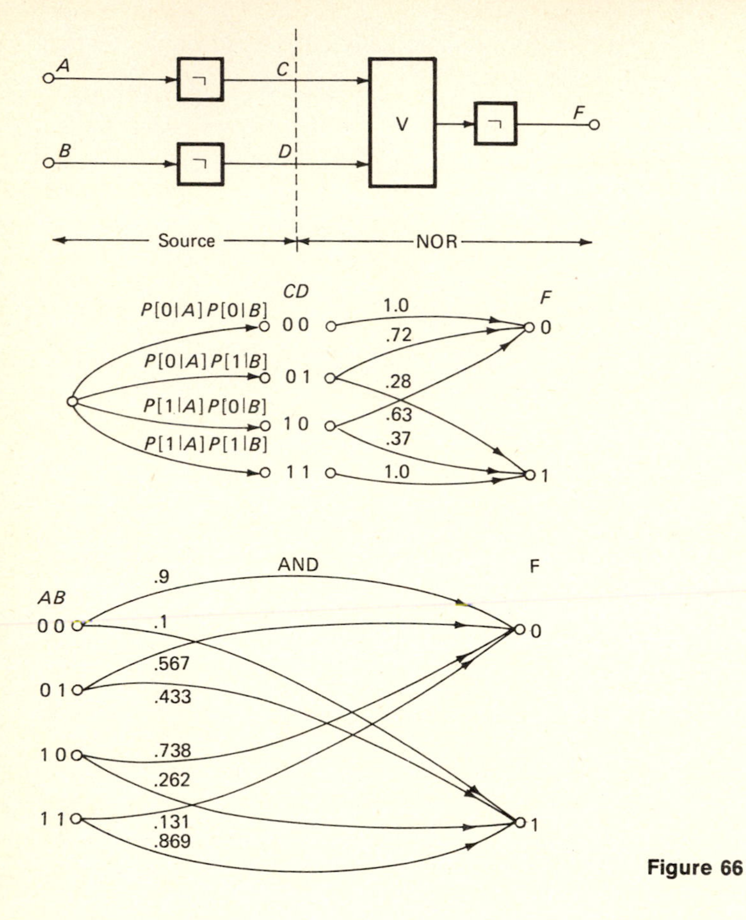

Figure 66

Repeated Independent Trials

A common procedure encountered in probabilistic systems is that of sequential repetition of an experiment for a number of trials or stages. For example, a coin may be tossed three times, or a system observed for 10 time intervals, or an experiment performed n times. All such systems have a basically similar structure, which will be studied now.

Repeated Coin Toss Consider tossing a single coin which is thick and bent so that the probabilities of landing on head, tail, and edge are

$$p[\mathrm{H}] = 0.5 \qquad p[\mathrm{T}] = 0.4 \qquad p[\mathrm{E}] = 0.1$$

If this same coin is thrown twice, all possible joint outcomes can be determined from the set product

$$\Omega = \{H, T, E\}^2 = \{\mathrm{HH, HT, HE, TH, TT, TE, EH, ET, EE}\}$$

Since the coin has no memory from one trial to another, the events of one trial can be assumed independent of the events of any other trial.

The probabilities of each joint event can be computed conveniently using a matrix as in Figure 67. First, we should realize that the probabilities on each trial are marginal probabilities. Then the joint probabilities can be determined by multiplying the corresponding marginal probabilities:

$$p[A_iB_j] = p[A_i] * p[B_j]$$

From this matrix we can now determine probabilities of various composite events. For example, the probability that the outcome on both trials is the same is

$$p[A_H B_H \vee A_T B_T \vee A_E B_E] = 0.25 + 0.16 + 0.01 = 0.42$$

Independent Arrows Three arrows (or missiles) A, B, C are fired in sequence at a target. Each arrow has the same probability of 0.6 of hitting the target independently of the others. Find the probability of at least one hit on the target.

By expressing this event E symbolically in three rather different ways, we have the following three different methods all arriving at the same answer:

$$\begin{aligned} p[E] &= p[\text{Any of the three } A \text{ or } B \text{ or } C \text{ is successful}] \\ &= p[A_S \vee B_S \vee C_S] \\ &= p[A_s] + p[B_s] + p[C_s] - p[A_sB_s] - p[A_sC_s] \\ &\qquad - p[B_sC_s] + p[A_sB_sC_s] \\ &= 0.6 + 0.6 + 0.6 - (0.6)(0.6) - (0.6)(0.6) - (0.6)(0.6) + (0.6)^3 \\ &= 0.936 \end{aligned}$$

$$\begin{aligned} p[E] &= p[\text{Success on first trial or else second or else third}] \\ &= p[A_S \vee A_F B_S \vee A_F B_F C_S] \\ &= (0.6) + (0.4)(0.6) + (0.4)(0.4)(0.6) \\ &= 0.936 \end{aligned}$$

$P[A_i B_j] = p[A_i]p[B_j]$

$i \backslash j$	H	T	E	
H	.25	.20	.05	.5
T	.20	.16	.04	.4
E	.05	.04	.01	.1
	.5	.4	.1	

Figure 67

$$
\begin{aligned}
p[E] &= p[\text{No failures on all three}] \\
&= p[\overline{\bar{A}_F \bar{B}_F \bar{C}_F}] \\
&= 1 - p[\bar{A}_F]\, p[\bar{B}_F]\, p[\bar{C}_F] \\
&= 1 - (0.4)(0.4)(0.4) \\
&= 0.936
\end{aligned}
$$

Repeated Conditional Trials

Although the previous repeated trials involved independent events, many of the methods can apply to nonindependent events. In such "conditioned" cases we would use the extended probabilistic product principle

$$
p[E_1 E_2 E_3 \cdots E_n] = p[E_1] * p[E_2 \mid E_1] * p[E_3 \mid E_2 E_1] * \cdots * p[E_n \mid E_{n-1} \cdots E_1]
$$

Such dependent repeated trials are conveniently described by a tree diagram.

Dependent Missiles A burst of three missiles (or arrows) is fired in a slow sequence, so that the outcome of one trial is known before another trial is attempted. Knowing this outcome may influence future trials. For example, if there is a success on a trial, the probability of another success on the following trial may decrease due to self-satisfaction or decrease in motivation, etc. Let us assume that the probability is halved in such a case. Similarly, if there is a failure on one trial, the probability of another failure on the next trial is halved. Construct the resulting probabilistic system if the probability of success on the first trial is 0.4. Determine the probability that the first hit occurs on the second or third trial.

Such a system, with its changing probabilities, is best visualized on a tree diagram as given in Figure 68. Each path from the "trunk" node to any "leaf" node is a possible outcome. Its probability is determined by taking the product of all the probabilities along this path; a very convenient method of applying the product principle. The probability that the first success occurs on the second or third trial is easily determined by inspection to be

$$
\begin{aligned}
p[A_F B_F C_S \vee A_F B_S C_S \vee A_F B_S C_S] &= 0.153 + 0.273 + 0.147 \\
&= 0.573
\end{aligned}
$$

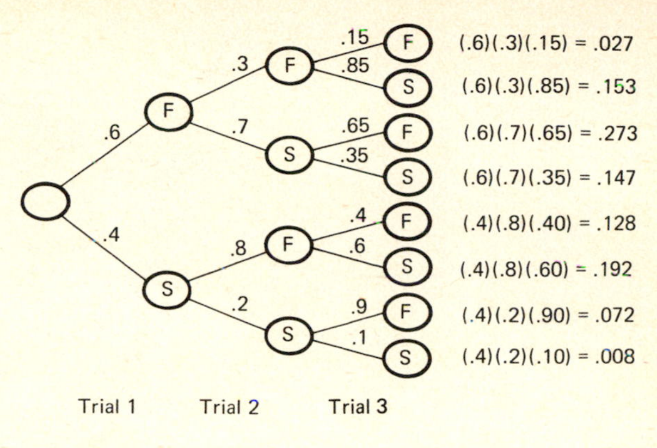

Figure 68

EXERCISE SET 18

1. Suppose a pair of ideal dice is modified by adding an extra dot to the face having four dots. Determine the matrix describing such a system. Find the probability of getting the sum of "7 or 11" and compare it to the ideal case.

2. For the example involving independent arrows:

(*a*) Determine all the basic events and their probabilities.

(*b*) Determine the probability of exactly two hits.

(*c*) Determine the probability that the first hit occurs on the second or third trial.

3. For the example involving dependent missiles (Figure 68):

(*a*) Find the probability of x hits, where $x = 0, 1, 2, 3, 4$.

(*b*) If there was only one hit, find the probability it was due to the third missile.

Ans.: $0.153/0.554 = 0.276$

(*c*) If there was at least one hit, find the probability it was due to the third missile.

5. RANDOM VARIABLES, DENSITY FUNCTIONS, EXPECTED VALUE

Random Variables

Until now we have been concentrating on events, rather than numbers. For example, a probabilistic system of three relays A, B, C can be described by eight basic events as shown in two ways in Figure 69. However, we are often not interested in the events themselves but rather in some numerical outcome which is a function of these events. For example, we may be interested in

$N =$ the number of relays operated,

or

$C =$ the amount of current taken by the relays (if relay A takes 2 amperes and the others only 1 ampere).

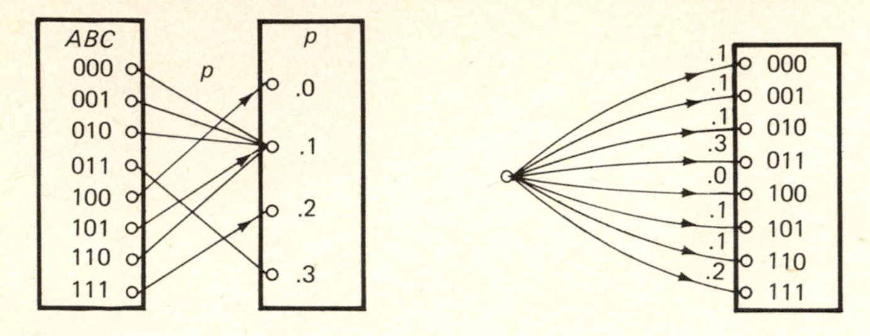

Figure 69

Any such numerical-valued function of events is called a *random variable*. Actually a random variable is neither random nor a variable. It is badly named but continues to remain so!

Some practical random variables are costs, profits, ages, utilities, magnitudes, speeds, tallies, voltages, and distances; all of them can be measured as real numbers. For the above relay system the two random variables N and C are shown graphically in Figure 70.

Other random variables which could be associated with this system are shown in Figure 71. This shows that a random variable could be any arbitrary real-valued function with no practical significance to these relays. Note that $Y = X^2$ illustrates that any real-valued function of a random variable is also a random variable.

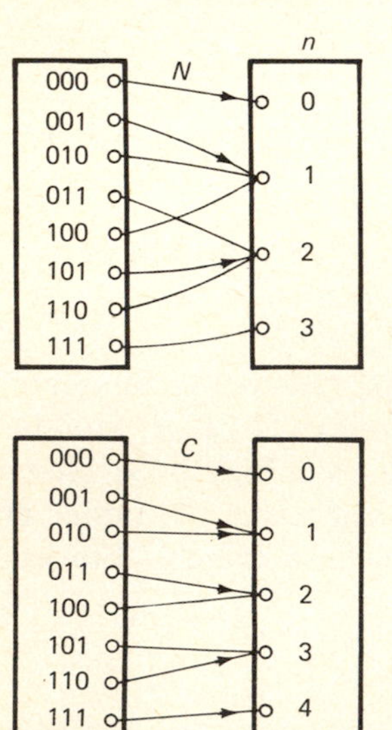

Figure 70

ABC	W	X	Y	Z
000	0	0	0	926
001	1	0	0	0.5
010	0	1	1	-33
011	1	2	4	10^9
100	0	3	9	1/3
101	1	1	1	0
110	1	0	0	2^{-5}
111	1	0	0	1

Figure 71

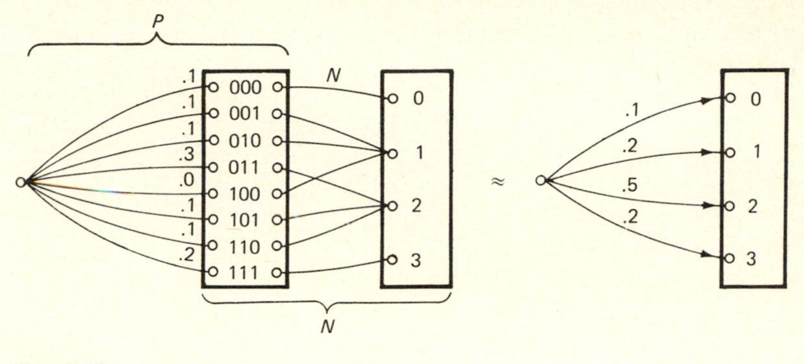

Figure 72

Probability Density Functions

The random variable N can be combined with the relay system as shown in Figure 72. The basic events are said to "*induce*" probabilities onto the random variable. For example, the probability that N has value 2 corresponds to the event

$$011 \vee 101 \vee 110$$

which occurs with probability

$$p[N = 2] = 0.3 + 0.1 + 0.1 = 0.5$$

The function which determines the probability of all values of the random variable is known as a *probability density function* (pdf). This probability density function is also known as the *probability mass function*. It can be drawn in other ways, as shown in Figure 73, but all these ways tend to suppress the functional dependence of the random variable on the basic events.

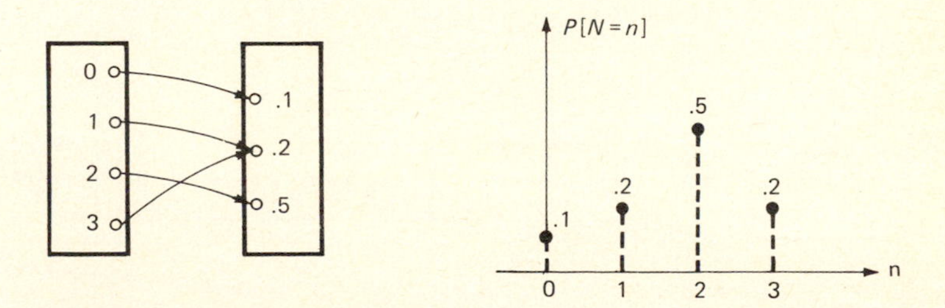

Figure 73

A probability density function $f(r_i)$ is a function which associates with each value r_i of a random variable R the probability that this value will be assumed. It is often denoted by $P[R = r_i]$ and has the following properties:

$$f(r_i) \geq 0$$

$$\Sigma f(r_i) = 1$$

$$p[a \leq R \leq b] = \sum_{a}^{b} f(r_i)$$

Since in many cases we are not interested in basic events but in numbers, we consider any function satisfying the above three properties as a random variable. For example, the source of Figure 74 emitting three values of voltage (0, 10, and − 10 volts) can be described directly by a pdf. In this case it is convenient to draw the pdf on its side (at the right) so that the vertical voltage scales will coincide. This provides an intuitive indication of thc "density" of the random voltage.

Expected Value

The concept of average value is commonly used in everyday life. In this section the intuitive concept of average value will be formalized to the probabilistic concept called expected value.

The average value of m numbers is computed by summing all the numbers and dividing by m. The resulting number, the average, indicates a central value of these numbers.

For example, consider the three-valued probabilistic source of Figure 74. The average value of the typical voltage sequence can be computed as

$$\frac{1}{T} \sum_{t} Y_t = \frac{0 + 10 + 0 + 0 - 10 + 10 + 10 + 0 - 10 + 0}{10}$$

$$= +1 \text{ volt}$$

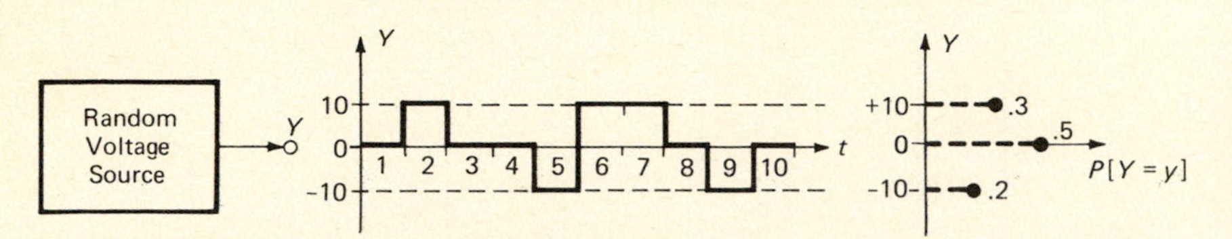

Figure 74

This average value of 1 volt can also be read directly on an average reading voltmeter (of the d'Arsonval type).

Since only three values appear, they can be grouped according to their frequency of occurrence as

$$(-10) * (\tfrac{2}{10}) + (0) * (\tfrac{5}{10}) + (+10) * (\tfrac{3}{10})$$

Recognizing that the above fractions represent probabilities, in the relative-frequency sense, we get

$$(-10) * P[Y=-10] + (0) * P[Y=0] + (+10) * P[Y=+10]$$

More generally this can be expressed as

$$\sum_i y_i P[Y=y_i] = \sum_i p_i y_i$$

This form enables the computation to be performed directly from the pdf, simply by summing the products of the random variable and its probability. This generalized average value is formally called the expected value.

Definition The *expected value* of a random variable R having values $r_1, r_2, \cdots, r_n$ occurring with probabilities $p_1, p_2, \cdots, p_n$ is denoted by symbol $E[R]$ and evaluated by

$$E[R] = \sum_i^n p_i r_i$$

The expected value of a random variable is a sort of center of gravity, a single number which indicates the value about which the random values are scattered. The expected value is often called the mean value or average value and is denoted in various ways such as

$$E[R] \qquad M[R] \qquad A[R] \qquad \tilde{R} \qquad \mu_R$$

We will generally use the symbol $E[R]$, but occasionally μ will be convenient.

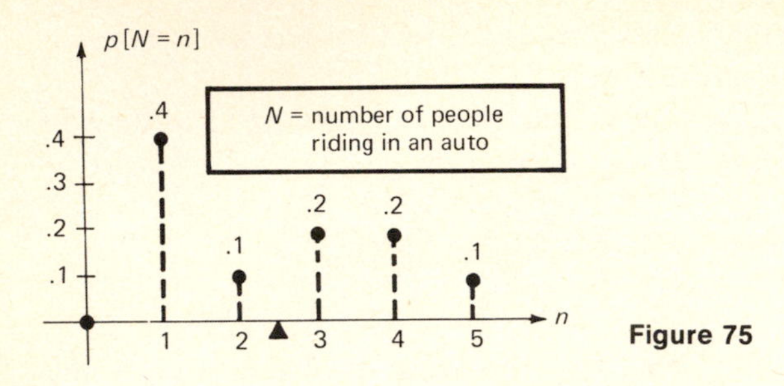

Figure 75

Example The pdf of Figure 75 describes the number of people riding in a car on a particular freeway during the rush hour in the middle of a typical week. The average number of people per car can be computed very simply from

$$E[N] = \Sigma n * p[N = n]$$

$$= (1)(0.4) + (2)(0.1) + 3(0.2) + (4)(0.2) + (5)(0.1)$$

$$= 2.5 \text{ people per car}$$

Of course, only in the most morbid situation can a car contain exactly this expected value of 2.5 people!

Example A machine produces items with probability of defect of 0.1. It costs \$50 to manufacture each item, which sells for \$60 if nondefective (and \$0 if defective). Find the expected profit or gain $E[G]$.

Solution If the item is not defective, the gain is

$$G[N] = 60 - 50 = 10 \text{ dollars per item}$$

If the item is defective, the gain is

$$G[D] = 0 - 50 = -50 \text{ dollars per item (a loss)}$$

In this case the random variable G has values $+10$ and -50 occurring with probabilities 0.9 and 0.1, thus having an expected value of

$$E[G] = p[N]\ G[N] + p[D]\ G[D]$$

$$= (0.9)(10) + (0.1)(-50)$$

$$= 4 \text{ dollars gain per item}$$

EXERCISE SET 19

1. Compute and sketch the pdf for the random variables Y and $(Y - X)^2$ of Figure 71 and also C of Figure 70. (Use the probabilities of Figure 69.)

2. For a system of two "honest" dice, sketch the pdf of the random variable S, where S is the sum of the dots on the two dice. Compute the expected value of S.

3. If the previous events A, B, and C of Figure 69 are independent, occurring with probability p, the resulting pdf of the number of occurrences N is known as the *binomial pdf* $b(n)$. Sketch this binomial pdf for $p = 0.4$, $p = 0.5$, and $p = 0.7$ and comment. Compute also the expected value of N in each case. Show that the expected value is $3 * p$.

4. If the previous events A, B, C of Figure 69 are independent, occurring with probabilities 0.3, 0.4, and 0.8, respectively, sketch the pdf for N (the number of events occurring). Determine the expected value of N. *Ans.*: 1.6

5. Consider the following game involving two ideal dice:

If the sum is 2 or 3 or 4, you win $4.
If the sum is 7 or 11 or 12, you win $3.
If the sum is 5 or 6, you lose $2.
If the sum is 8 or 9 or 10, you lose $3.

First, guess your average winning per game. Then, compute the expected value of your winnings. Compare!

Some Particular Probability Density Functions

So far, when considering random variables and density functions we have been quite general. However, there are some very particular density functions which are sometimes useful. We will consider some of these briefly in this section.

Binomial Probability Density Function. One probability density function which is often encountered is the binomial pdf. It arises when there are two outcomes such as 0 and 1, or defective and nondefective, or success and failure. It applies to systems consisting of n such binary events which are independent, and it gives the probability that exactly x of the n events occur. For example, it describes the following situations.

1. A coin is flipped three times; the probability of a head is 0.6; what is the probability that exactly two heads show up?
2. In a production lot of 10 items, the probability of a defect is 0.2 (independent of all others); find the probability that the number of defects is two.
3. A radioactive source is observed for 20 time intervals. The probability of emission during any time interval is 0.05 (independent of any other intervals). What is the probability of no emissions in the 20 intervals?

These three problems will be solved in three different ways. The first problem is sufficiently simple to enumerate in detail; the second problem will be solved after deriving a general result; the third problem will be solved by resorting to tables of compiled values for the binomial pdf.

In general, n independent events $E_1, E_2, \cdots, E_n$ having two values 0 and 1 generate a probability space $\{0, 1\}^n$, which is illustrated by the table-of-combinations of Figure 76 for $n = 3$.

The probability of the individual events is

$$p[E_i = 1] = p \qquad \text{for all } i$$

$$p[E_i = 0] = q = 1 - p$$

The probability of each combination of events is found by direct multiplication. From this table we can study the random variable X which represents the number of occurrences of value 1 in the n events.

$$p[X = 0] = q^3$$

$$p[X = 1] = 3pq^2$$

$$p[X = 2] = 3p^2q$$

$$p[X = 3] = p^3$$

For the value $p = p[E_i = 1] = 0.6$, the pdf can be graphed as shown in Figure 77*a*. The probability of exactly two heads is then seen to be

$$p[X = 2] = 3p^2q = 3(0.6)^2(0.4) = 0.432$$

The second problem involving 10 items can be treated similarly except that it requires a table-of-combinations of $2^{10} = 1{,}024$ rows. Since this would be a tedious task to evaluate, let us attempt to generalize the previous result.

E_1	E_2	E_3	$p[E_1E_2E_3]$	X
0	0	0	q^3	0
0	0	1	pq^2	1
0	1	0	pq^2	1
0	1	1	p^2q	2
1	0	0	pq^2	1
1	0	1	p^2q	2
1	1	1	p^2q	2
1	1	1	p^3	3

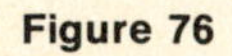

Figure 76

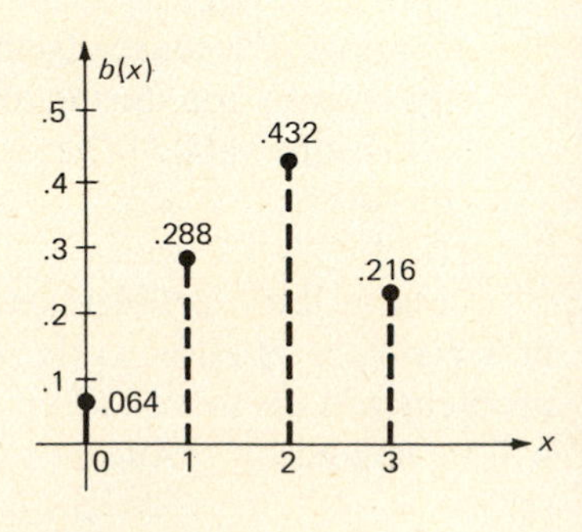

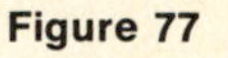

Figure 77

The probability of any x events occurring (each with probability p) and $n - x$ events not occurring is

$$p^x(1-p)^{n-x}$$

But the number of times that x events occur and $n - x$ events do not occur is simply the number of (distinguishable) arrangements of x elements of one type (occurring) and $n - x$ of another type (not occurring) which is

$$\frac{n!}{(x)!\,(n-x)!}$$

Thus the probability of exactly x events occurring out of n is given by the general expression

$$p[X=x] = \frac{n!}{(x)!\,(n-x)!}\,p^x(1-p)^{n-x}$$

This binomial density function is often denoted by $b(x)$ or sometimes $b(x; n, p)$.

Now returning to the previous problem involving $n = 10$ items occurring with probability 0.2, the probability that the number of defects is exactly $X = 2$ is given by substituting as

$$p[X=2] = b(2;10,0.2) = \frac{(10)(9)(8)!}{(2)(1)(8)!}(0.2)^2(0.8)^8 = 0.30$$

Although such computations are simpler to evaluate than the 2^n rows in a table-of-combinations, they still are somewhat tedious as n becomes large. Therefore the values of the binomial density function have been tabulated in many reports and books. The short table in Figure 78 is sufficient for most of our problems.

The third problem above, involving 20 observations of a radioactive sample with probability of emission of 0.05 in each interval, is described by the binomial density function $b(x;\ 20,\ 0.05)$. The probability of no emissions in the 20 intervals can be determined directly from the tables as

$$b(0;20,\ 0.05) = 0.359$$

Also the most likely number of emissions is 1, which occurs with probability 0.377.

Selected Values of the Binomial Density Function $b(x;n,p)$

x	$p=.1$, $n=5$	$p=.1$, $n=7$	$p=.1$, $n=9$	$p=.1$, $n=10$	$p=.2$, $n=10$	$p=.5$, $n=10$	$p=.05$, $n=20$	$p=.1$, $n=20$	$p=.15$, $n=20$
0	.590	.478	.387	.349	.107	.001	.359	.122	.039
1	.328	.372	.387	.387	.268	.010	.377	.270	.137
2	.073	.124	.172	.194	.302	.044	.189	.285	.229
3	.008	.023	.045	.057	.201	.117	.060	.190	.243
4	.001	.003	.007	.011	.088	.205	.013	.090	.182
5	.000	.000	.001	.002	.026	.246	.002	.032	.103
6		.000	.000	.000	.006	.205	.000	.009	.045
7		.000	.000	.000	.000	.117		.002	.016
8			.000	.000	.000	.044		.000	.005
9			.000	.000	.000	.010			.001
10				.000	.000	.001			.000

Figure 78

EXERCISE SET 20

1. An ideal coin is flipped 10 times. Find the probabilities of the following events:

(*a*) All 10 trials result in heads.

(*b*) Exactly 5 heads come up. *Ans.*: 0.246

(*c*) The number of heads is between 4 and 6 (inclusively).

(*d*) There are less than 5 heads.

(*e*) There are more than 5 heads.

(*f*) There are more than 8 heads.

(*g*) There are less than 3 or more than 7 heads.

2. Some components (transistors, nuts, etc.) are subjected to an endurance test in a hostile environment in which the probability of each component's failing is 0.15 (independent of any other components). If 20 such components are tested, find the probability of the following events:

(*a*) All components fail.

(*b*) No components fail.

(*c*) Exactly three components fail.

(*d*) At most three fail.

(*e*) More than five fail.

(*f*) Exactly six fail, given that more than five fail.

3. The method of improving the reliability of level networks of Figure 61 used the majority of an odd number of symmetric identical components. If the probability of error of each component is 0.1, find the probability of error of a system having three, five, seven, and nine components.

4. The quality-control procedure for evaluating a fuel is as follows. Five independent samples are selected from each batch of fuel and tested. The remainder of the batch is then rejected if one or more samples are inferior. What is the probability of accepting the batch if the probability of its being of inferior quality is 0.1? If the acceptance policy is changed to read "two or more samples are inferior," find the probability of accepting the batch.

Ans.: 0.59, 0.92

5. Describe two realistic situations which can be modeled by the binomial pdf.

Poisson Probability Density Function Under certain conditions the binomial pdf can be approximated by other density functions. For example, if the probability p is small (approaching 0) and the number of independent events n is large (approaching infinity) but the product pn is constant, then the binomial pdf can be approximated by the *Poisson density function*

$$p(x) = \frac{e^{-m} m^x}{x\,!}$$

The constant $m = pn$ is the expected value of the random variable, and $e = 2.71$ is the base of the natural logarithms. Figure 79 shows the Poisson pdf for $m = 3$.

Example On a complex system (computer, spaceship) the probability of making an improper electric connection is 10^{-6}; however, the number of such connections is 3 million, and so the average is $m = pn = (10^{-6})(3 * 10^6) = 3$ errors per system. The probability of having no errors is $p(0) = 0.050$ from Figure 79.

Example The average number of emissions per second from a block of radioactive material is 3. If the Poisson density function describes this emission behavior (as it often does), and if the lethal dose is 5 emissions per second, find the probability of a lethal dose of radioactivity. The solution is $p(5) + p(6) + \cdots$, an infinite sum which is more easily evaluated as

$$\begin{aligned} p[X \geq 5] &= 1 - [p(0) + p(1) + p(2) + p(3) + p(4)] \\ &= 1 - [0.815] = 0.185 \end{aligned}$$

Example There are 90 raisins in a loaf of bread which has been cut into 30 slices. What is the probability of getting the average, 3 raisins, in any particular slice of this bread?

Solution It is a temptation to say $p(3) = 0.224$ is the answer. However it is subject to some error since $n = 90$ is not too large, and $p = 0.033$ is not too small. Also the raisins may not be equiprobably or independently distributed because of insufficient stirring or

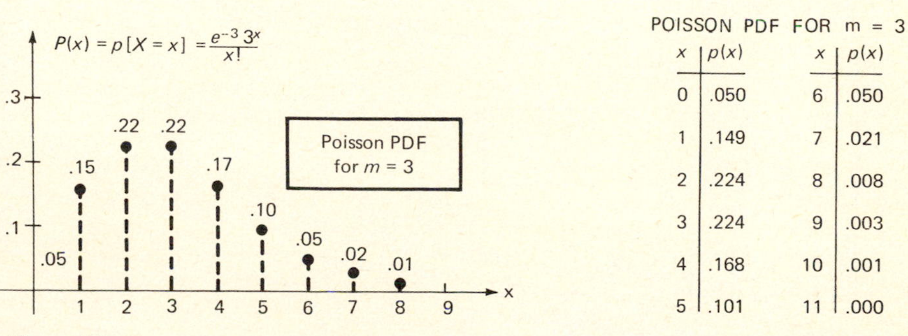

POISSON PDF FOR m = 3

x	p(x)	x	p(x)
0	.050	6	.050
1	.149	7	.021
2	.224	8	.008
3	.224	9	.003
4	.168	10	.001
5	.101	11	.000

Figure 79

because of settling during baking. Furthermore, raisins may be cut in two as the bread is sliced.

EXERCISE SET 21

1. On a typical day 100,000 cars use a freeway. The probability of each car's having an accident is 0.00003. Determine the probability of less than three accidents a day.

2. Suppose a typical typewritten letter consists of 1,000 characters. If each character is in error with probability 0.005, independent of any other characters, determine the probability of less than 3 errors in a letter.

†3. Prove that the binomial pdf yields the Poisson pdf for the conditions given in this section.

Geometric Probability Density Function. Consider a system (car, transistor, etc.) having a constant probability p of failing during a month, independent of any previous month. The probability of (the first) failure on the tth trial (month) is given by the geometric pdf

$$g(x) = p[T = t] = p(1 - p)^{t-1}$$

which corresponds to the occurrence of $t - 1$ successful months before the first failure. The diagram of Figure 80 shows that the probabilities decrease as the time t increases. This geometric density function also describes lengths of telephone calls, the lifetime of batteries, and even the number of attempts to get an automatic dispenser to finally accept a coin.

The expected value of the number of trials can be determined from the sum

$$E[T] = \sum_{1}^{\infty} tg(t) = \sum_{1}^{\infty} tp(1 - p)^{t-1}$$

$$= 1p + 2p(1 - p) + 3p(1 - p)^2 + 4p(1 - p)^3 + \cdots$$

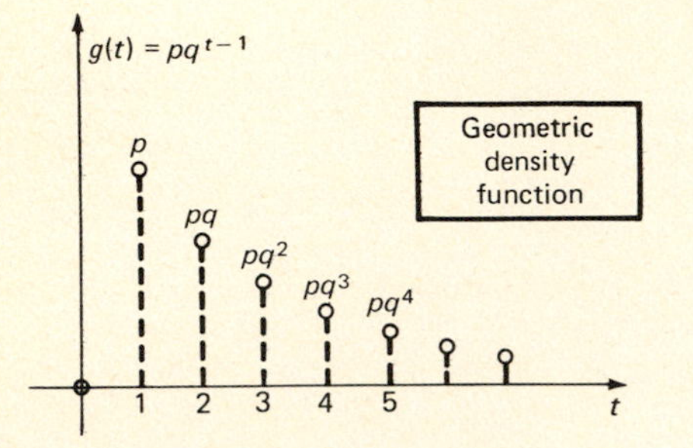

Figure 80

Notice that this is an infinite sum, but it can be proved that it converges by using the equality

$$\sum_{1}^{\infty} tx^{t-1} = \frac{1}{(1-x)^2}$$

The average number of trials is now

$$E[T] = p * \sum_{1}^{\infty} tq^{t-1} \qquad \text{where } q = 1 - p$$

$$= p * \frac{1}{(1-q)^2}$$

$$= \frac{1}{p}$$

Thus, for a geometric pdf, the average number of trials is the inverse of the probability of occurrence of the event on each trial. This average, or mean time before failure, is often denoted as MTF.

EXERCISE SET 22

1. A battery guaranteed for 3 years has probability 0.2 of failing each year regardless of its age. Determine its probability of failing within the guarantee period. Determine also its MTF. Repeat the problem if the probability of succeeding halves after each year.

2. An automatic dispensing machine rejects a good coin with probability 0.6 and rejects a counterfeit coin with probability 0.9. Find the average number of trials to accept a good coin and to accept a counterfeit coin.

†**3.** Prove the equality which was used to find $E[T]$ from the hint below:

$$\frac{d}{dx} \sum_{t=1}^{\infty} x^t = \frac{d}{dx} \quad \frac{1}{1-x}$$

Expected Value of a Function of Random Variables

Recall that a real-valued function of random variables is also a random variable. Thus if $Z = f(X, Y)$ is a function of the random variable X, having values $x_1, x_2, \cdots, x_m$, and the random variable Y, having values $y_1, y_2, \cdots, y_n$, then Z may have as many as mn different values z_{ij} corresponding to all possible combinations of values of X and Y. If each combination has probability $p(x_i, y_j)$, the expected value of random variable Z is

$$E[Z] = \sum_{k}^{mn} z_k p(z_k)$$

which can also be written in the form

$$E[f(X, Y)] = \sum_{j}^{n} \sum_{i}^{m} p(x_i, y_j)\, f(x_i, y_j)$$

The following example will be done using both of these forms.

Sums of Random Variables

A crummy communication system is sometimes modeled with a channel which is assumed to add (arithmetically) some noise to the signal to produce the output as in Figure 81. Of course the noise is assumed independent of the signal. Let us find the expected value of the output voltage Z corresponding to the given binary signal source X when perturbed by noise Y. There are three methods for computing this expected value.

Method 1. In this case the function and the joint probability become

$$f(x_i, y_j) = [x_i + y_j] \qquad \text{and} \qquad p(x_i, y_j) = p(x_i) * p(y_j)$$

The table-of-combinations of Figure 82 is most convenient for evaluating this expected value

$$\sum_{j}^{n} \sum_{i}^{m} p(x_i) * p(y_j) * [x_i + y_j]$$

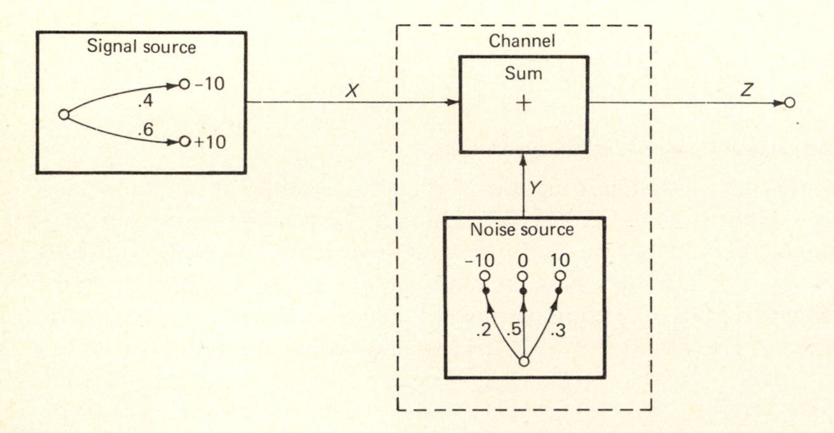

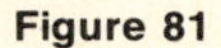
Figure 81

X	Y	$p(X)$	$p(Y)$	$p(X,Y)$	$f(X,Y)$	$p(X,Y)f(X,Y)$
-10	-10	.4	.2	.08	-20	-1.6
-10	0	.4	.5	.20	-10	-2.0
-10	10	.4	.3	.12	0	0.0
10	-10	.6	.2	.12	0	0.0
10	0	.6	.5	.30	10	3.0
10	10	.6	.3	.18	20	3.6
					$E[X+Y]=$	3.0

Figure 82

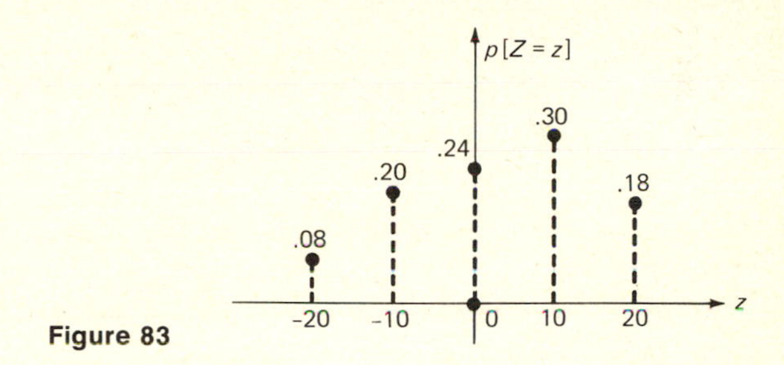

Figure 83

Method 2. The table-of-combinations of Figure 82 may also be used to determine the probability density function for the new random variable Z (where $Z = X + Y$) as in Figure 83. Then the expected value of Z could be computed directly as

$$
\begin{aligned}
E[Z] &= \sum_k p_k * z_k \\
&= (0.08)(-20) + (0.20)(-10) + (0.30)(10) + (0.18)(20) \\
&= 3.0 \text{ volts}
\end{aligned}
$$

Method 3. Although the above two methods are general and apply to all functions of random variables, there are other alternative methods which may be more convenient in particular cases. This will be shown in the next section after proving some theorems which we will need.

Properties of Expected Value

The expected value is an important concept having useful properties. In this section we shall study some of these properties, in the form of theorems. Then these properties will be applied to random voltage sources.

Theorem

$E[k * X] = k * E[X]$ for k a constant $(k \neq f(X))$

Proof

$$E[f(X)] = \sum_{i} p(x_i) f(x_i) \quad \text{by definition}$$

$$\begin{aligned} E[k * X] &= \sum_{i} p(x_i) * kx_i && \text{by substitution} \\ &= k \sum p(x_i) * x_i && \text{by distribution} \\ &= k * E[X] && \text{by definition} \end{aligned}$$

This theorem can be interpreted to indicate that if a random variable (such as a voltage) is scaled, either amplified or attenuated, the expected value is scaled by this same proportion. Such a property is sometimes known as *homogeneity*.

Theorem 2

$E[X + Y] = E[X] + E[Y]$ for any X, Y

Proof

$$E[f(X,Y)] = \sum_{i} \sum_{j} p(x_i, y_j) f(x_i, y_j) \quad \text{by definition}$$

$$\begin{aligned} E[X + Y] &= \sum_{i} \sum_{j} p(x_i, y_j)[x_i + y_j] && \text{by substitution} \\ &= \sum_{i} \sum_{j} p(x_i, y_j) x_i + \sum_{i} \sum_{j} p(x_i, y_j) y_j && \text{by distribution} \\ &= \sum_{i} x_i \sum_{j} p(x_i, y_j) + \sum_{i} \sum_{j} p(x_i, y_j) y_j && \text{by Theorem 1} \\ &= \sum_{i} x_i p(x_i) + \sum_{i} \sum_{j} p(x_i, y_j) y_j && \text{by definition (marginal)} \\ &= \sum_{i} x_i p(x_i) + \sum_{j} \sum_{i} p(x_i, y_j) y_j && \text{by commutativity} \\ &= \sum_{i} x_i p(x_i) + \sum_{j} y_j \sum_{i} p(x_i, y_j) && \text{by Theorem 1} \end{aligned}$$

$$= \sum_i x_i p(x_i) + \sum_j y_j p(y_j) \qquad \text{by marginal probability}$$

$$= E[X] + E[Y] \qquad \text{by definition}$$

This theorem indicates that the expected value is additive in general. The additivity is not a consequence of any other property such as independence. Thus, if two random voltages are added together, the resulting average voltage is the sum of the individual average voltages. This can be computationally convenient as an alternative way of solving some problems.

Example In the previous section we encountered noise added to a signal. We found the expected value of the sum by two methods. Now we can also determine the expected value by summing the expected value of the signal and the expected value of the noise. The expected value of the signal voltage X is

$$E[X] = \sum_i x_i \, p(x_i) = (-10)(0.4) + (10)(0.6) = 2.0 \text{ volts}$$

The expected value of the noise voltage Y is

$$E[Y] = \sum_j y_j \, p(y_j) = (10)(0.2) + (10)(0.3) = 1.0 \text{ volts}$$

The expected value of the total signal-plus-noise voltage is

$$\begin{aligned} E[X+Y] &= E[X] + E[Y] \\ &= 2.0 + 1.0 \\ &= 3 \text{ volts} \end{aligned}$$

Notice that this method is much more convenient computationally than the previous two methods.

Some other useful theorems are summarized below, for reference purposes. You should convince yourself of their validity, either intuitively or formally (by proving them).

Theorem 3

$$E[k] = k$$

Theorem 4

$$E[X + k] = E[X] + k$$

Theorem 5

$$E[X - Y] = E[X] - E[Y]$$

Theorem 6

$$E[XY] = E[X] * E[Y] \qquad \text{if } X \perp Y$$

Theorem 7

$$E[k_1 X + k_2 Y] = k_1 E[X] + k_2 E[Y]$$

EXERCISE SET 23

1. If the signal and noise of Figure 81 are combined so that the product of their two values results at the output, i.e.,

$$Z = X * Y$$

determine the pdf and expected value of the new output Z.

2. Two equiprobable and independent binary signals having values 0 and 10 volts (corresponding to 0 and 1, respectively) are connected to an OR element. Determine the pdf and expected value of the output of this OR element.

3. If the signal and noise of Figure 81 are combined so that the maximum of their two values results at the output, i.e.,

$$Z = \max[X, Y]$$

determine the pdf and expected value of the new output Z. *Ans.*: 6.4 volts

4. Nuts and bolts are combined to form nut-bolt assemblies in an almost independent manner as described by the matrix of Figure 84. The number of flaws in a bolt is denoted A; the number of flaws in a nut is denoted B. The gain $G[A_i B_j]$ (profit) associated with each nut-bolt assembly is also given.

(*a*) Compute the expected gain (profit) of a nut-bolt assembly. *Ans.*: 12

(*b*) Compute the expected total number of flaws in the assembly.

5. Prove Theorems 4 and 6.

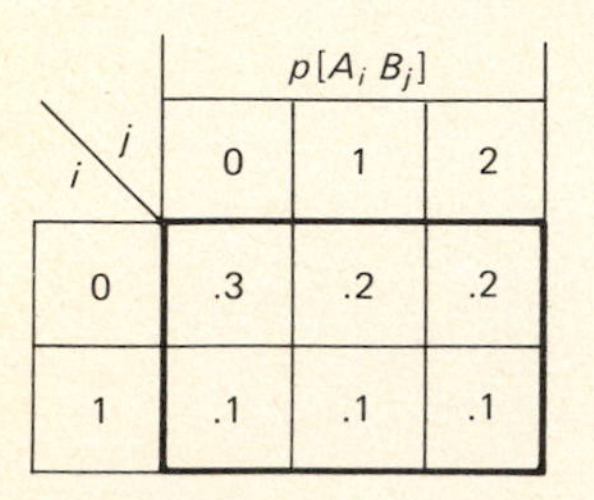

$p[A_i\, B_j]$: $i \backslash j$	0	1	2
0	.3	.2	.2
1	.1	.1	.1

$G[A_i\, B_j]$: $i \backslash j$	0	1	2
0	30	20	- 5
1	10	0	-10

Figure 84

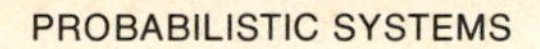

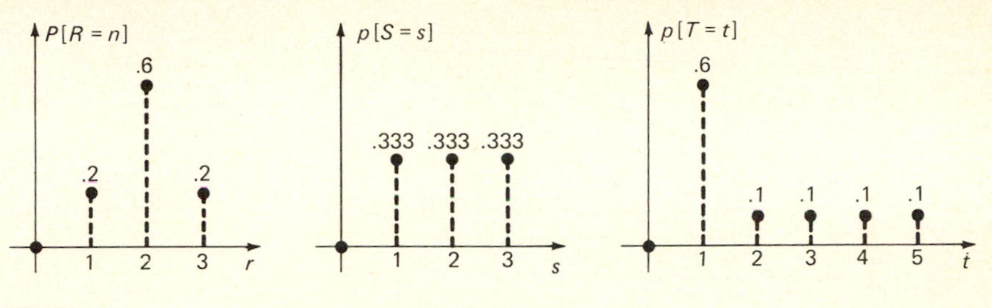

Figure 85

*6. VARIANCE, COVARIANCE, DEPENDENCE

It is often convenient to describe some attribute of a probabilistic system by a single number. For example, the central value of a random variable is described by the expected value. In this section we will briefly introduce other measures such as the variance, standard deviation, covariance, and correlation coefficient. We will also introduce in greater detail the concept of dependence, which is also known as mutual information or transinformation.

Variance

Consider the three random variables of Figure 85. They all have the same expected value of 2.0, but the values of R vary slightly, the values of S vary more, and the values of T vary considerably from the expected value. A measure of this variability, spread, dispersion, or variance can be developed as follows.

Consider the pdf of Figure 86 describing random variable X having values X_i and expected value $E(X)$. The deviation of each value X_i from the expected value is

$$X_i - E(X)$$

Each such deviation can be squared to exaggerate the difference and to make this value positive, yielding

$$[X_i - E(X)]^2$$

This is done because the amount of deviation is considered more important than its direction or polarity. The deviations can then be averaged by taking the expected value to yield

$$E[(X_i - E(X))^2] = \sum_i p(X_i)(X_i - E(X))^2$$

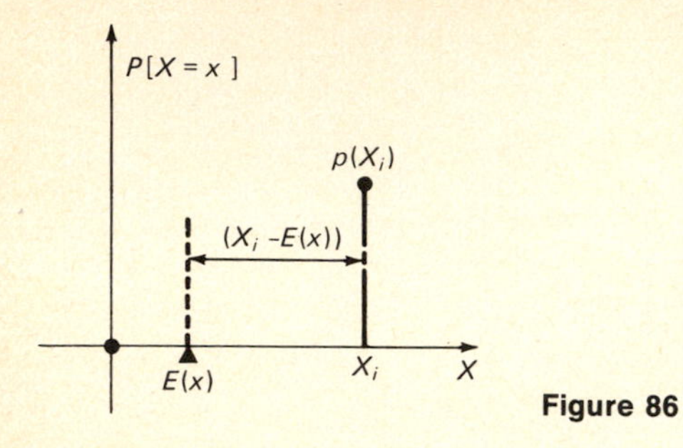

Figure 86

This measure is referred to as the *variance* of X and denoted

$$V[X] = E[(X - E(X))^2]$$

Example The random voltage Y of Figure 74 has values

$$y_1 = -10 \qquad y_2 = 0 \qquad y_3 = +10$$

occurring with probabilities

$$p_1 = 0.2 \qquad p_2 = 0.5 \qquad p_3 = 0.3$$

The expected value is computed as

$$\begin{aligned} E(Y) &= \sum_{i=1}^{3} y_i \, p_i \\ &= (-10)(0.2) + (0)(0.5) + (10)(0.3) \\ &= 1 \text{ volt} \end{aligned}$$

The variance can be computed as

$$\begin{aligned} V[Y] &= \sum_{i=1}^{3} p_i \, [(y_i - 1)^2] \\ &= (0.2)(-10 - 1)^2 + (0.5)(0 - 1)^2 + (0.3)(10 - 1)^2 \\ &= 49 \end{aligned}$$

Similarly the variance of the three random variables of Figure 85 can be computed as

$$V[R] = 0.40 \qquad V[S] = 0.667 \qquad V[T] = 2.00$$

Notice that this numerical measure is in agreement with our original intuitive indication of variability which was determined by inspection.

Standard Deviation

Sometimes the square root of the variance is chosen to indicate the measure of the spread dispersion or variability. This *root mean square deviation* is commonly known as the *standard deviation* and is denoted $S[X]$ or σ (sigma)

$$S[X] = \sqrt{V[X]}$$

The standard deviation has the same units as the random variable. For the previous random voltage the standard deviation is $\sqrt{49} = 7$ volts, which can be shown as an average deviation as in Figure 87.

Properties of Variance and Standard Deviation

A number of properties involving variance are computationally and conceptually convenient. For example, the variance need not be computed directly from its definition

$$V[X] = E[(X - E(X))^2]$$

but can be computed from an equivalent expression

$$V[X] = E[X^2] - [E[X]]^2$$

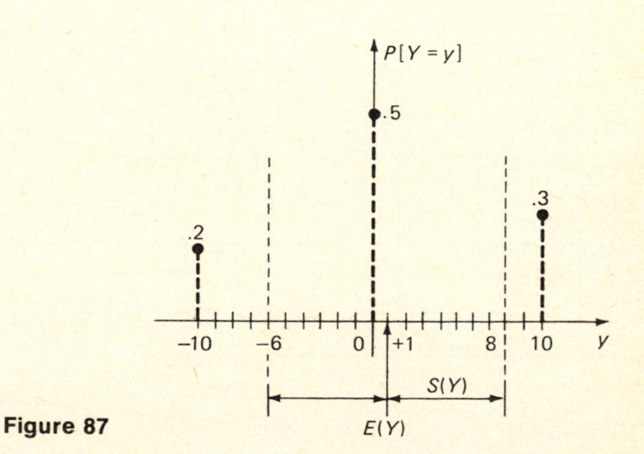

Figure 87

This can be proved by using some properties of expected value as follows. Note that $E(X)$ is a constant, written as μ for convenience.

$$\begin{aligned} V[X] &= E[(X-\mu)^2] \\ &= E[X^2 - 2\mu X + \mu^2] \\ &= E[X^2] - 2\mu E[X] + E[\mu^2] \\ &= E[X^2] - 2\mu^2 + \mu^2 \\ &= E[X^2] - [E[X]]^2 \end{aligned}$$

The algorithms corresponding to these two methods are given in Figure 88. The second method is computationally faster since it involves only one iteration loop compared to the first method, requiring two loops.

***Covariance and Correlation Coefficient**

It is often desirable to have some indication or measure of the degree of association, relation, or dependence between two random variables. One attempt at such a measure is the covariance $C[X,Y]$, defined as

$$C[X, Y] = E[(X - E(X)) * (Y - E(Y))]$$

which can also be written (see Exercise 4*d*)

$$C[X,Y] = E[XY] - E[X] * E[Y]$$

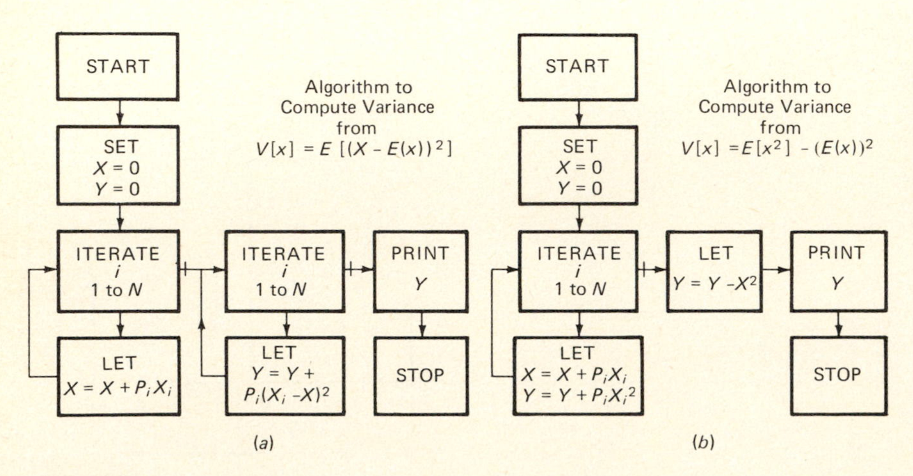

Figure 88

Two random variables X and Y are defined to be *uncorrelated* if their covariance $C[X,Y]$ is zero. This also means that for uncorrelated random variables

$$E[XY] = E[X] * E[Y]$$

Notice that this relation between expected values also holds for independent random variables. Thus, if two random variables are independent, they are uncorrelated. However, the converse does not hold; there are unusual cases where two random variables are uncorrelated but they are not independent (see Problem 1 in Exercise Set 24). Correlation is not necessarily a measure of dependence, but it does indicate some degree of association.

The covariance is sometimes normalized and called the *correlation coefficient* ρ (rho):

$$\rho = \frac{C[X,Y]}{S[X]\,S[Y]}$$

It has values only within the range -1 to $+1$, indicating the degree of association as well as its direction (direct or inverse).

EXERCISE SET 24

1. For the two random variables A and B of Figure 89:
   (*a*) Determine the variances $V[A]$ and $V[B]$.
   (*b*) Determine the covariance $C[A,B]$ and comment!

A \ B	-1	0	+1
-1	.2	0	.2
0	0	.2	0
+1	.2	0	.2

Figure 89

2. A typical input-output sequence of some system is given as

Time t	1	2	3	4	5	6	7	8	9	10	11	12	13	14	15	16	17	18
Input X	0	2	1	2	2	0	2	1	1	0	0	0	1	2	1	2	0	1
Output Y	1	2	1	1	0	0	1	1	0	1	2	1	2	2	0	1	0	2

Determine the covariance between the following and comment:

(*a*) $X(t)$ and $Y(t)$ (*b*) $X(t)$ and $Y(t+1)$

(*c*) $X(t)$ and $X(t+1)$ (*d*) $X(t)$ and $X(t+2)$

3. Construct an algorithm (in flow-diagram form) to compute the covariance $C[X, Y]$. How could you modify this algorithm to compute the variance $V[X]$?

4. Prove the following results:

(*a*) $V[kX] = k^2 V[X]$

(*b*) $V[X+Y] = V[X] + V[Y] + 2C[X,Y]$

(*c*) $C[A * X, B * Y] = A * B * C[X, Y]$

(*d*) $C[X, Y] = E[XY] - E[X] * E[Y]$

Degree of Dependence (Mutual Information)

The variance, covariance, and correlation coefficient provide some measure of variables which have real values. However, not all variables have real values. For example, the coloring of hair is a variable having values such as brunette, red, black, and blonde. Also the color of eyes may be classified as blue or brown. Of course we could always impose a measure of color such as the wavelength of the reflected light, but this is just as arbitrary as the nonmeasurable values, and this is being too precise about an inherently fuzzy concept. Notice that even a precise concept of sex, having values male and female, is nonmeasurable.

Despite the nonmeasurability of the colors of hair and eyes, we should still be able to measure the dependence or relationship between them. Similarly we may wish to relate the sex (male or female) of people to the opinion (like or dislike) about some product. Neither sex nor opinion is measurable. However, the relationship between such nonmetric variables can be measured as follows.

Suppose that the probability of occurrence of one event is $p[X_i]$, the probability of a second event is $p[Y_j]$, and the probability of their joint occurrence is $p[X_i Y_j]$. If the two events were independent, the probability of joint occurrence would be $p[X_i]\, p[Y_j]$. Comparing the two numbers $p[X_i Y_j]$ and $p[X_i]\, p[Y_j]$ would provide an indication of the dependence of these two events.

A mathematical comparison of these two probabilities could be given by the difference, the ratio, or the logarithm of the ratio. Actually taking the logarithm of the ratio leads to a most meaningful measure

$$D[X_i Y_j] = \log \frac{p[X_i, Y_j]}{p[X_i]\, p[Y_j]}$$

This function has the convenient property that when the events are independent, they have a dependence measure of 0.

Such a measure can be computed for each combination $X_i Y_j$ in a system, and then the overall dependence $D[X{:}Y]$ is determined by averaging

$$D[X:Y] = \sum_i \sum_j p[X_i Y_j]\; D[X_i Y_j]$$

$$= \sum_i \sum_j p[X_i Y_j] \log \frac{p[X_i Y_j]}{p[X_i]\, p[Y_j]}$$

This measure has an interesting interpretation and great application in information theory, where it is known as *mutual information*. When the base of the logarithm is 2, this is a measure of the number of bits of information. The measure is also used in psychology and biology.

Let us compute the degree of dependence between the input and output of the original crummy relay system. First the joint probability matrix must be obtained; it is repeated in Figure 90. Then the individual measures $D(X_i Y_j)$, labeled D_{ij}, are computed for each input-output combination:

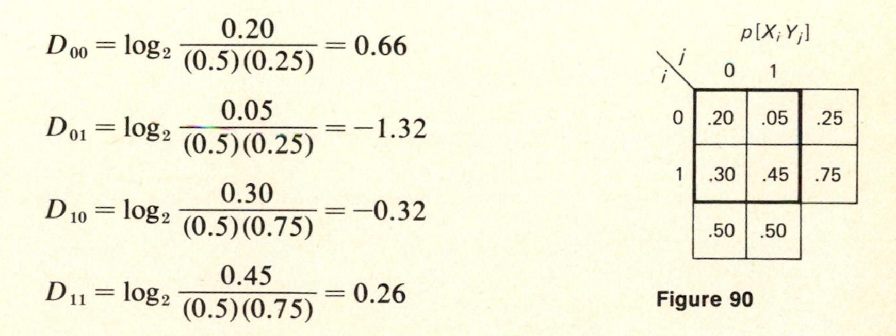

$$D_{00} = \log_2 \frac{0.20}{(0.5)(0.25)} = 0.66$$

$$D_{01} = \log_2 \frac{0.05}{(0.5)(0.25)} = -1.32$$

$$D_{10} = \log_2 \frac{0.30}{(0.5)(0.75)} = -0.32$$

$$D_{11} = \log_2 \frac{0.45}{(0.5)(0.75)} = 0.26$$

Figure 90

These individual measures are then averaged to yield

$$D[X:Y] = \sum_i \sum_j p_{ij} D_{ij} = 0.2D_{00} + 0.05D_{01} + 0.30D_{10} + 0.45D_{11}$$

$$= 0.09$$

The degree of dependence can now be compared to other systems. It should especially be compared to an ideal system as shown in Figure 91. The maximum dependence in this case is

$$D_{\max} = 0.25 \log \frac{0.25}{(0.25)(0.75)} + 0.75 \log \frac{0.75}{(0.75)(0.75)} = 0.811$$

The ratio of the actual degree of dependence to this maximum degree of dependence could be used as a measure of efficiency of this information system (source and relay). In this case it is

$$\text{Information efficiency} = \frac{D[X:Y]}{D_{\max}} = \frac{0.092}{0.811} = 0.113$$

This system is about 11 percent efficient, i.e., about a tenth as good as it could be with this source. For other sources the efficiency could be lower or higher.

In summary, the degree of dependence is a measure of variables which are not real-valued. It also allows for comparison of very different systems. For example, it could indicate how the eye- and hair-coloration dependence compares to the dependence between smoking and cancer. But perhaps more importantly it can be used to compare existing systems with optimal systems, and the resulting efficiency can be used for evaluation of information systems. More detailed consideration of this measure can be found in the field of information theory.

EXERCISE SET 25

1. Determine the degree of dependence between the color of hair and eyes from the data of Figure 92.
2. Compute the efficiency of the previous crummy relay system when connected to a source having probability $p[X_1] = 0.4$
3. Find the degree of dependence between smoking and cancer using the data of Figure 16.

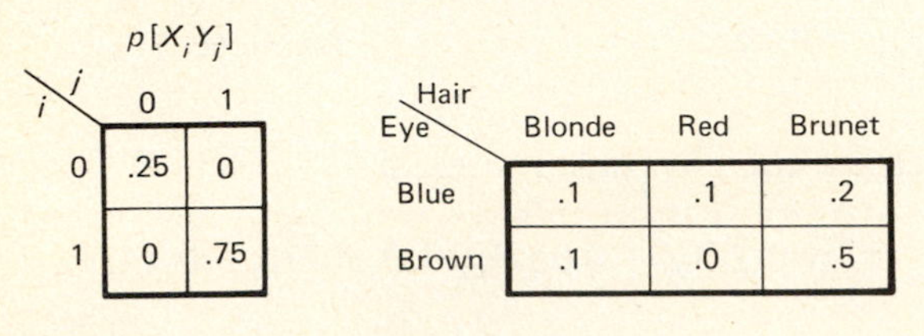

$p[X_i Y_j]$

i \ j	0	1
0	.25	0
1	0	.75

Eye \ Hair	Blonde	Red	Brunet
Blue	.1	.1	.2
Brown	.1	.0	.5

Figure 91 Figure 92

4. Compute the degree of dependence of the weather on the previous day's barometer reading using the data of Figure 96. How does this compare to the dependence between smoking and cancer?

5. For the survey of Figure 11, determine the degree of dependence between:
(*a*) Opinion and age (*b*) Opinion and sex
(*c*) Age and sex (*d*) Age-sex combination and opinion

6. Prove the following computationally convenient result, justifying each step:

$$D[X:Y] = \sum_i \sum_j p(x_i, y_j) \log p(x_i, y_j) - \sum_i p(x_i) \log p(x_i) - \sum_j p(y_j) \log p(y_j)$$

7. Extend this measure to determine the degree of dependence between the three relays A, B, C of Figure 12. Comment.

8. Construct an algorithm (in flow-diagram form) for computing the degree of dependence between any two variables when given by a joint-probability matrix.

7. DECISION MAKING

Decision making is the process of optimizing the behavior of a system. Intuitively this process involves the following sequence of steps:

1. Listing all relevant inputs (actions or alternatives)
2. Enumerating all outputs (outcomes or consequences)
3. Relating the inputs to outputs (assigning probabilities)
4. Determining the gain for each input-output combination
5. Evaluating the gain for each input value
6. Selecting the input which yields the optimal gain

This intuitive process will be formalized and extended after the following example.

Example A system is to manufacture some item. Three types of materials can be used: aluminum, costing \$8 per item; brass, costing \$2 per item; and copper, costing \$5 per item. If aluminum is used, no defective items are produced, but with brass the probability of defect is 0.2 and with copper the probability of defect is 0.6. A nondefective item sells for \$20, and a defective item must be scrapped at a loss of \$5. The problem is to determine which of the three materials should be used for the maximum gain.

Solution This system can be modeled by the sagittal diagram of Figure 93. The input X has values A, B, C, corresponding to the three materials. The output Y has two values D, N corresponding to defective and nondefective. For each input-output combination there is an arrow labeled with both a probability p_{ij} and a gain G_{ij}. The gain G_{ij} for each input-output combination X_iY_j is determined by taking the selling price S_j less the cost of the material C_i

$$G_{ij} = S_j - C_i$$

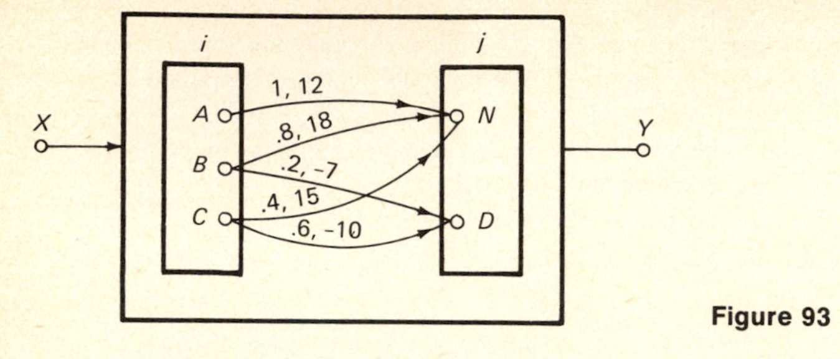

Figure 93

For example, the gain associated with a defective item of brass is

$$G_{BD} = S_D - C_B = -5 - 2 = -7 \text{ dollars} \quad \text{(a loss)}$$

For each input we can evaluate the expected gain. When the input is A (aluminum), the output quality is always good, and so the gain is \$12 per item. When the input is B (brass), the expected gain is

$$(0.8)(18) + (0.2)(-7) = 13 \text{ dollars per item}$$

When the input is C (copper), the expected gain is

$$(0.4)(15) + (0.6)(-10) = 0 \text{ dollars per item}$$

The maximum of the three gains, 12, 13, and 0, is 13, which occurs when the input is B; the optimal input material is brass.

Let us now generalize this process of decision making. A probabilistic system is given by a sagittal relation specifying the forward conditional probabilities p_{ij}

$$p_{ij} = p[(Y = j) \mid (X = i)]$$

With each input-output combination there may be associated a random variable G_{ij} called the *gain*

$$G_{ij} = G[(Y = j) \text{ and } (X = i)]$$

The expected value of the gain given any input X_i can be computed from

$$E[G \mid X_i] = \sum_j p[Y_j | X_i] G(Y_j \cdot X_i)$$

This is known as the *conditional expected value* and abbreviated

$$E[G_i] = \sum_j p_{ij} G_{ij}$$

The maximum gain, denoted G^*, is determined by computing the conditional expected gain associated with each input value i and simply selecting the maximum, i.e.,

$$G^* = \max_{\text{all } i} \left[E(G \mid X_i) \right]$$

$$G^* = \max_{i} \left[\sum_{i} p_{ij} G_{ij} \right]$$

The optimal input, denoted X_i^*, is the input (or inputs) associated with this maximum value of gain G^*.

Example A manufacturing system can be operated at two speeds, S (slow) and F (fast), corresponding to three and five items per hour. However, the probability of a defect also increases with speed, with corresponding values 0.1 and 0.3. If a manufactured item of good quality yields a gain of 20 units whereas one of defective quality yields a loss of 10 units, which of these two inputs would you choose? Why?

Solution The goal in this case is not simply to maximize the gain per item, but to maximize the gain per hour which is determined from the product

$$\frac{\text{Gain}}{\text{Hour}} = \frac{\text{gain}}{\text{item}} * \frac{\text{items}}{\text{hour}}$$

The corresponding sagittal diagram is shown in Figure 94. The optimal gain is

$$\begin{aligned} G^* &= \max \ [G_S, G_F] \\ &= \max \ [(0.9)(60) + (0.1)(-30), (0.7)(100) + (0.3)(-50)] \\ &= \max \ [51, 55] \\ &= 55 \end{aligned}$$

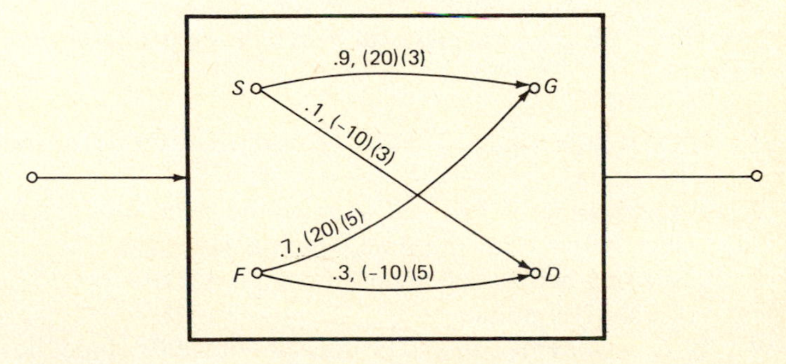

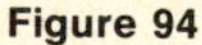
Figure 94

This optimal gain of 55 units per hour corresponds to the fast input. However, it is significant to notice that the slower input yielded almost as much gain. The differences between these gains may be negligible; one may be just as satisfactory as the other. In fact, the slower speed may be preferable since it may result in less machine breakdown. The slower speed may also result in better working conditions, less noise, less tension, and fewer accidents.

We should realize that we often make simplifying assumptions about the model, we approximate some costs, and we usually use a single simple goal. Therefore when we get outcomes that are close, we should stop to reconsider before blindly choosing the maximum outcome. At this final stage, we can consider some less easily measured goals, such as convenience or beauty.

Decision Making Using Information Instruments

An information instrument in a decision-making system is any device or system which relates relevant events and which may aid in achieving the goal. For example, if a decision is to be made today but depends on the weather tomorrow, some information instruments could be a barometer, a newspaper forecast, a radio forecast, the activity of animals, the aches and pains of some person, or any combinations of them. Although none of these instruments deterministically indicates the next day's weather, we may still be able to use them to help in achieving our goal. The method for using such instruments will be illustrated by the following example.

Consider an individual, called a *controller*, who wishes to control a system such as a tourist resort, construction firm, lumber mill, or hamburger stand, all of which are highly dependent on the weather. For a simple example consider a hamburger stand near a recreation area where the main activity occurs on Sunday, weather permitting. Although the controller cannot determine the weather, he can control buying the necessary foods, such as bread or vegetables, which are unfortunately perishable. He must, however, make this decision on the day *before*, Saturday, to order the food in sufficient time for Sunday.

From observation and experience the controller makes the following estimates:

1. If the weather is good and he buys the extra food, he makes a profit of 300 units.
2. If the weather is bad and he buys the extra food, he loses 200 units since he pays for the food which is not used.
3. If the weather is good and he did not buy extra food, he makes a profit of only 200 units.
4. If the weather is bad and he does not buy extra food, he makes a small profit of 100 units.

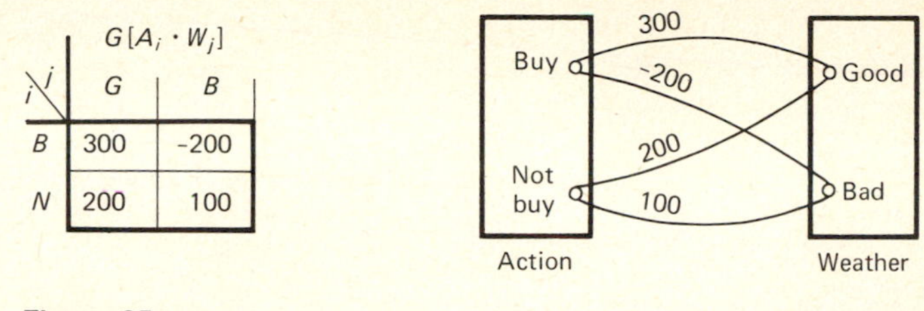

$i \backslash j$	G	B
B	300	-200
N	200	100

Figure 95

This information can be summarized by a gain function, which is described by both a *payoff* matrix and by a sagittal function-on-a-relation in Figure 95. Since the decision to buy is made on the day previous to the day of selling, it must be made without certain knowledge of the weather. To aid in making this decision, an information system such as a barometer may be used to attempt to predict the weather. A probabilistic barometer is described in Figure 96.

A controller (or control policy) is any function which specifies an action for each value of the information instrument. The controller which optimizes the average gain is known as the *optimal controller*. The optimal controller for this system is given in Figure 97. Its policy or strategy is expressed verbally as "buy only when the barometer is rising." This optimal controller was determined by the following method, which considers the optimal action for each barometer reading.

Case 1 Barometer is falling, B_F. The probabilities of resulting good and bad weather are then 0.5 and 0.5. If the action is then to buy, the average gain is

$$G_H = \sum_j p(W_j \mid B_F)\, G[W_j \cdot A_B] = (0.5)(300) + (0.5)(-200) = 50 \text{ units}$$

If the action is not to buy, the resulting gain is
$G_N = (0.5)(200) + (0.5)(100) = 150$ units

The maximum gain G^* occurs when the decision is not to buy.

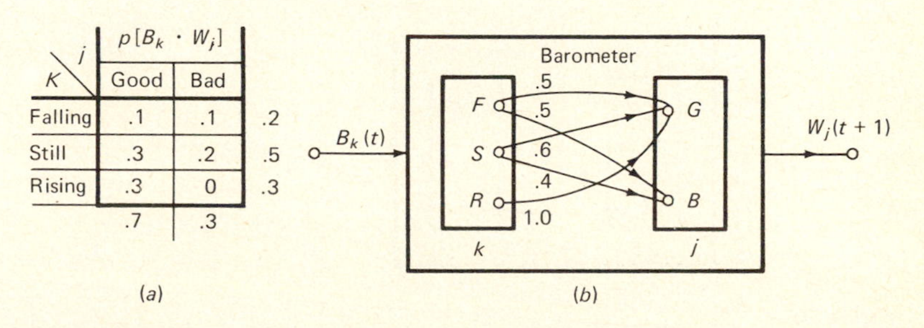

$K \backslash j$	Good	Bad	
Falling	.1	.1	.2
Still	.3	.2	.5
Rising	.3	0	.3
	.7	.3	

(*a*) (*b*)

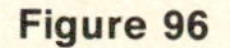

Figure 96

This process may be generalized so that the optimal gain associated with each barometer reading B_k is

$$G^*(B_k) = \max_i \left[\sum_j p(W_j \mid B_k) G(W_j \cdot A_i) \right]$$

This general result can be applied to the remaining two cases.

Case 2 Barometer remains still, B_S.

$$G^*(B_s) = \max\left[\sum_j p(W_j \mid B_s)G(W_j \cdot A_B), \sum_j p(W_j \mid B_S)G(W_j \cdot A_N)\right]$$

$$= \max[(0.6)(300) + (0.4)(-200), (0.6)(200) + (0.4)(100)]$$

$$= \max[100, 160]$$

$$= 160 \text{ units}$$

with optimal action of not buying.

Case 3 Barometer is rising, B_R.

$$G^*(B_R) = \max[(300)(1), (200)(1)]$$

$$= \max[300, 200]$$

$$= 300 \text{ units}$$

with optimal action of buying.

This completes the description of the optimal controller, which is summarized conveniently on the given sagittal diagram of Figure 97.

The overall optimal gain for this system can be determined by averaging the individual optimal gains $G[B_k]$ over all possible inputs (barometer readings).

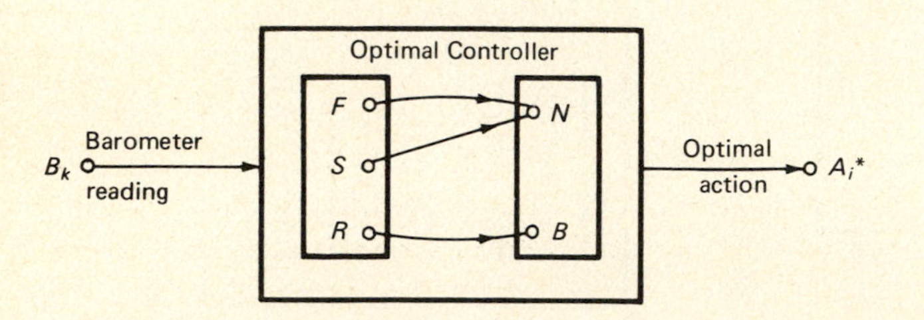

Figure 97

$$E[G^*] = \sum_k p[B_k]\ G^*(B_k)$$

$$= p[B_F]\, G^*(B_F) + p[B_S]\, G^*(B_S) + p[B_R]\, G^*(B_R)$$

$$= (0.2)(150) + (0.5)(160) + (0.3)(300)$$

$$E[G^*] = 200 \text{ units of gain}$$

With the proper substitutions we can obtain the following forboding result:

$$E[G^*] = \sum_k p[B_k] \max_i \left[\sum_j p(W_j \mid B_k) G(W_j \cdot A_i)\right]$$

Extreme Information Instruments

Oracles. To evaluate how well some probabilistic information instrument behaves, it could be compared to an ideal instrument, or *oracle*. For example, let us evaluate the optimal gain of the previous system when the weather can be predicted perfectly. If the weather is good (both predicted and actual), the optimal gain is

$$\max[300, 200] = 300 \text{ units}$$

and if the weather is bad, the optimal gain is

$$\max[100, -200] = 100 \text{ units}$$

Since good weather occurs with probability 0.7 and bad weather with probability 0.3 (from a previous joint matrix), the overall average optimal gain is

$$(0.7)(300) + (0.3)(100) = 240 \text{ units}$$

Now comparing the crummy barometer and its gain of 200 with the perfect oracle and its gain of 240 gives us an idea of the efficiency of the barometer.

Demons. The worst possible system occurs when the information instrument (a demon) knows the future weather perfectly but indicates just the opposite. In addition, the controller does not observe that the instrument is always wrong! Thus 70 percent of the time when the weather will be good the instrument indicates it will be bad; the controller chooses not to buy (maximizing -200 and 100), thereby making 200 units of gain.

Similarly, 30 percent of the time when the weather will be bad the instrument indicates it will be good; the controller chooses to buy (maximizing 300 and 200) and so loses 200 units. The average gain is

$$(0.70) * (200) + (0.30) * (-200) = 80 \text{ units}$$

Even in this incredibly bad case the gain is still 80 units. With some systems you just can't lose!

Decision Making Using Combinations of Instruments

Intuitively it appears that a combination of information instruments may be more effective than any one instrument used alone. This is illustrated by the following example, consisting of the previous barometer B and a television forecast T, applied to the previous weather-dependent system.

The table of joint probabilities in Figure 98*a* describes the two instruments. Now the two predictions can be treated as one combination, and all the previous methods apply. For example, if the barometer is still and the television forecast indicates bad weather, the conditional probabilities of good and bad weather are 0.333 and 0.667, respectively. If the decision is to buy, the expected gain is

$$(0.333)(300) + (0.667)(-200) = -33.33$$

and if the decision is not to buy, this expected gain is

$$(0.333)(200) + (0.667)(100) = 133.33$$

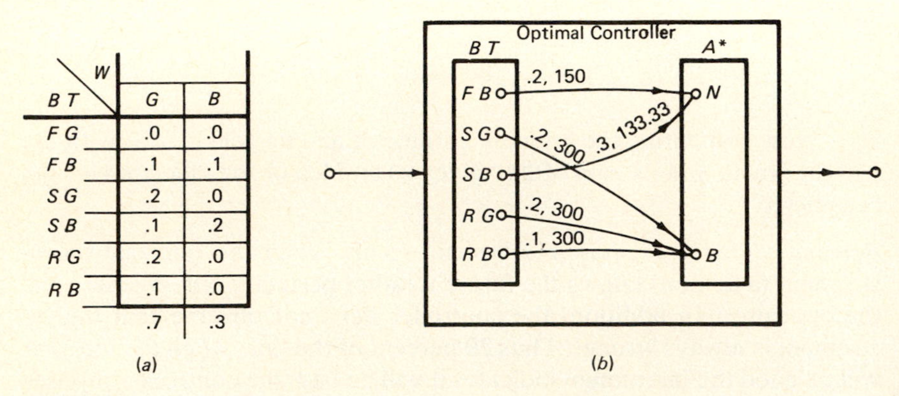

BT \ W	G	B
F G	.0	.0
F B	.1	.1
S G	.2	.0
S B	.1	.2
R G	.2	.0
R B	.1	.0
	.7	.3

(*a*)

(*b*)

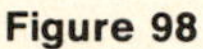

Figure 98

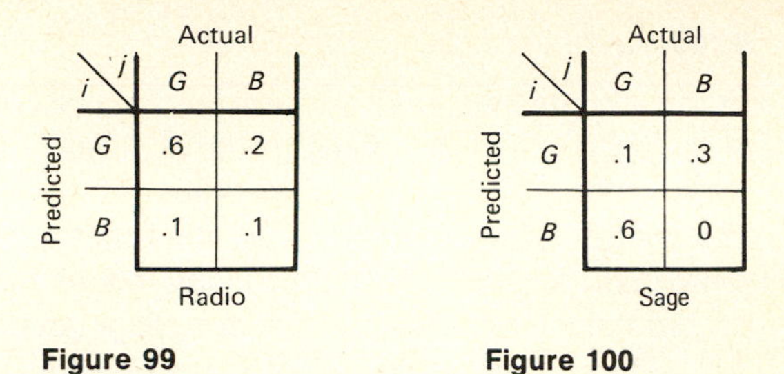

Figure 99 Figure 100

The optimal gain G^*_{SB} is 133.33 for optimal action of not buying. Continuing in this way yields the optimal controller of Figure 98*b*. The overall gain for this system is

$$E[G] = (0.2)(150) + (0.2)(300) + (0.3)(133) + (0.2)(300) + (0.1)(300)$$
$$= 220 \text{ units}$$

This gain of 220 units can be compared to the 200 units if the barometer is used alone and 210 units (check it) if the television forecast is used alone.

EXERCISE SET 26

1. If a radio station predicts the weather in a manner given by the joint-probability matrix of Figure 99, find the optimal controller and the average optimal gain for this instrument.
Ans.: two controllers yield 170

2. If the strategy of the controller is always to buy, what is the average optimal gain?

3. A coin is flipped to determine whether or not to buy. What is the average optimal gain using this very common information instrument? Comment on the optimal controller.
Ans.: 170

4. If a coin is used with the control policy "if heads, then buy; if tails, then don't buy," find the average gain.

5. An old sage predicts the weather from his various aches and pains, as described in Figure 100. Show that his predictions are so bad that he is one of the best information instruments.

6. How many controllers were possible for the barometer? Rank them according to their overall average gains. *Ans.*: 8

7. Construct an algorithm to determine the optimal controller for any given information instrument.

SUMMARY

In this chapter we have encountered many concepts. We have assigned probabilities in many ways (relative frequency, classical, equiprobable, etc.). We have used many types of probability (conditional, joint, marginal, forward, reverse). We also met a number of density functions (binomial, geometric, Poisson) and some measures or indicators (expected value, variance, covariance, dependence).

The viewpoint was quite general. We did not emphasize independent systems or mutually exclusive systems because they hold only in particular cases. Independence may be mathematically convenient but often physically meaningless. Similarly, there was very little emphasis on specialized density functions, their properties, and other measures. Instead the emphasis was on the use of the general concepts for optimizing, synthesizing, evaluating, and decision making in spite of the uncertainty or unreliability.

Despite the many concepts and great generality of this chapter it has been quite brief (compared to most introductions to probability). This is mainly because it was limited to digital or discrete systems. However, all these concepts can be extended to continuous systems, where the random variables can have any value in a continuous range. For example, the properties of expected value and variance can be extended to continuous systems by replacing each summation by an integral.

Time was not a very significant concept in this chapter. The probabilities were not dependent on time in any essential way. The concept of time will be introduced in the following two chapters. Either chapter 5 or 6 may be considered next.

PROBLEMS:

1. Justify your solution to the following problems:

(*a*) Can mutually exclusive events also be independent?

(*b*) If events A and B are independent, must the events $\bar{A}$ and $\bar{B}$ be independent also?

Ans.: yes

(*c*) If events A and B are mutually exclusive, must the events $\bar{A}$ and $\bar{B}$ be mutually exclusive also?

(*d*) Can the variance of a random variable be negative? *Ans.*: no

2. Determine the probability of the event $AB \vee \bar{C}$ in its simplest form for each of the following conditions:

(*a*) A is mutually exclusive of B.

(*b*) A is independent of B.

(*c*) A is a subevent (or subset) of B.

(*d*) A is a subevent of B, and A is independent of C.

3. One type of missile hits its target with probability 0.4. How many missiles should be fired (independently) for the probability of hitting the target to be at least 0.8?

4. An insurance company estimates that men of age w have probability p of living through 1 more year. What is the size of yearly premium x that the company should charge for amount y of insurance if their expected profit is to be z?

5. Sketch the pdf corresponding to the experiment of throwing a single die if the probability of any side's landing up is proportional to the number of dots on this side. Find the expected value of the number of dots.

6. Determine and sketch the pdf which describes the random variable D, the number of days of activity in a week (of 7 days):

(*a*) If the probability of activity is directly proportional to this number of days, i.e.,

$$p[D = d] = k_1 d$$

(*b*) If the probability of activity is inversely proportional to this number of days, i.e.,

$$p[D = d] = \frac{k_2}{d}$$

Evaluate both constants k_1 and k_2.

7. In a medical study of 100 patients some actually had a particular disease, some believed they had the disease, and some were classified (diagnosed) as having it. These three categories A (actual), B (believed), and C (classified) and their numbers are on the given set diagram.

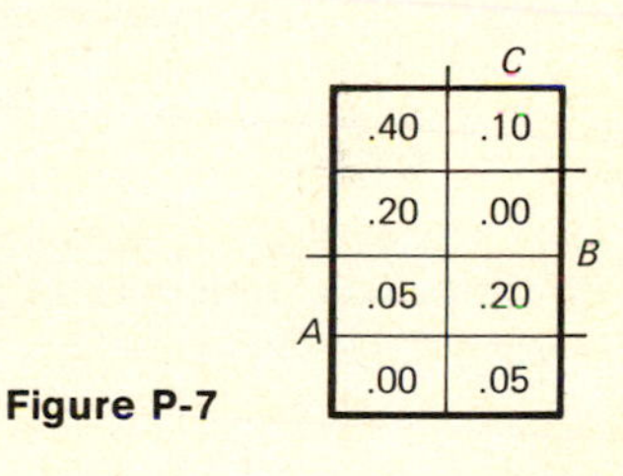

Figure P-7

(*a*) What is the probability of incorrect diagnosis?
(*b*) What is the probability that a person does not have the disease, given that he is diagnosed as having it?
(*c*) What portion of these people are hypochondriacs (believe they have the disease when they actually do not)?
(*d*) What portion do the doctors feel are hypochondriacs?
(*e*) What other conclusions can you draw from these data?

8. The given diagram (a set diagram in disguise) describes the fraction of the time that the various combinations of temperature T, pressure P, and rate R occur. Show that the rate is not dependent on the pressure and that the rate is not dependent on the temperature but that it is dependent on the temperature-pressure combination.

Temp.	Press.	Rate: Low	Rate: High
Low	Low	.01	.09
Low	High	.05	.15
High	High	.03	.17
High	Low	.11	.39

Figure P-8

9. The given set diagram describes the weather at various towns T_i on the same day.

$T_i = 1$ corresponds to good weather
$T_i = 0$ corresponds to bad weather

For example, the probability of bad weather at all four towns on the same day is $p[\bar{T}_1\bar{T}_2\bar{T}_3\bar{T}_4] = 0.05$; the probability of good weather at all four towns is $p[T_1T_2T_3T_4] = 0.10$.

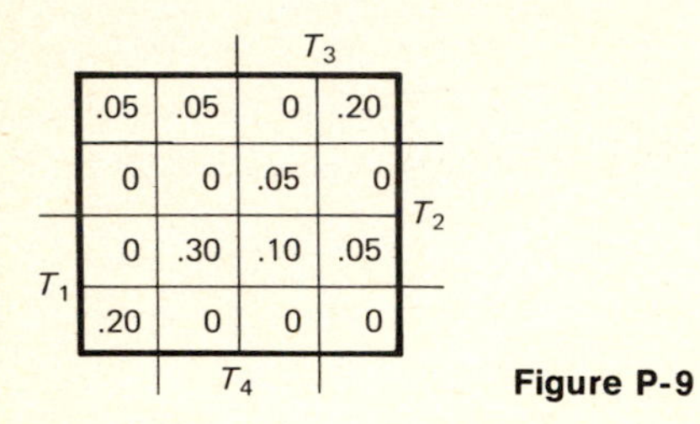

Figure P-9

(*a*) Which town has the best weather?
(*b*) Which two towns are most independent weatherwise?
(*c*) How is the weather in towns T_2 and T_4 related?
(*d*) What can you say about the weather in towns T_1 and T_3?
(*e*) Show *how* knowing the weather in towns T_1, T_3, and T_4 completely determines the weather in town T_2.

10. Four events A, B, C, and D are sampled simultaneously for 100 times and the results summarized on the given brief table-of-combinations (where 1 indicates occurrence of an event and 0 indicates nonoccurrence).

A	B	C	D	n(ABCD)
0	0	1	0	30
0	1	0	0	30
1	0	0	1	20
1	1	1	0	20
all the rest				0

Figure P-10

(*a*) Draw the corresponding set diagram.
(*b*) Which of the four events A, B, C, D occur most often? *Ans.*: B and C
(*c*) Which events are independent?
(*d*) Which events are mutually exclusive?
(*e*) How many possible combinations may be checked to see whether any of these four events is a subset of any other event? *Ans.*: 12
(*f*) Event D is a subset of which set?
(*g*) Show how B is a function of A and C.
(*h*) Is it true that if D occurs, A must occur? *Ans.*: yes
(*i*) Is it true that if A occurs, D must occur?

11. The given diagram shows the probabilistic relation between some diseases d_i and their resulting predominant symptom s_j. Disease d_1 occurs four times as often as disease d_2.

(*a*) If a person has symptoms s_1 or s_3, find the probability of disease d_1.
(*b*) Why are the s_j labeled as predominant symptoms rather than any symptoms?

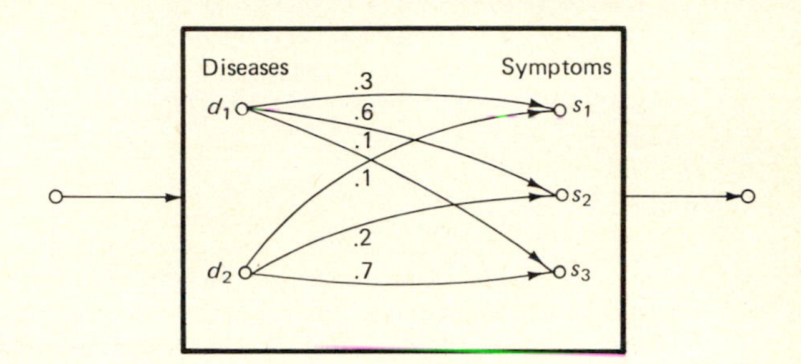

Figure P-11

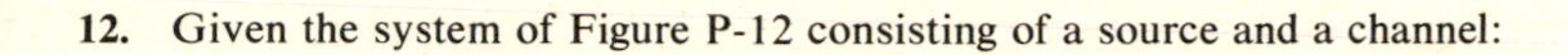

12. Given the system of Figure P-12 consisting of a source and a channel:

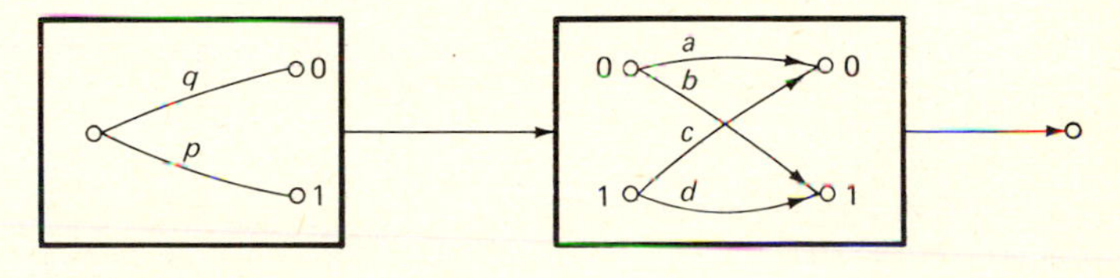

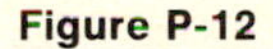

Figure P-12

(*a*) What is the relation between different probabilities: a, b, c, d, p, q?
(*b*) Write the forward, joint, and reverse probability matrices.
(*c*) What is the relation between different probabilities if the system is symmetric?
(*d*) What is the relation between the probabilities if the output is independent of the input?
(*e*) Can the system be symmetric and independent?
(*f*) Under what conditions is the probability of error of the system of Figure 1 the same as the probability of the system obtained by encoding the source with an ideal NOT encoder?

13. Prove the following results:
(*a*) $P[(E \vee F) \mid G] = p[E \mid G] + p[F \mid G] - p[EF \mid G]$.
(*b*) $p[EF] \leq p[F] \leq p[E \vee F] \leq p[E] + p[F]$.
(*c*) $p[E] \leq p[\bar{F}]$ if EMF.

14. Prove the following results concerning expected value.
(*a*) $E[X - E(X)] = 0$.
(*b*) $E[X^2 - X] = 0$ if X has only values 0 and 1.
(*c*) $E[XY] \leq [E(X^2) + E(Y^2)]/2$.
(*d*) $E[X] = (N+1)/2$ if X is the discrete uniform pdf given by $p[X = x] = 1/N$ for $x = 1, 2, 3, \ldots, N$.

15. Prove the following results concerning variance:
(*a*) $V[k] = 0$.
(*b*) $V[X + k] = V[X]$
(*c*) $V[X + Y] = V[X] + V[Y]$ if $X \perp Y$.
(*d*) $V[X - Y] = V[X] + V[Y]$ if $X \perp Y$.

16. Prove the following results concerning covariance:

(*a*) $C[X, X] = V[X]$

(*b*) $C[X + Y, Z] = C[X, Z] + C[Y, Z]$

(*c*) $C[X + Y, X + Y] = C[X, X] + 2C[X, Y] + C[Y, Y]$

17. Under what conditions do the following results hold?

(*a*) $p[E \vee F] = p(E) + p(F) - p[E] * p[F]$

(*b*) $p[E \mid F] = 1$

(*c*) $E[X^2] = (E[x])^2$

(*d*) $V[X^2] = (V[x])^2$

(*e*) $\text{cov}[X, Y] \geq V[x] + V[Y]$

18. Under severe environmental conditions the probability of a device's failing in a year is 0.8. Given 10 such devices under these conditions (using the tables of Figure 78):

(*a*) Find the probability that more than half of them fail in a year.

(*b*) Find the probability that none of them fail in a year.

(*c*) Find the probability that more than half of them fail in a year given that they do not all fail.

(*d*) Find the average number of failures in a year.

19. The probability that a system (bulb, car) will fail destructively during the xth month of use is given by the geometric pdf

$$g[x] = (0.2)(0.8)^{x-1} \qquad \text{for } x = 1, 2, 3, \ldots$$

Three such systems are in use simultaneously. Find the probability that:

(*a*) All the systems fail in the first month.

(*b*) No systems fail in the first month.

(*c*) Exactly one system fails each month.

20. A set of 10 numbers is to be added. Each number is given to one decimal, i.e., with one digit to the right of the decimal. Before adding the numbers each number is rounded off to the nearest integer, thus introducing an error (random variable). Sketch the pdf of the error. Find the mean and standard deviation of the error in each number. Find the mean and standard deviation of the error in the sum of the 10 numbers. Do this problem for the following two rounding-off processes.

(*a*) When 0.5 is encountered, always choose the next higher integer.

(*b*) When 0.5 is encountered, half the time choose the next higher integer and half the time choose the next lower integer.

21. The three given transmission lines are to be connected in cascade to form one long line. In what order should they be interconnected for the minimum probability of error when connected to a source having probabilities $p[X_0] = 0.4$ and $p[X_1] = 0.6$? Do you need to try all six possible combinations of interconnection?

$i \backslash j$	0	1
0	.8	.2
1	.3	.7

(*a*)

	0	1
0	.9	.1
1	.4	.6

(*b*)

	0	1
0	.8	.2
1	.3	.7

(*c*)

Figure P-21

22. A man drives to work each morning using the same freeway route. His experience is summarized by the given payoff matrix, which describes the (average) time to get to his place of work $T(S_i, A_j)$ as a function of the state of the freeway (S_i which may be bad or good) and his action A_j (departing times indicated by *E*arly, *N*ormal, *L*ate). For example, if the state of the freeway is good and the man leaves early, the time to get to work is 20 minutes. If this person wishes to minimize his driving time, find his optimal action under the following circumstances:

(*a*) On Mondays the probability of good traffic is 0.6.
(*b*) On all other weekdays this probability is 0.8.
(*c*) During a rainstorm this probability is 0.5.

$T[S_i A_j]$

$i \backslash j$	E	N	L
G	20	30	40
B	50	40	30

Figure P-22

23. The sagittal diagram of Figure P-23 describes a system for the manufacture of bolts. The inputs are two types of raw materials A and B. To make items from material A costs $C(A) = 3$ units per item, whereas B costs $C(B) = 2$ units per item. The output values j represent the number of defects in the manufactured item.

If an item has 0 defects, it can sell for $G(0) = 10$ units.
If an item has 1 defect, it can sell for $G(1) = 5$ units.
If an item has 2 defects, it can sell for $G(2) = 0$ units.

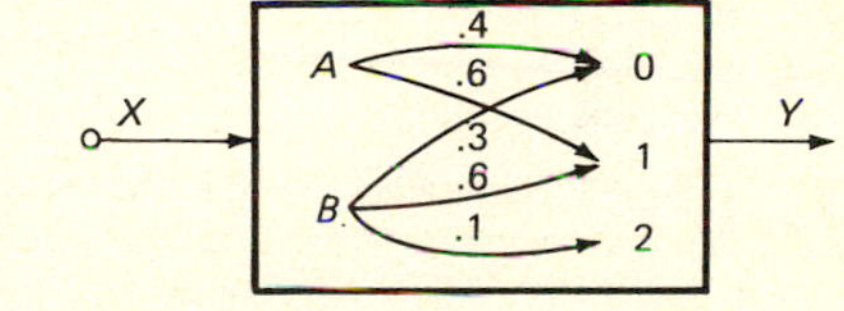

Figure P-23

(*a*) Find the optimal input to minimize the average number of defects.
(*b*) Find the optimal input to maximize the average profit.
(*c*) Which input would you select? Why?
(*d*) If both materials are randomly fed into the system, with B occurring three times as often as A, find the pdf of the number of defects produced.

24. It is required to control the system of Figure 95 (by buying or not buying) using two information instruments X and Y in combination to predict the weather. Specify the optimal controller and overall optimal gain in the three cases:

(*a*) Only the first instrument X is used.
(*b*) Only the second instrument Y is used.
(*c*) Both instruments X and Y are used.

X Y \ W	G	B
G G	.4	.0
G B	.2	.1
B B	.0	.1
B G	.1	.1

Figure P-24

25. The given figure describes the "parity" error-detecting communication system considered in Chapter 2. Recall that the encoder was designed to send an even number of 1s over the four channels and that the decoder was to indicate an error if an odd number was received. Determine the probability of error of this system if each of the channels has probability of error of $p = 0.1$ (and if the errors occur independently). Write a general result, in terms of the binomial pdf, for the case of n channels.

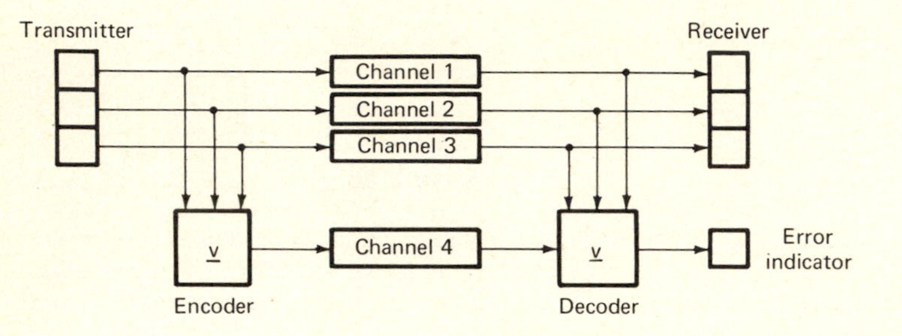

Figure P-25

26. Construct an algorithm for the following problem. You are given at most N crummy relays specified by their conditional probabilities of error p_{01} and p_{10}. They are to be connected to make a more reliable system specified by the equivalent conditional probabilities P_{01} and P_{10}. If the crummy relays are to be connected in a hammock configuration, determine the smallest number of relays and their configuration (L and W) to meet or just exceed the given specifications.

27. Consider the majority vote system of Figure 61:

(*a*) If there are three identical symmetric channels each with probability of error p, and if the MAJORITY element has probability of error q, determine the probability of error of this system.

(*b*) If there is an odd number n of identical symmetric channels each with probability of error p (and an ideal MAJORITY element), determine the probability of error of the system in terms of the binomial density function $b(x)$.

*(*c*) If there are three identical nonsymmetric channels described by Figure 29, construct the equivalent system.

28. *Chebyshev's theorem* states: The probability that *any* random variable deviates from its expected value by more than k standard deviations is less than $1/k^2$. Symbolically

$$p[|x - E(x)| > k * S(x)] < \frac{1}{k^2}$$

Prove this theorem. If a random voltage has an expected value of 1 volt and a standard deviation of 7 volts find the maximum probability of it deviating from this expected value by more than 10 volts. Compare this to the particular random voltage of Figure 87.

Ans.: 0.5 v.s. 0.2.

29. *Project.* Investigate a practical probabilistic system (involving computation, communication, control, or decision making), determine (or estimate) the relevant probabilities, and use them to gain some insight leading to recommendation or action.

30. *Project.* Investigate the philosophical aspects of the various probability assignments: Classical, Von Mises, Subjective, etc.

Chapter 5

Sequential Systems

Dynamic, Deterministic Digital Systems

In the instantaneous deterministic model of Chapter 1 the output at any instant was assumed to be a function only of the input at the same instant. However, in a more general case, the output may also depend on the past history or sequence of inputs. Thus some knowledge of the past may be stored in the system and may influence the output. A model describing such sequential systems will be developed here. This sequential model may describe the behavior associated with time delays, order, hysteresis, irreversibility, feedback, and memory. Briefly this chapter is about the concept of time viewed in a digital or discrete way.

Sequential systems (often called automata or finite-state machines) originally evolved from digital-computer technology but now find use also in psychology, economics, sociology, linguistics, ecology, and biology. However, our examples and motivation will be related to computation, communication, and control systems.

1. MODELING AND REPRESENTATION

A sequential system, like all systems, can be viewed as a black box. Inputs will be applied to the system at discrete time intervals labeled 0, 1, 2, 3 . . . . The intervals are usually equally spaced, but they need not be. At any time t, when an input is applied, the output is observed at this *same* instant. Before applying another input, the system is allowed sufficient time to respond to the previous input. This process is continued in such a sequential way, yielding an input-output sequence as shown in Figure 1.

Modeling a Relay

Any relay, when examined closely, will be observed to operate at one value of voltage (or current) and release at another, lower value of voltage. This phenomenon defines three ranges of input:

$X = 0$ represents any voltage insufficient to operate the relay.
$X = 2$ represents any voltage sufficient to operate the relay.
$X = 1$ represents any voltage insufficient to operate the relay, but once the relay is operated, this value is sufficient to hold it operated.

The output values are the usual two positions of the contact:

$Y = 0$ representing an open contact
$Y = 1$ representing a closed contact

The relation between the input and output is given in Figure 2. This shows that as the input voltage is increased and then decreased, different paths are followed.

Behavior

A representative input-output sequence is shown in both a continuous and digital representation in Figure 3.

Suppose initially (at time $t = 0$) the relay is unoperated. If the intermediate input voltage $X = 1$ is applied, the contact will remain open ($Y = 0$) and the relay will remain in the unoperated position. Now if at

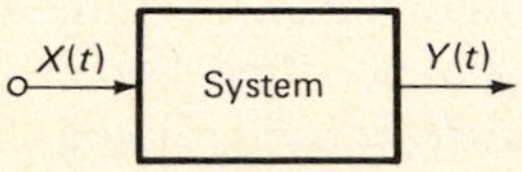

$X(t)$ → System → $Y(t)$

Time	0	1	2		t	$t+1$	
Input	$X(0)$	$X(1)$	$X(2)$		$X(t)$		
Output	$Y(0)$	$Y(1)$	$Y(2)$		$Y(t)$		

Figure 1

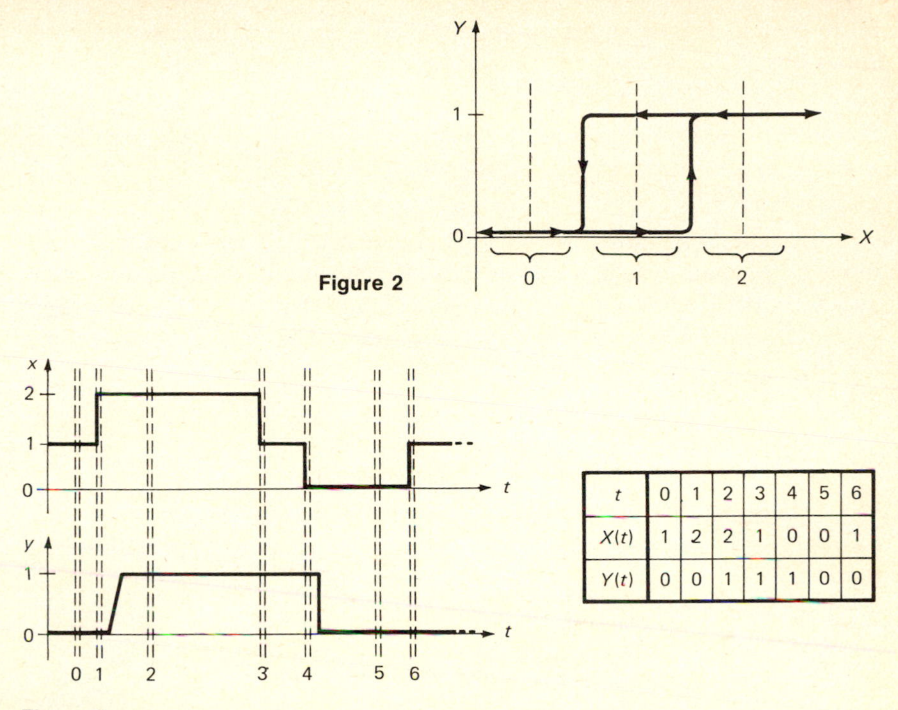

Figure 2

Figure 3

t	0	1	2	3	4	5	6
$X(t)$	1	2	2	1	0	0	1
$Y(t)$	0	0	1	1	1	0	0

time $t = 1$ the input has value $X = 2$, the output at this instant is still $Y = 0$ but the relay will operate after some time delay. If at the next time instant ($t = 2$) the input value of $X = 2$ is maintained, the output is $Y = 1$.

Now (at time $t = 3$) if the input is returned to $X = 1$, the relay output still remains closed ($Y = 1$) and the relay will not change in the next interval to an unoperated condition. If (at time $t = 4$) the input value $X = 0$ is applied, the output will still be $Y = 1$ until the relay releases. For the remaining inputs of 0 and 1 there are no transitions, and the output remains open ($Y = 0$).

Similarly you can show that if initially the relay was operated, this same input sequence 1221001 yields the output sequence 1111100. Notice that inputs are changed only at the sampling intervals.

At each time interval there is an input-output pair which could be designated as $\langle X(t), Y(t)\rangle$. For example, at $t = 3$ the input-output pair is $\langle 1, 1 \rangle$, whereas at time $t = 6$ it is $\langle 1,0\rangle$. This indicates that at different times for the same input $X = 1$, the output is not uniquely determined since it is sometimes 1 and sometimes 0.

The output may be specified uniquely if it is also known which of the two input values 0 or 2 was most recently applied. Any such entity which is required to determine the output is called a state. In this case the states could be:

$S = 0$ if the most recent input (other than 1) was 0
$S = 2$ if the most recent input (other than 1) was 2

In general the *state* $S(t)$ of a system is a label associated with each input-output pair in such a way as to make the output $Y(t)$ uniquely determined by the label (state) and the input $X(t)$ at that time. The state then essentially converts a relation into a function. The *output function* f has the form

$$Y(t)=f\,[S(t),\,X(t)]$$

There may be more than one choice of state for any system. For example, in this case the states correspond physically to the conditions of the relay coil:

$S = 0$ if the coil is unenergized
$S = 2$ if the coil is energized

This is a simpler and preferred choice of state compared to the original choice. It also makes it clear that the output is a function of the state only (in this case) since the state (condition of the coil) directly determines the output (position of contact).

It is important to realize that the states may change in time. In general, the next state $S(t + 1)$ is a function of the present state $S(t)$ and the present input $X(t)$. The *next-state function* g has the form

$$S(t + 1) = g[S(t), X(t)]$$

For example, at time $t = 4$, the relay is energized ($S = 2$) and when an input of $X = 0$ is applied, the state will change to unenergized ($S = 0$). Formally this could be written $g[2,0] = 0$.

The behavior of a sequential system could be described by a *state diagram* in which the states are represented by circles and the changes or transitions between states are represented by arrows labeled with the corresponding input-output values. The form of each transition is given in Figure 4. A sequential system consists of a state diagram showing all possible transitions. The state diagram for the relay example is given in Figure 5.

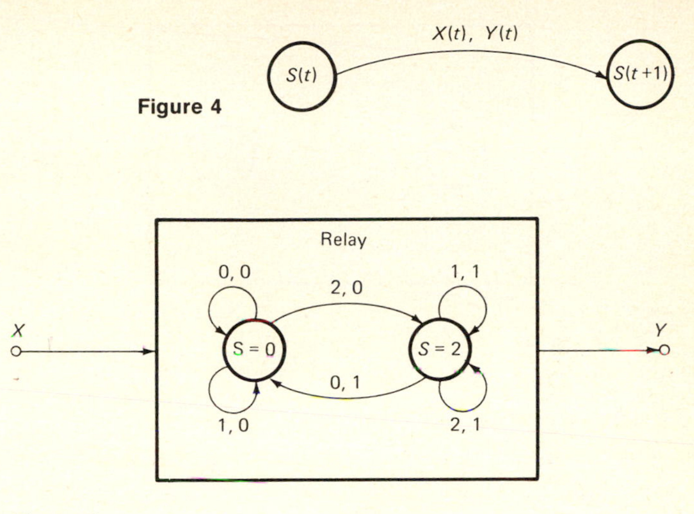

Figure 4

Figure 5

Now the output sequence for any particular input sequence can be determined by following through the transitions in the "leap frog" manner of Figure 6*a*. This can also be redrawn as the multiple transition of Figure 6*b*. It indicates graphically that starting in state $S = 0$ and applying input sequence 1221001 results in output response 0011100 and final state $S = 0$. The complete input-output state sequences can also be written in the following tabular form, called a *protocol*.

Time	0	1	2	3	4	5	6	7	...	t	$t+1$	...
Input	1	2	2	1	0	0	1			$X(t)$		
State	0	0	2	2	2	0	0			$S(t)$	$S(t+1)$	
Output	0	0	1	1	1	0	0			$Y(t)$		

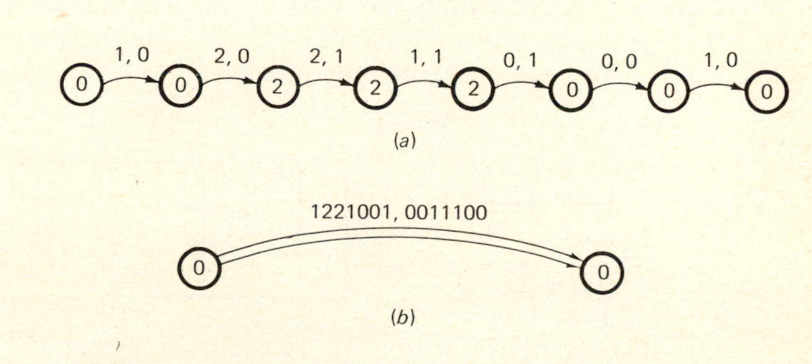

Figure 6

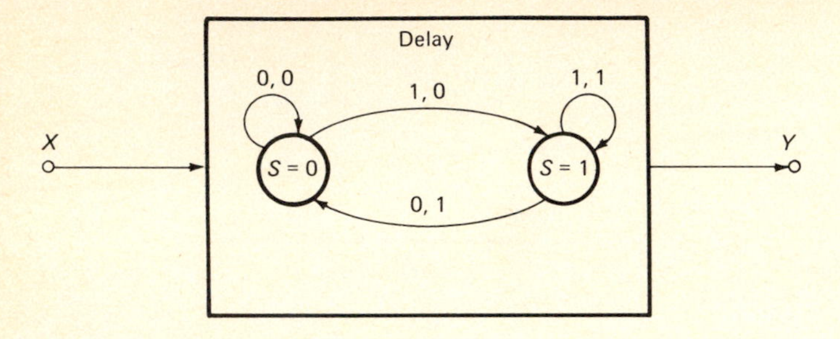

Figure 7

Remodeling the Relay

If the relay is operated only at the two extreme values 0 and 2 (which will be relabeled 0 and 1), the state diagram of Figure 7 results. A typical input-output sequence is shown in both the continuous and digital representation in Figure 8. Notice that in the digital representation the output is simply the input delayed by 1 time unit. This system is called a *formal delay element*. It differs from an *ideal delay element*, which delays the continuous waveform by some fixed amount. Later we will consider the ideal delay as well as a very useful clocked delay. All three delay elements will have the same diagrammatic representation of Figure 7; however, time will be interpreted differently (as digital, continuous, or clocked).

Relays are too restrictive to abstract immediately into the most general sequential system. Let us first consider one more example before considering the general model.

Ternary Counter

A ternary counter (or divide-by-3 or scaler) has an output of 1 for every three inputs of value 1; thus the output sequence has one-third as many pulses as the input. The state diagram of Figure 9 describes this behavior. The value of the state indicates the number of pulses accumulated since

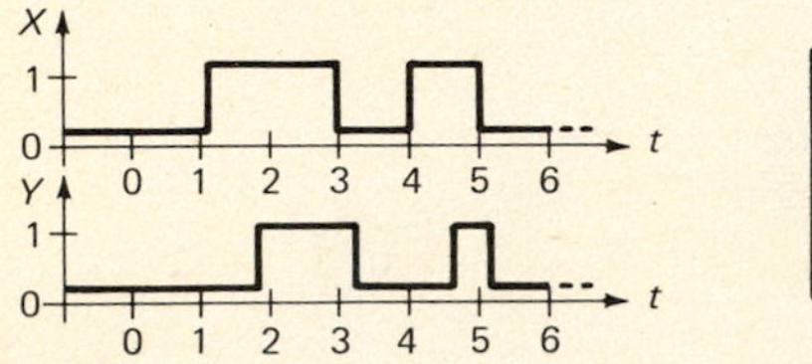

t	0	1	2	3	4	5	6	
X	0	1	1	0	1	0		$X(t-1)$
Y	0	0	1	1	0	1		$Y(t)$

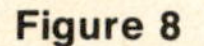

Figure 8

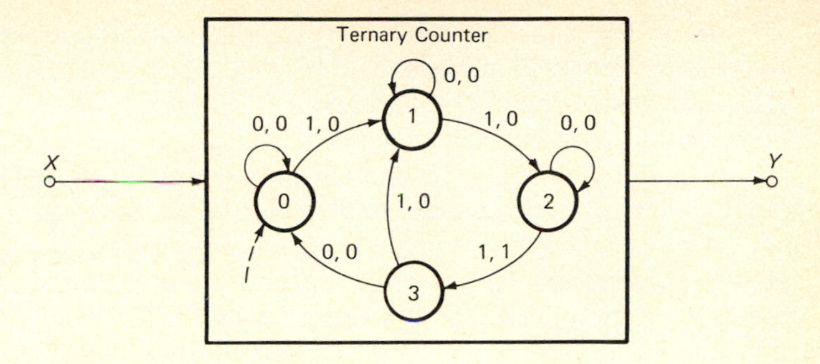

Figure 9

the last output pulse. The dotted arrow into state 0 indicates that it is the *initial* or starting state. A typical input-output sequence is

Time	0	1	2	3	4	5	6	7	9				
Input	0	1	1	0	1	1	1	1	1	0	1	0	1
Output	0	0	0	0	1	0	0	1	0	0	0	0	1

Although the state diagram is conceptually convenient for visualizing the behavior of a system, there are other useful representations. Figure 10*a* illustrates the table-of-combinations representation, and Figure

$S(t)$	$X(t)$	$Y(t)$	$S(t+1)$
0	0	0	0
0	1	0	1
1	0	0	1
1	1	0	2
2	0	0	2
2	1	1	3
3	0	0	0
3	1	0	1

$S(t)$ —$X(t)$, $Y(t)$→ $S(t+1)$

(*a*)

$X(t)$ \ $S(t)$	$Y(t)$		$S(t+1)$	
	0	1	0	1
0	0	0	0	1
1	0	0	1	2
2	0	1	2	3
3	0	0	0	1

(*b*)

Figure 10

10*b* shows the transition-table representation. The correspondence between these representations should be clear. Try it, for example, to show that state 2 with input 1 yields output 1 and next state 3. The table-of-combinations will be useful for synthesizing and implementing, whereas the transition table will be useful for optimizing. Later in this chapter we will synthesize, optimize, and implement this ternary counter.

There are other possible representations (such as regular expressions and matrices), but they will not be introduced here.

Other examples of sequential systems for computation, communication, and control will be considered after the following generalization of this model.

A General Model of a Sequential System

A sequential system may now be defined in general as follows.

> ***Definition*** A *sequential system*, denoted $\langle X, Y, S, f, g \rangle$, consists of three sets X, Y, S and two functions f, g, where:

1. $X = \{X_1, X_2, \ldots, X_p\}$ is a finite set of input symbols
2. $Y = \{Y_1, Y_2, \ldots, Y_q\}$ is a finite set of output symbols
3. $S = \{S_1, S_2, \ldots, S_n\}$ is a finite set of states
4. f: $S \times X \rightarrow Y$ is an output function
5. g: $S \times X \rightarrow S$ is a transition function

These two functions f and g are often called the *input-output-state equations*. They may be rewritten to show explicitly the dependence on time as

$$Y(t) = f\,[S(t),\ X(t)]$$

$$S(t+1) = g\,[S(t),\ X(t)]$$

They are also sometimes abbreviated as

$$Y = f\,[S, X]$$
$$S' = g\,[S, X]$$

This model of a sequential system is also known as the *Meally model* or the *Shannon model*. Another model, the Moore model, is sometimes used, but it will not be considered here. In general it requires more states than the Meally model.

Sequential Behavior

Notice that these input-output-state equations f and g relate only individual input symbols X and output symbols Y at any one interval of time. These functions can be extended to relate input sequences or *tapes*, such as

$$T = X(0)\ X(1)\ X(2) \cdots X(n)$$

and output or *response sequences*, such as

$$R = Y(0)\ Y(1)\ Y(2) \cdots Y(n)$$

If the initial state $S(0)$ and input sequence T are given, the response sequence is uniquely determined by iteration of the input-output-state equations as follows:

$$Y(0) = f[S(0), X(0)]$$
$$S(1) = g[S(0), X(0)]$$
$$Y(1) = f[S(1), X(1)] = f[g[S(0), X(0)], X(1)]$$
$$S(2) = g[S(1), X(1)] = g[g[S(0), X(0)], X(1)]$$
$$Y(2) = f[S(2), X(2)] = f[g[g[S(0), X(0)], X(1)], X(2)]$$

....

$$\begin{aligned} Y(n) &= f[S(n), X(n)] \\ &= f[g[g[\cdots [g[g[S(0), X(0)], \\ &\quad X(1)], X(2)] \cdots], X(n)] \end{aligned}$$

$$\begin{aligned} S(n+1) &= g[S(n), X(n)] \\ &= g[g[g[g \cdots [g[g[S(0), X(0)], \\ &\quad X(1)], X(2)] \cdots], X(n)] \end{aligned}$$

Although the iterations look hideous, they correspond directly to the leapfrog operations of Figure 6*a*.

This iterative procedure effectively defines two new output and state functions F and G which apply to sequences only:

$$Y(0)\ Y(1) \cdots Y(n) = F[S(0), X(0), X(1), X(2), \ldots, X(n)]$$
$$S(n+1) = G[S(0), X(0), X(1), X(2), \ldots, X(n)]$$

The *input-output-state equations for sequences* can be written as

$$R = F\,[S(0), T\,]$$

$$S^n = G\,[S(0), T\,]$$

Example In Figure 6, the initial state is $S(0) = 0$; applying input tape $T = 1221001$ results in response sequence

$R = F\,[0{,}1221001\,] = 0011100$ and the final state $S^7 = G\,[0.1221001\,] = 0$

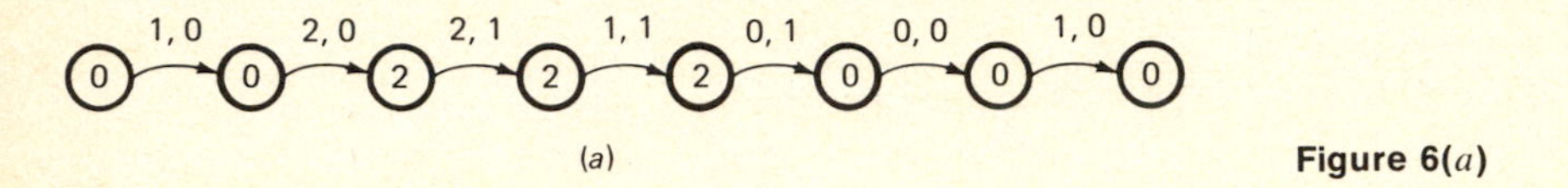

(a) **Figure 6(*a*)**

Similarly, if the initial state is $S(0) = 2$, this same input tape $T = 1221001$ results in another response

$R = F\,[2{,}1221001\,] = 1111100$

This indicates that the output response is not a function of the input sequence alone but depends also on the initial state.

EXERCISE SET 1

1. Represent the original relay by a transition table and a table-of-combinations.

2. Determine the output response sequences for the following input tapes to the original relay in state 0:

(*a*) 21001 (*b*) 10012
(*c*) 10000 · · · (*d*) 20000 · · ·

3. Is time reversible? In other words, if the input sequences are reversed, do the output responses also reverse?

4. For the ternary counter, construct an alternative choice of states and an alternative diagram to relate the input and output sequence. *Ans.*: See Figure 38

2. APPLICATION TO COMPUTATION, COMMUNICATION, CONTROL

Many systems used for computation, communication, and control can be modeled as sequential systems. Some examples follow, illustrating the behavior (as opposed to structure) of some practical systems. Later we will consider the structure for synthesizing, optimizing, and implementing such systems.

Synchronous and Asynchronous Systems

Most of the following examples are *synchronous* systems, in which the time intervals (samples) are predetermined by some clock which usually marks equal intervals. The inputs are applied and the outputs are observed only at these clock times.

An *asynchronous* system has no clock, but the application and removal of the inputs define the time intervals. Thus a sequence of operations can be performed by having the end of one operation signaling the start of the next operation (without waiting for any clock control).

Synchronous systems are usually slower than asynchronous systems, but they are also usually simpler in design and structure. Most large systems such as computers are synchronous. Our emphasis is also on the synchronous systems.

Application to Computation: The Binary Sequential Adder

The sequential system of Figure 11 adds two numbers A and B which are coded as binary sequences. The least significant binary digits are read first into the adder. The sequence generated at the output is the binary representation of the sum. The states in this case indicate the amount of carry (if any) to the next stage of the computation. A sample sequence is also shown to illustrate the addition of the numbers 12 and 6. Notice that this simple sequential adder can add numbers of any arbitrary magnitude, whereas parallel adders are limited in magnitude. We will construct such a binary sequential adder in a later section.

*Application to Communication: Comma Coding

Consider the communication system of Figure 12, which is to transmit four symbols A, B, C, and the blank □ but only a binary channel is available. (More generally there would be 26 letters and a blank.) Each symbol can be encoded as a sequence of binary digits, transmitted along the

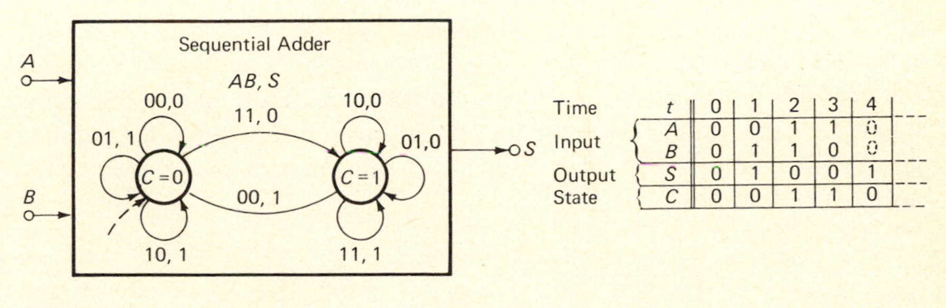

Figure 11

Figure 12

Symbol	Probability	Code
A	.4	0
B	.3	10
C	.2	110
□	.1	1110

Figure 13

binary channel, and then decoded into the original symbols. If some symbols appear more frequently than others, they can be assigned a shorter code length, as indicated in Figure 13. This is called a *comma code* since the 0 serves as a form of punctuation (comma) indicating the end of a symbol. For example, the message

ABACA □ BA

is encoded into the binary sequence

010011001110100

After being transmitted through the channel this binary sequence can be converted into the original message by the decoder in Figure 14. Note the distinction between the binary symbol 0, the blank symbol □, and the empty nonsymbol ϵ. Notice also that one of the next states is not completely specified. Such incompletely specified systems will not be considered in detail here.

Application to Control: Coin-Operated Dispenser

Consider a simple coin-operated dispenser of some item as given in Figure 15. The cost of the item dispensed is 15 cents but only dimes and nickels are accepted as inputs. The output is the item and 5 cents change if necessary. In this case the states indicate the amount of money accumulated for an item. Notice that the inputs are all the combinations of nickels and dimes; i.e., combination 01 indicates an input of no nickel but one dime. Similarly each output is a combination *PC* indicating whether the product *P* or change *C* is output. In some cases not all possible combinations can occur. For example, if only one slot is provided, both a nickel and a dime cannot be inserted at the same instant and so the combination 11 is a don't-care situation.

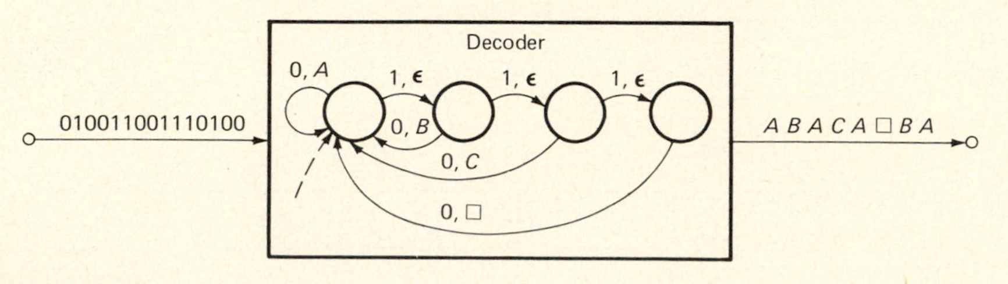

Figure 14

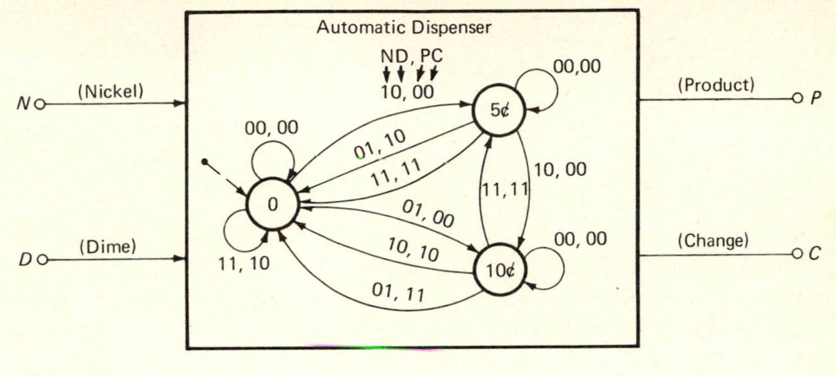

Figure 15

EXERCISE SET 2

1. Construct the state diagram of a binary sequential *subtractor* similar to the adder of Figure 11. (Assume $A \geq B$.)

2. Construct the state diagram of a sequential *comparator* which compares two numbers A and B coded as binary sequences and indicates which number has the larger magnitude:

(*a*) If the least significant digits are entered first.

(*b*) If the most significant digits are entered first.

3. In the table of Figure 13 code the blank □ as the sequence 111 and then draw the corresponding decoder.

4. If the sequences of Figure 13 are reversed (so that C is sent as 011), draw the corresponding decoder.

5. Construct a state diagram for a *sequential lock* which produces an output of 1 whenever the subsequence 1012 appears in any sequence. *Hint*: Check that the lock opens for sequence 101012.

6. It is required to construct a system to control the water supply in some artificial environment. The decision to water or leave dry is dependent on the present temperature (high or low) and the previous 2 days' moisture (wet or dry). On a day of high temperature, the system is to provide water unless both previous 2 days were wet. On a day of low temperature the system is to water only when the previous 2 days were dry. Construct a state diagram describing the behavior of this system. *Hint*: There are four states.

3. ANALYSIS AND SYNTHESIS OF SEQUENTIAL SYSTEMS

In this section we will study interconnections of instantaneous systems and ideal delay elements to see how they can be analyzed as sequential systems. Then later we will synthesize systems.

Ideal Delay

An ideal delay element as shown in Figure 16 delays the input by exactly 1 time unit. The output value at time t equals its input value at the previous time $t-1$ and may be described by either

$$q(t) = p(t-1) \qquad \text{or} \qquad q(t+1) = p(t)$$

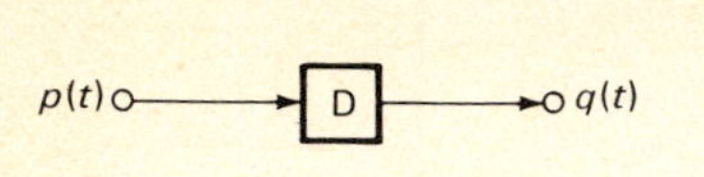

t	0	1	2	3	4	5	6
$p(t)$	0	1	1	0	1	0	
$q(t)$		0	1	1	0	1	0

Figure 16

Networks having Ideal Delays

Any interconnection of instantaneous systems and delay elements such as the feedback form of Figure 17*a* is a sequential system. The output $Y(t)$ can be expressed as

$$Y(t) = f_1[q(t), X(t)]$$

which is of the required form for the output function

$$Y(t) = f[S(t), X(t)]$$

This suggests that the switching function f_1 is the output function f and also that the variable q (at the output of the delay element) is that state S.

$$q(t) = S(t)$$

The variable p at the input of the delay element is the next state variable since

$$p(t) = q\,(t + 1) = S(t + 1)$$

From the diagram the second equation is

$$p(t) = f_2[S(t), X(t)]$$

which has the form of the transition equation

$$S(t + 1) = g[S(t), X(t)]$$

We have thus seen that by assigning the state variable as the output of the delay, the network has the form of a sequential system. The final sequential system is redrawn and relabeled in Figure 17*b*.

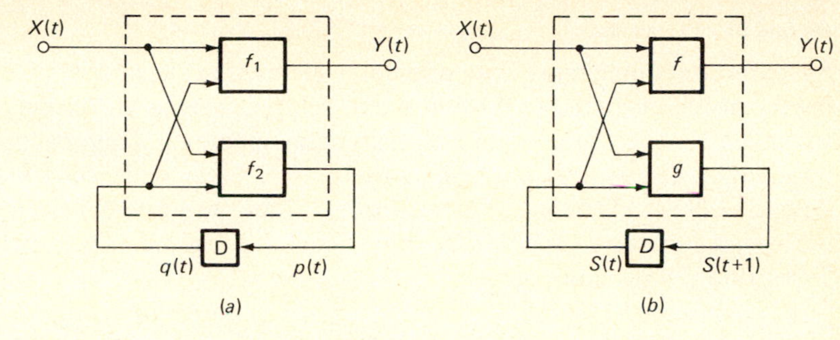

Figure 17

As an example, let us construct the sequential system corresponding to the network of Figure 18. The input-output-state equations can be algebraically expressed as

$$Y(t) = S(t) \cdot \overline{X(t)}$$
$$S(t+1) = \overline{S(t)} \vee \overline{X(t)}$$

A table-of-combinations would now be most convenient to evaluate these equations for all possible values. This table could then be converted to a state diagram by drawing one transition arrow for each row of the table-of-combinations. The table-of-combinations and corresponding state diagram are shown in Figure 19.

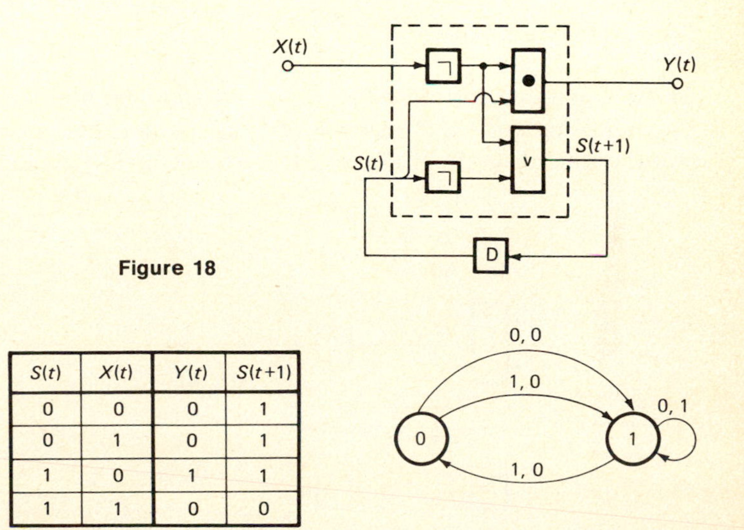

Figure 18

$S(t)$	$X(t)$	$Y(t)$	$S(t+1)$
0	0	0	1
0	1	0	1
1	0	1	1
1	1	0	0

Figure 19

Multiple Delay Elements

Systems containing more than one delay element are analyzed by extending the previous method. If there are m delay elements, we assign m state variables S_i, one to the output of each delay element. Then each combination of m values is a state. Thus there are 2^m states in a binary system having m delay elements. For example, if there are two delay elements, we assign two state variables S_1 and S_2 each having the two values 0 and 1, resulting in the four states

$$(0,0) \qquad (0,1) \qquad (1,0) \qquad (1,1)$$

which are more conveniently written as 00, 01, 10, and 11 (and occasionally abbreviated to 0, 1, 2, and 3, their decimal equivalents). This will be used in the next example.

In Chapter 1 we assumed that the switching elements had no delays and that they could not be connected so that outputs could influence inputs (no feedback). Now we can remove both these constraints. For example, Figure 20 shows OR and NOT components each having one unit of delay (shown following the element). They are interconnected in a feedback manner which has an interesting behavior.

The input-output-state equations for this system are

$$Y(t) = S_1(t)$$

$$S_1(t+1) = X(t) \vee S_2(t)$$

$$S_2(t+1) = \overline{S_1(t)}$$

These equations can now be converted into the table-of-combinations shown in Figure 21*a*. This table-of-combinations can then be converted into the state diagram of Figure 21*b*. The behavior of such a system can

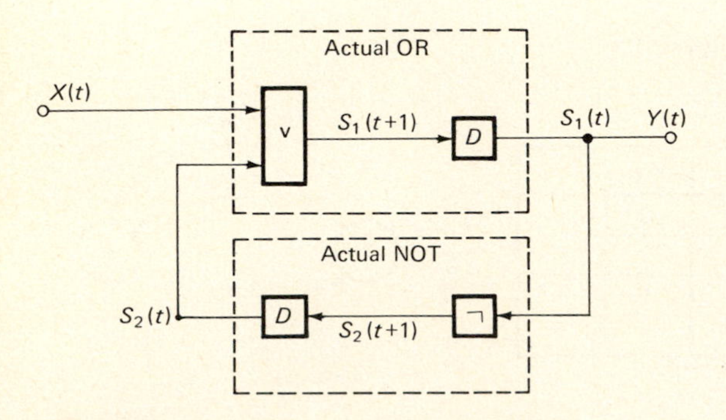

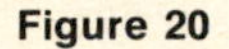
Figure 20

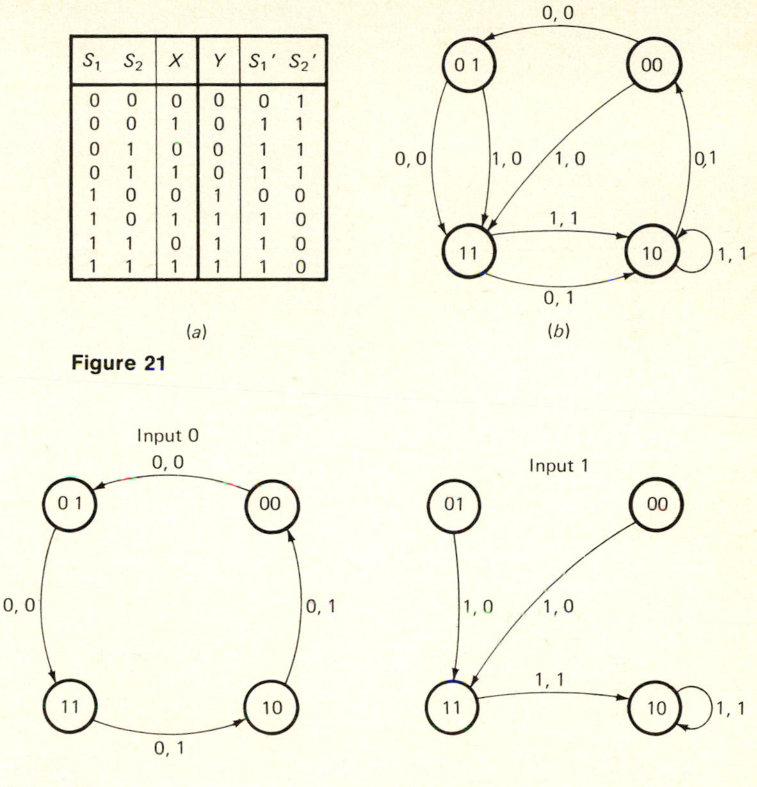

S_1	S_2	X	Y	S_1'	S_2'
0	0	0	0	0	1
0	0	1	0	1	1
0	1	0	0	1	1
0	1	1	0	1	1
1	0	0	1	0	0
1	0	1	1	1	0
1	1	0	1	1	0
1	1	1	1	1	0

Figure 21

Figure 22

be seen more readily if the state diagram is decomposed into two diagrams, one with input value 1, the other with input value 0. This is done in Figure 22. Notice that when input value 0 is applied, the output oscillates, yielding the sequence 001100110011 · · · . When this input is changed ($X = 1$), the oscillation settles to one final state and the output remains at value 1. This system acts as an oscillator which is controlled by input X.

EXERCISE SET 3

1. Using the instantaneous system of Figure 18 but interconnecting the delay in different ways yields different sequential systems. Draw the state diagram corresponding to the interconnections of Figure 23.

2. For the given network of Figure 24:

(*a*) Describe its behavior in words.

(*b*) Redraw it in the form of Figure 17*b* to emphasize the feedback.

(*c*) Construct the corresponding state diagram.

Ans.: Output Y at any instant has value 1 if both inputs have value 1 and exactly one of immediately preceding inputs had value 1.

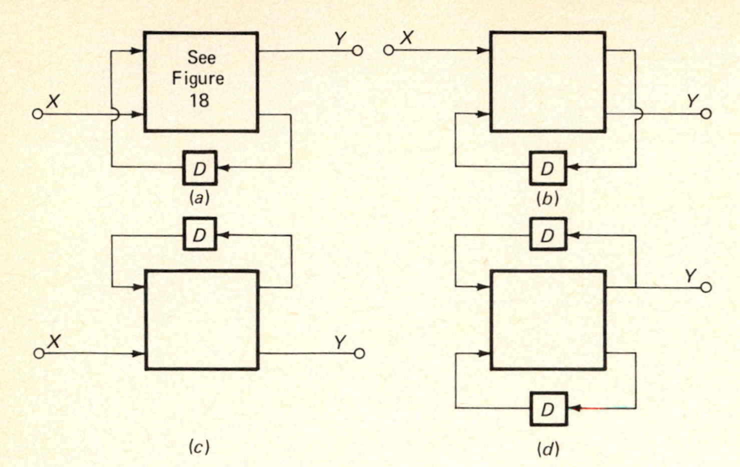

Figure 23

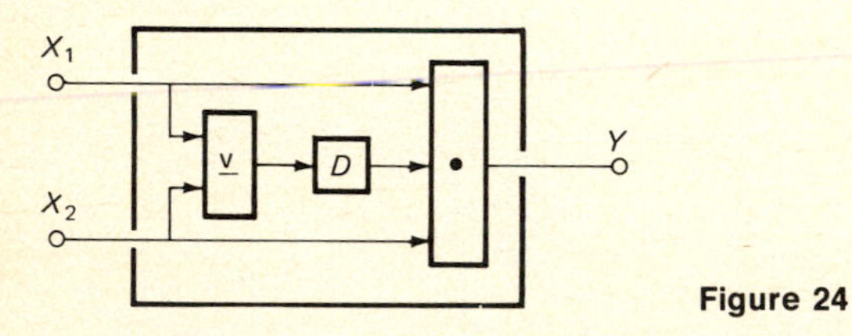

Figure 24

3. Put the network of Figure 25*a* into standard feedback form of Figure 25*b* and construct the state diagram.

4. Draw a state diagram corresponding to the following equations, and comment.

$$S_1' = X\bar{S}_1 \vee \bar{X}S_2$$

$$S_2' = X\bar{S}_2 \vee \bar{X}S_1$$

$$Y = (S_1 \vee S_2)X$$

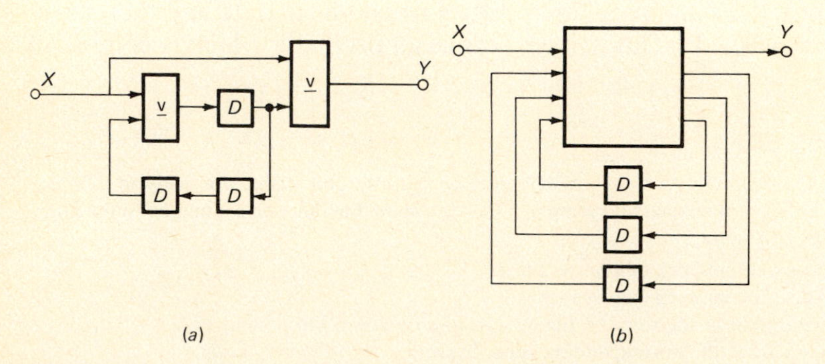

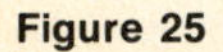

Figure 25

5. Construct the state diagram of the system of Figure 20 if:
  (*a*) The delay of the NOT is negligible (compared to the OR).
  (*b*) The delay of the OR is negligible.
  (*c*) The delay of the NOT is twice that of the OR.
  (*d*) Neither element has delay. *Ans.*: see Figure 88

Synthesis: From State Diagrams to Networks

Until now we have analyzed systems of instantaneous elements and delay elements as sequential systems. We can easily reverse this process to synthesize or construct a given sequential system from these components. As an example, let us synthesize the binary sequential adder. The state diagram (repeated in Figure 26) is first converted to a table-of-combinations representation for convenience. From this table-of-combinations the output function (describing the sum S) is

$$\begin{aligned} S &= \bar{A}\bar{B}C \vee \bar{A}B\bar{C} \vee A\bar{B}\bar{C} \vee ABC \\ &= \bar{A}(\bar{B}C \vee B\bar{C}) \vee A(\bar{B}\bar{C} \vee BC) \\ &= \bar{A}(B \veebar C) \vee A\overline{(B \veebar C)} \\ &= A \veebar B \veebar C \end{aligned}$$

Similarly the next state function (describing the carry C') is

$$\begin{aligned} C' &= \bar{A}BC \vee A\bar{B}C \vee AB\bar{C} \vee ABC \\ &= (\bar{A}B \vee A\bar{B})C \vee AB(\bar{C} \vee C) \\ &= (A \veebar B)C \vee AB \end{aligned}$$

Notice that these two equations also describe the parallel adder of Chapter 1. The resulting network is shown in Figure 27. It consists of one full adder with the carry delayed. This result is what intuitively would be expected.

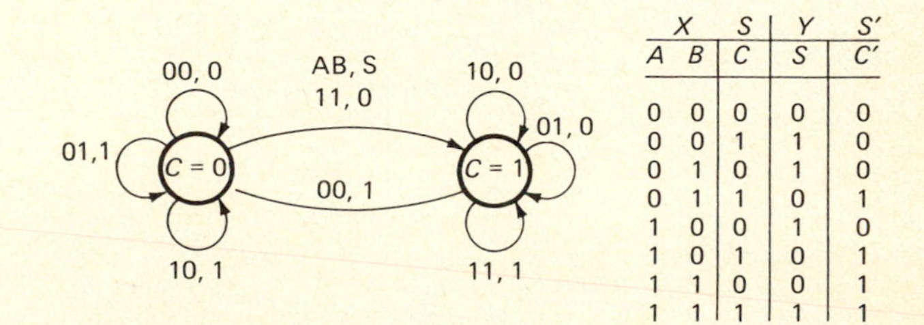

X		S	Y	S'
A	B	C	S	C'
0	0	0	0	0
0	0	1	1	0
0	1	0	1	0
0	1	1	0	1
1	0	0	1	0
1	0	1	0	1
1	1	0	0	1
1	1	1	1	1

Figure 26

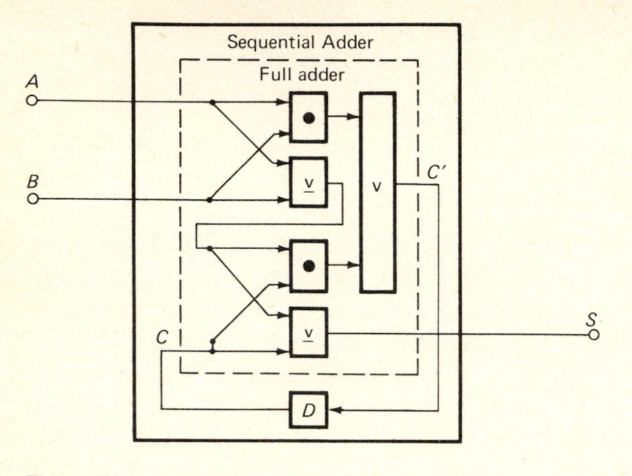

Figure 27

Some additional aspects of synthesis are illustrated by attempting to synthesize a ternary counter. The table-of-combinations for the ternary counter is repeated in Figure 28*a*. The four state values can be coded as combinations of two state variables S_1 and S_2, each having two values. One possible state assignment is shown in Figure 28*b*. The corresponding binary table-of-combinations is given in Figure 28*c*. From the binary table-of-combinations we can specify the required instantaneous network as

$$Y = S_1 \bar{S}_2 X$$

$$S_1' = \bar{S}_1 S_2 X \vee S_1 \bar{S}_2 \bar{X} \vee S_1 \bar{S}_2 X$$

$$S_2' = \bar{S}_1 \bar{S}_2 X \vee \bar{S}_1 S_2 \bar{X} \vee S_1 \bar{S}_2 X \vee S_1 S_2 X$$

S	X	Y	S'
0	0	0	0
0	1	0	1
1	0	0	1
1	1	0	2
2	0	0	2
2	1	1	3
3	0	0	0
3	1	0	1

(*a*)

S	S_1	S_2
0	0	0
1	0	1
2	1	0
3	1	1

(*b*)

S_1	S_2	X	Y	S_1'	S_2'
0	0	0	0	0	0
0	0	1	0	0	1
0	1	0	0	0	1
0	1	1	0	1	0
1	0	0	0	1	0
1	0	1	1	1	1
1	1	0	0	0	0
1	1	1	0	0	1

(*c*)

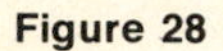

Figure 28

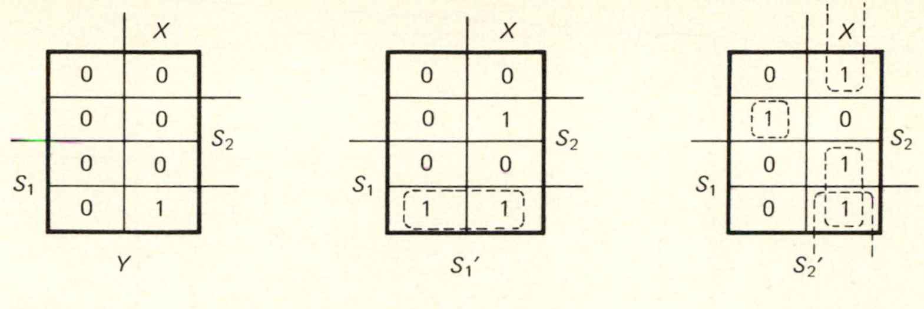

Figure 29

These input-output-state equations may now be optimized by any of the methods which apply to the Boolean algebra. For example the corresponding three set (Karnaugh) diagrams are given in Figure 29. The resulting optimized input-output-state equations are given below and shown implemented on the diagrams of Figure 30.

$$Y = S_1 \bar{S}_2 X$$

$$S_1' = S_1 \bar{S}_2 \vee \bar{S}_1 S_2 X$$

$$S_2' = S_1 X \vee \bar{S}_1 S_2 \bar{X} \vee \bar{S}_2 X$$

State Assignment

It is important to realize that the above coding (state assignment) and its corresponding network are not unique. In this case there are 24 different ways of coding the states as binary combinations. This is because the first state can be assigned as any of the four combinations but, once it is assigned, the second state can be assigned only one of the remaining three combinations. Continuing this by the product principle we get the total number of different assignments as

$$4 * 3 * 2 * 1 = 24$$

There is, unfortunately, no simple, systematic method of making an optimum state assignment. The state assignment problem remains one depending on the designer's creativity, experience, and luck. However, there are useful methods for optimizing the number of states (before making any state assignment). But before considering this we need more analysis, which follows in the next section.

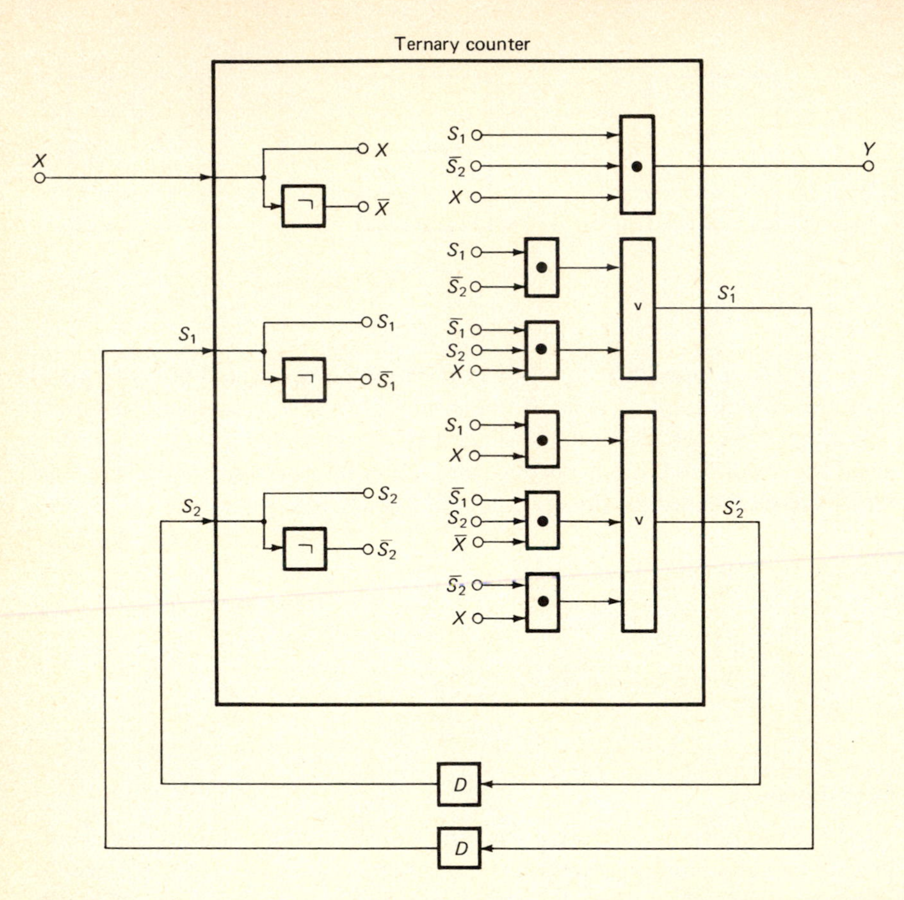

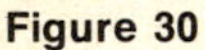
Figure 30

EXERCISE SET 4

1. For a system having 25 states, how many binary state variables are necessary? How many are necessary for 1 million states?

2. If a sequential system having x states is to be synthesized using components from a y-valued system, estimate the total number of state assignments possible.

3. Construct an algorithm to select systematically all possible binary state assignments for a system having n states.

4. Construct, optimize, and compare the networks for a ternary counter corresponding to the three state assignments of Figure 31. Can you use the fact that Figure 31a corresponds to a Karnaugh map?

5. For the automatic dispenser of Figure 15 construct and optimize a network using the state assignment

$$0 = 00 \qquad 5 = 01 \qquad 10 = 11$$

S	S_1	S_2
0	0	0
1	0	1
2	1	1
3	1	0

(a)

S	S_1	S_2
0	1	1
1	0	0
2	1	0
3	0	1

(b)

S	S_1	S_2
0	0	1
1	1	0
2	0	0
3	0	1

(c)

Figure 31

General Behavioral Properties

In this section we will briefly consider three general properties which hold for all sequential systems. They will be useful for the state optimization in the next section. These properties can also be extended to other physical systems.

Causality. In all physical systems the response at some time is not dependent on inputs after this time; outputs do not anticipate inputs. Otherwise you would have perfect prediction of the future and could use it to avoid this predicted future, which is a contradiction. So we cannot use feedback from the future.

This property can be expressed formally by considering an input tape T_1 followed by an input tape T_2 yielding the output response R_1R_2. For a causal system response R_1 should depend only on T_1 and not on the future inputs T_2; that is,

$$F[S, T_1T_2] = R_1R_2 \qquad \text{implies} \qquad F[S,T_1] = R_1$$

The Response Separation Property. This property indicates that the state at any time (present) relates the past responses to the future responses. The response to a sequence T_1 followed by T_2 can be determined by finding the corresponding responses R_1 and R_2 separately if we know the state S' common to both sequences. Symbolically the property is given as

$$F[S,T_1T_2] = F[S,T_1]\,F[S',T_2] \qquad \text{where } S' = G[S,T_1]$$

For example, the original relay sequence of Figure 6 is modified as shown in Figure 32. In this case the intermediate state $S' = 0$ "unhinges" the past sequence 111 from the future sequence 00.

$$\begin{aligned} F[2,21001] &= F[2,210]\,F[G(2,210),01] \\ &= 111F[0,01] \\ &= 11100 \end{aligned}$$

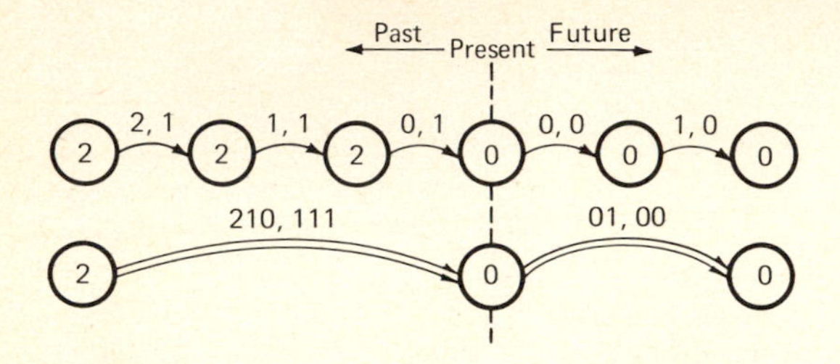

Figure 32

The State Separation Property. Similar to the previous separation property, this property indicates that the present state unhinges future states from past states by the expression

$$S^n = G[S, T_1 T_2] = G[G[S, T_1], T_2]$$

This is shown diagrammatically in Figure 33. For example, in the previous relay sequence of Figure 32

$$\begin{aligned} S^5 = G[2, 21001] &= G[G(2, 210), 01] \\ &= G[0, 01] \\ &= 0 \end{aligned}$$

The two separation properties can be extended to three (and more) segments as follows

$$F[S, T_1 T_2 T_3] = F[S, T_1] F[S', T_2] F[S'', T_3]$$

$$G[S, T_1 T_2 T_3] = G[G[G[S, T_1], T_2], T_3]$$

Notice the different structures of these two functions. The output functions F follow one another (are concatenated), whereas the state functions G are nested (embedded) within one another.

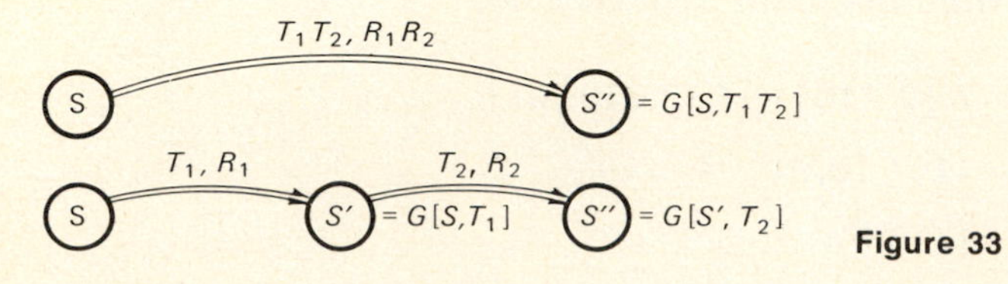

Figure 33

***Particular Properties of Sequential Systems (a Digression)**

In addition to the general properties which hold for all sequential systems, there are also some particular properties which may hold for only some systems. We will now consider some restrictive properties of finite memory and linearity. This may be an interesting digression, but it is unnecessary for anything that follows.

Finite Memory. A sequential system is defined to have finite memory length if the output can be written as a function of a finite number of preceding time intervals in the form

$$Y(t) = [X(t), X(t-1), X(t-2), \ldots, X(t-m), Y(t-1), Y(t-2), \ldots, Y(t-n)]$$

The length of the memory is the largest of these two time intervals, m or n:

$$L = \max(m, n)$$

Notice that an instantaneous system has a memory of length 0 and a delay element has a memory of length 1.

A finite-state system may have infinite length of memory. Such an infinite machine is shown in Figure 34. This can be seen by considering the input tape

$$T = 000 \cdots 0001$$

The corresponding output response is

$$R = 222 \cdots 222?$$

The last output, labeled ?, can be either 2 or 3 but it cannot be determined from any number of previous inputs or outputs; it depends on the initial state.

Linearity. This particular property holds for a small number of systems. However, when it does hold, it is extremely convenient and powerful. Here we will briefly consider a linear system and some of its aspects.

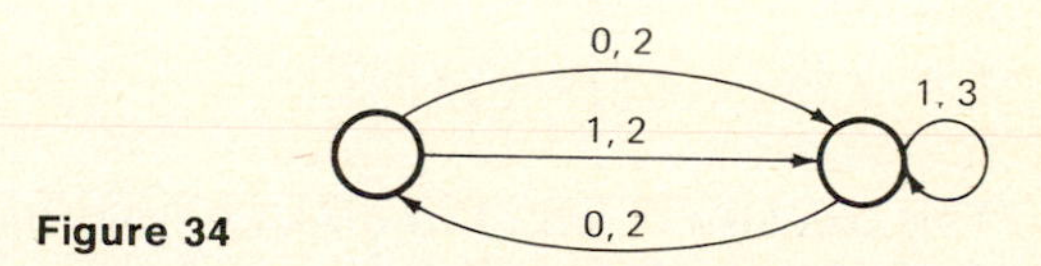

Figure 34

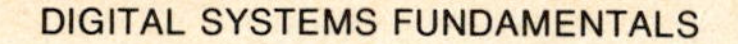

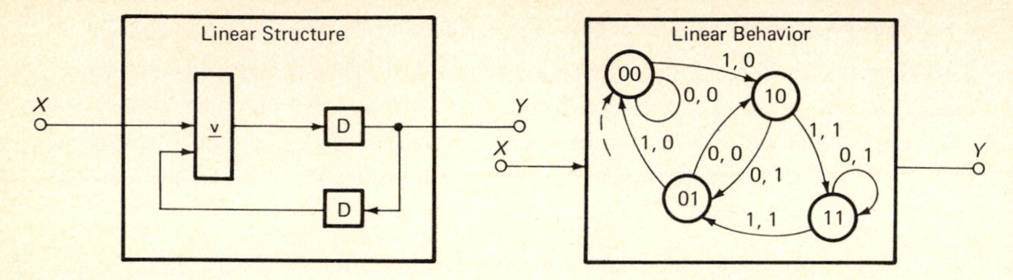

Figure 35

The corresponding results can be extended and compared to continuous systems which are described by other mathematical tools such as differential equations and transforms.

As an example, any (proper) interconnection of EXCLUSIVE-OR components and delay components such as in Figure 35 is linear.

One property of all linear systems is the *superposition property*; the response to the sum of a number of input sequences can be determined by summing the responses of each individual input sequence taken separately, provided that the system starts in the zero state initially. In this case the sum is the EXCLUSIVE-OR operation, and the zero state is 00. For example, the response to input sequence $T_1 = 1101$ is

$$R_1 = F[S_0, 1101] = 0111$$

and the response to input sequence $T_2 = 0100$ is

$$R_2 = F[S_0, 0100] = 0010$$

Then the response to the sum of those sequences

$$1101 \text{ and } 0100 = (1 \veebar 0)(1 \veebar 1)(0 \veebar 0)(1 \vee 0) = 1001$$

can be determined by summing the responses

$$0111 \text{ and } 0010 = (0 \veebar 0)(1 \veebar 0)(1 \veebar 1)(1 \veebar 0) = 0101$$

In other words we have determined algebraically that

$$F[S_0, 1001] = 0101$$

This can also be checked directly from the state diagram.

Another property of all linear systems is that they can be completely represented by their response to a single sequence known as the *impulse function.* In this case the impulse function is the sequence

$$\delta(t) = 100000 \cdots$$

The corresponding impulse response $H(t)$ of the system of Figure 35 is

$$H(t) = 010101010 \cdots$$

The response to any other input sequence can be determined by a process called *convolution*, defined as

$$Y(t) = \sum_{i=0}^{t} H(t-i)X(i)$$

This is essentially a process of weighting the past inputs by the impulse function to get the present output.

There are many other interesting aspects of linear systems, but they are too specialized to be considered here.

EXERCISE SET 5

1. Determine by inspection the length of memory of the systems of Figure 36, giving reasons for each choice.

2. Show that the system of Figure 18 has infinite memory.

3. For the system of Figure 37 write the output as a function of a finite number of past inputs and outputs. What is the memory length of this system?

4. For the linear system of Figure 35 write the output in terms of a finite number of preceding inputs and outputs. *Ans.*: $Y(t) = X(t-1) \veebar Y(t-2)$

5. Using the method of convolution, determine the response of the system of Figure 35 to the sequence 1101.

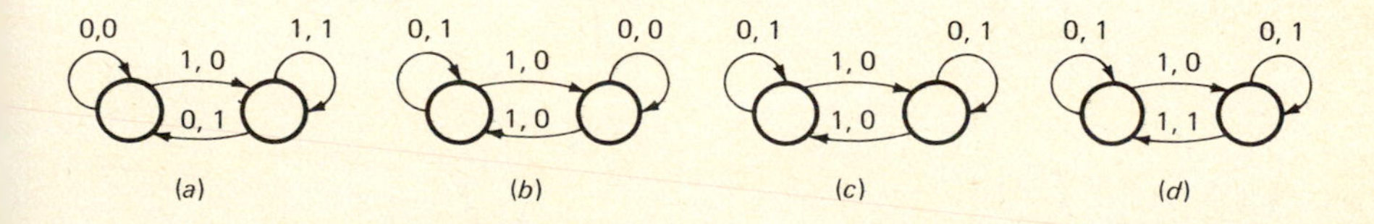

Figure 36

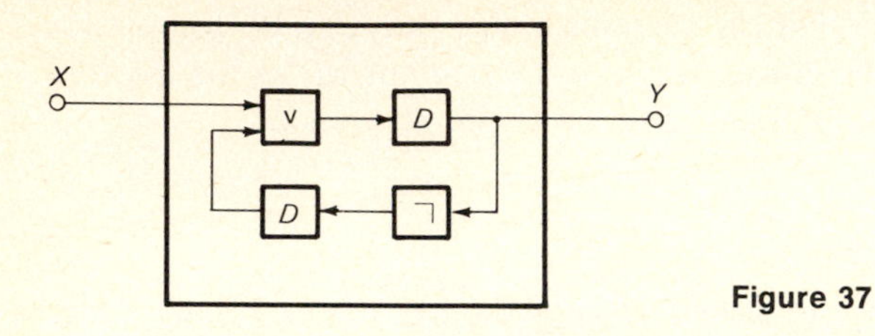

Figure 37

4. OPTIMIZATION OF SEQUENTIAL SYSTEMS

The concept of state has been thought of as an intermediate although important concept relating input and output sequences. The set of states describing a system need not be observable, or physical, or unique. Thus two state diagrams with different numbers of states may describe the same input-output behavior. For example, the two state diagrams of Figure 38 describe the behavior of the same ternary counter. *Try* to check this.

State Equivalence

Since the above two distinct systems differ in the number of states but exhibit the same behavior, it appears that some states in the larger machine are equivalent in some sense and could be removed. Thus two states could be considered equivalent if when started in either state, all possible input-output sequences are identical. We could define state S_i equivalent to state S_j if and only if for all input sequences T the response sequence $F[S_i,T]$ is identical to the response sequence $F[S_j,T]$. Symbolically,

$$S_1 \approx S_j \quad \leftrightarrow \quad F[S_i,T] = F[S_j,T] \qquad \text{for } \textit{all sequences T}$$

Recall that each input tape T is a sequence

$$T = X(0)X(1)X(2)X(3) \cdots X(n)$$

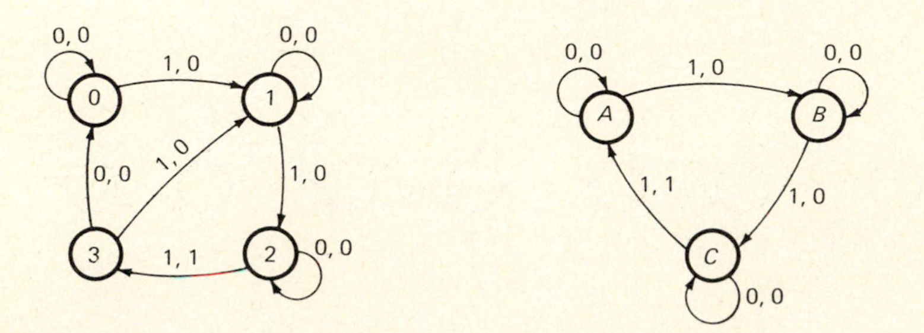

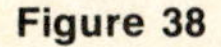
Figure 38

Thus to verify state equivalence experimentally, it is necessary to check all such possible sequences of all possible lengths—an infinite task.

There are, however, some finite methods for determining the equivalence of states according to such an infinite definition. They fall directly into two categories. The first method begins by assuming all states different and proceeds by *merging* pairs of states into one single state. The second method begins by assuming all states equivalent and proceeds by *partitioning* the states into equivalence classes. In general both methods will be applied to optimize any sequential system.

Optimizing States by Method of Merging

To develop insight into optimizing of sequential systems, let us consider the four-state ternary counter and attempt to see how it could be reduced to a three-state device. The transition table is the most convenient representation for this optimization. Examining the structure of the transition table of Figure 39 reveals that the first and last rows are identical, indicating that states 0 and 3 yield identical outputs and next states for all input symbols (but not sequences). The diagram of Figure 40 shows that if the initial state is either 0 or 3, the response to all sequences is also identical, and so the definition of state equivalence is satisfied for these two states 0 and 3. (This will be more formally proved shortly.) One of these two states (say state 3) can then be replaced by the equivalent state 0, resulting in the simpler merged transition table and state diagram of Figure 41. The proof of this method of determining state equivalence follows directly from the response separation property.

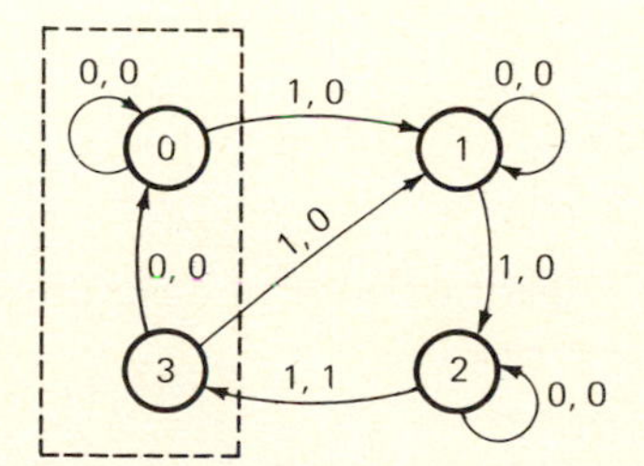

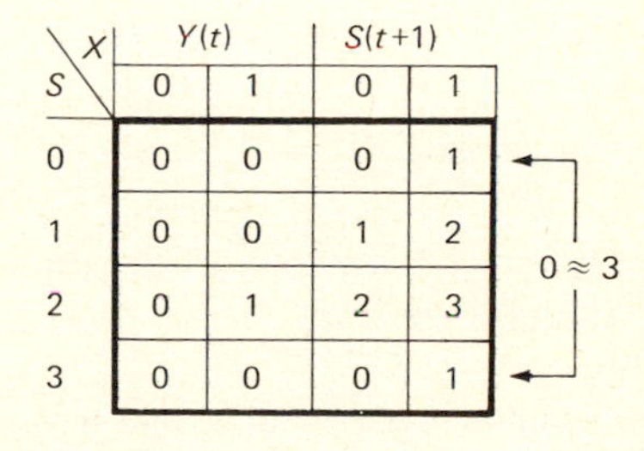

S \ X	Y(t) 0	Y(t) 1	S(t+1) 0	S(t+1) 1
0	0	0	0	1
1	0	0	1	2
2	0	1	2	3
3	0	0	0	1

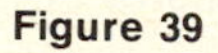
Figure 39

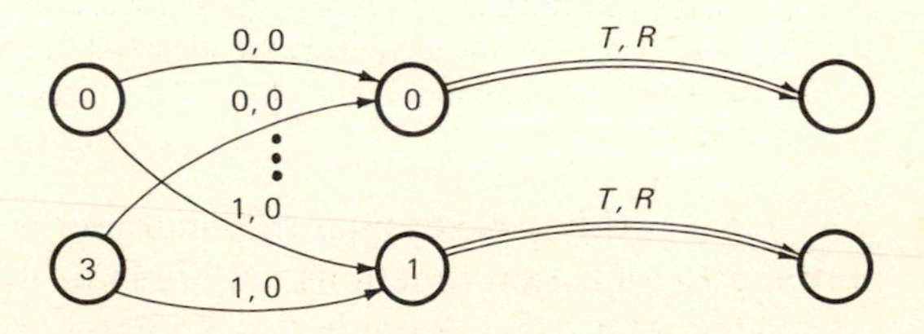

Figure 40

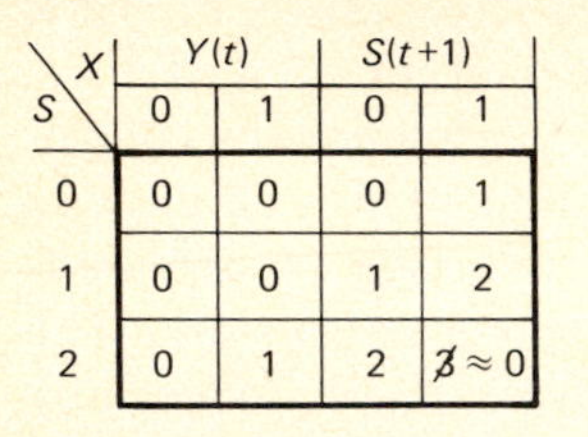

S \ X	Y(t) 0	Y(t) 1	S(t+1) 0	S(t+1) 1
0	0	0	0	1
1	0	0	1	2
2	0	1	2	$\not 3 \approx 0$

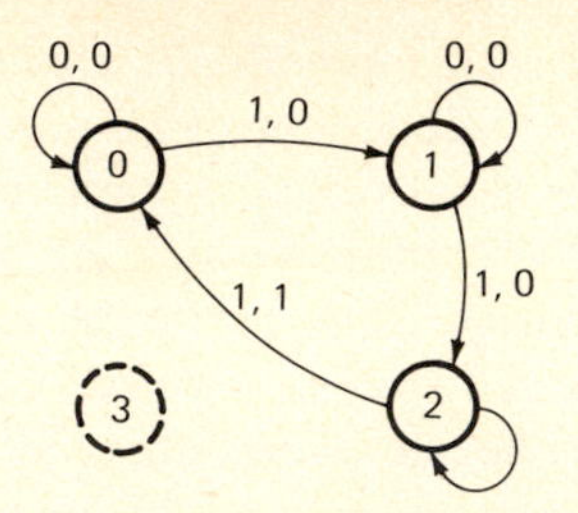

Figure 41

Theorem

Given two states S_i and S_j with properties

$$f(S_i,X) = f(S_j,X) \text{ and } g(S_i,X) = g(S_j,X) \qquad \text{for } \textit{all symbols X}$$

then the states S_i and S_j are equivalent.

Proof

$$\begin{aligned} F[S_i,T] &= F[S_i,X(0)X(1)X(2)\cdots X(n)] \qquad \text{for all } \textit{sequences T} \\ &= F[S_i,X(0)]\,F[g(S_i,X(0)),X(1)X(2)\cdots X(n)] \\ &= f(S_i,X(0))\;F[g(S_j,X(0)),X(1)X(2)\cdots X(n)] \\ &= f(S_j,X(0))F[g(S_j,X(0)),X(1)X(2)\cdots X(n)] \\ &= F[S_j,X(0)]F[g(S_j,X(0)),X(1),X(2)\cdots X(n)] \\ &= F[S_j,X(0)X(1)X(2)\cdots X(n)] \\ &= F[S_j,T] \end{aligned}$$

Therefore $\quad S_i = S_j$

Notice that this proof is simply a general algebraic way of describing a diagram such as Figure 40.

Example Let us consider optimizing the system of Figure 42*a* to illustrate some details of the merging method. First notice that only two rows are identical, indicating that state 2 is equivalent to state 4. Merging these two states yields the system of Figure 42*b*. But notice now that the merging has made rows 1 and 3 identical. Merging these yields the final optimal system of Figure 42*c*. This example illustrates the fact that continued merging is possible.

An algorithm describing this merging is shown in Figure 43. The output and transition functions are given as matrices $F(I,J)$ and $G(I,J)$ associated with states I and inputs J. The part of the algorithm at the left

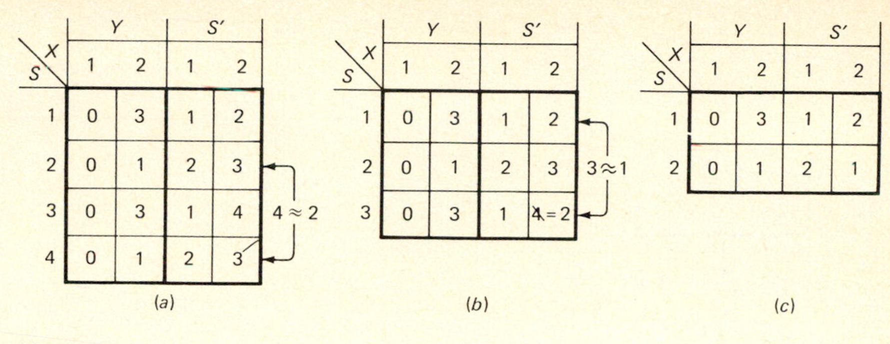

Figure 42

of Figure 43 checks for equivalence of states, the part at the right substitutes equivalent states. The one-dimensional array H indicates which states have merged and need not be compared with others.

Optimizing States by the Method of Partitioning

A second method of optimizing the number of states follows from negating the definition of *state equivalence*

$$S_1 \approx S_j \quad \leftrightarrow \quad F[S_i,T] = F[S_j,T] \qquad \text{for } \textit{all sequences } T$$

to obtain the concept of *state distinguishability*

$$S_i \not\approx S_j \quad \leftrightarrow \quad F[S_i,T] \neq F[S_j,T] \qquad \text{for } \textit{some sequence } T$$

Notice that the definitions apply to input sequences T but can be related to input symbols X from the previous theorem

$$S_i \approx S_j \leftrightarrow \begin{cases} f[S_i,X] = f[S_j,X] \\ \qquad \text{and} \\ g[S_i,X] = g[S_j,X] \end{cases} \qquad \text{for } \textit{all symbols } X$$

to yield the new result on state distinguishability

$$S_i \not\approx S_j \leftrightarrow \begin{cases} f[S_i,X] \neq f[S_j,X] \\ \qquad \text{or} \\ g[S_i,X] \neq g[S_j,X] \end{cases} \qquad \text{for } \textit{some symbol } X$$

The procedure starts by considering the class of all states as one large super state, which is successively split up or partitioned into smaller sets of states according to the definition of distinguishability. The splitting continues until no other states can be seen to be distinguishable.

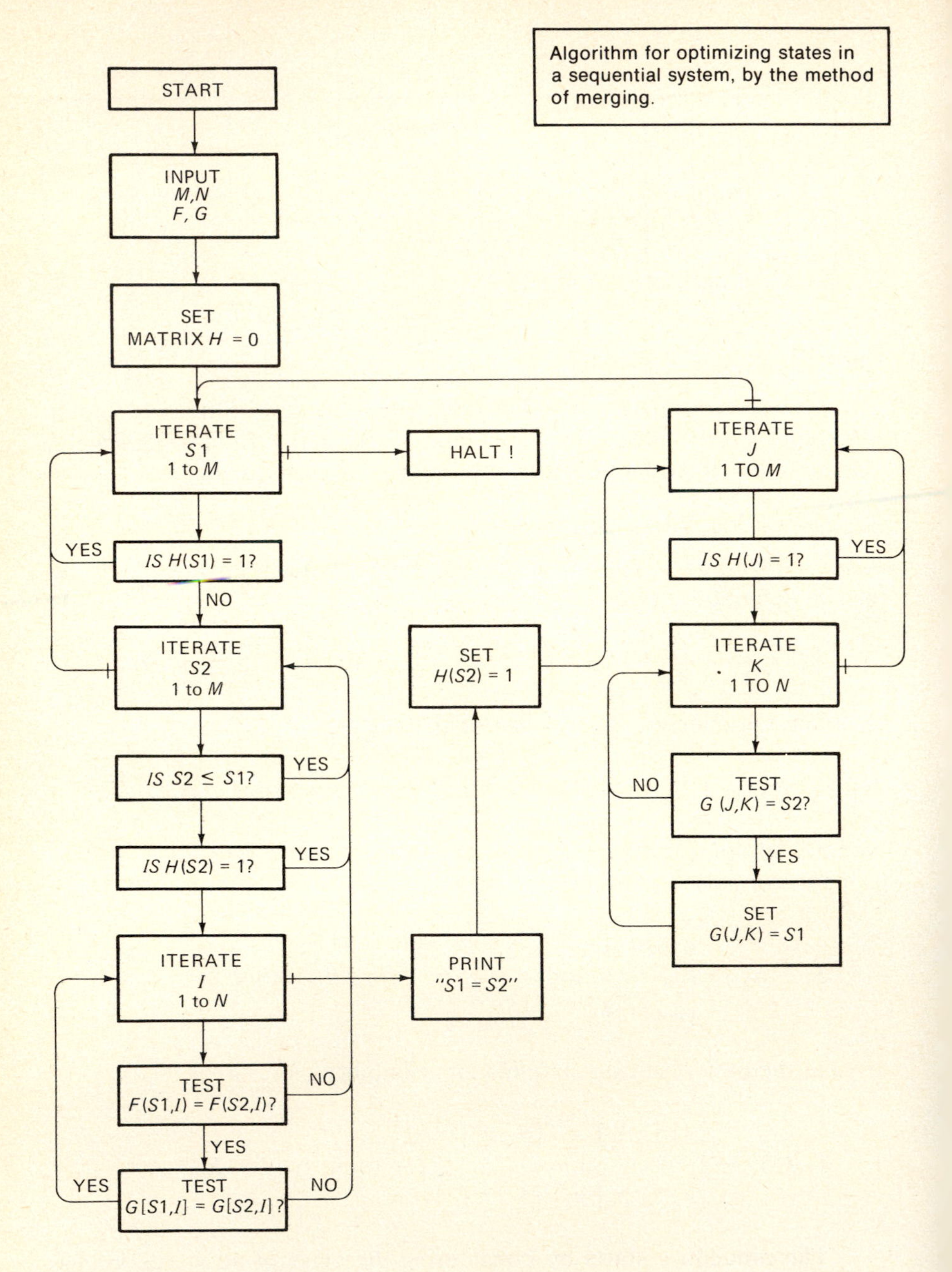
Algorithm for optimizing states in a sequential system, by the method of merging.
START
INPUT
M,N
F, G
SET
MATRIX H = 0
ITERATE
S1
1 to M
HALT !
YES
IS H(S1) = 1?
NO
ITERATE
S2
1 to M
IS S2 ≤ S1?
YES
IS H(S2) = 1?
YES
ITERATE
I
1 to N
TEST
F(S1,I) = F(S2,I)?
NO
YES
TEST
G[S1,I] = G[S2,I]?
YES
NO
SET
H(S2) = 1
PRINT
"S1 = S2"
ITERATE
J
1 TO M
IS H(J) = 1?
YES
ITERATE
K
1 TO N
NO
TEST
G (J,K) = S2?
YES
SET
G(J,K) = S1

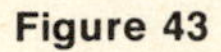
Figure 43

Then each remaining set of states is considered one state. The example of Figure 44 illustrates this procedure. Notice that no direct merging is possible. First, the five states could be partitioned into two "candidate states," called a and b, where

$$a = \{1,2\} \qquad \text{and} \qquad b = \{3,4,5\}$$

S \ X	Y: 0	Y: 1	S′: 0	S′: 1
1	0	1	3	1
2	0	1	4	1
3	0	0	5	4
4	0	0	5	3
5	0	0	4	2

Figure 44

These two states are distinguishable since they have different outputs for some inputs, i.e.,

$$f[a,1] \neq f[b,1]$$

This does not yet mean that all the substates of state b are equivalent; this must still be checked. The partition can be redrawn on a transition table, with labeled subscripts as in Figure 45. The next-state function g indicates that state 5 is distinguishable from states 3 and 4 since the (subscripted) next states for all inputs are not identical. This leads to the following refined partition of new candidate states:

$$a = \{1,2\} \qquad \text{and} \qquad c = \{3,4\} \qquad \text{and} \qquad d = \{5\}$$

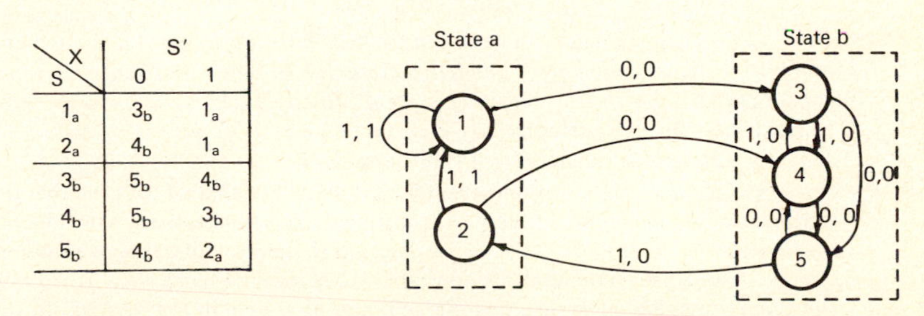

S \ X	S′: 0	S′: 1
1_a	3_b	1_a
2_a	4_b	1_a
3_b	5_b	4_b
4_b	5_b	3_b
5_b	4_b	2_a

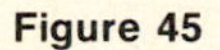
Figure 45

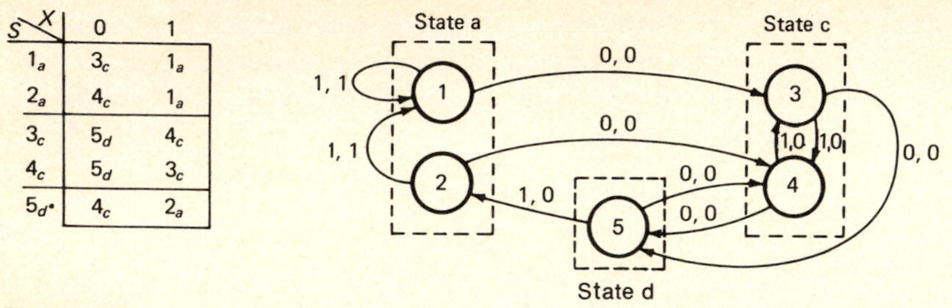

S \ X	0	1
1_a	3_c	1_a
2_a	4_c	1_a
3_c	5_d	4_c
4_c	5_d	3_c
5_d*	4_c	2_a

Figure 46

Finally the relabeled transition table of Figure 46 shows all three candidate states as distinguishable and the substates of each candidate state as equivalent. No further partitioning is possible; optimizing is complete. Now the optimal system can be redrawn in terms of the three distinguishable states *a*, *c*, and *d*, as in Figure 47.

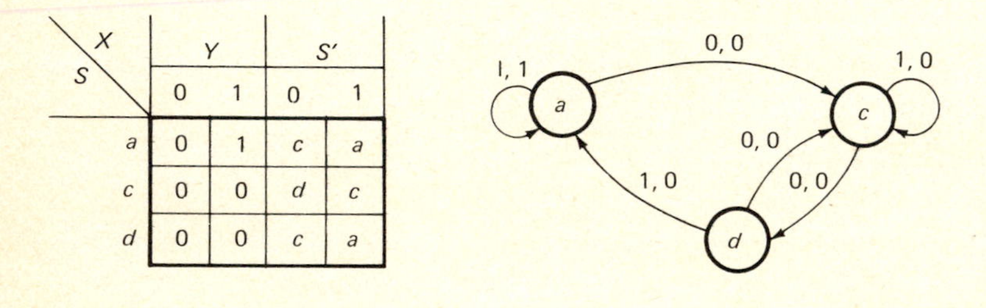

S \ X	Y 0	Y 1	S′ 0	S′ 1
a	0	1	*c*	*a*
c	0	0	*d*	*c*
d	0	0	*c*	*a*

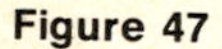

Figure 47

Example The 10 decimal digits can be coded by their positional notation, as sequences of four binary digits (called *binary coded decimal*, or BCD). Construct and optimize a state diagram for an error-detecting system which checks sequences to determine whether they represent the properly coded digits 0 to 9. It indicates an error (output 1) if any other value (from 10 to 15) has appeared.

Let us assume that the most significant digits are applied first and that the system and the sequences are initially synchronized. It is conceptually convenient to indicate all possible sequences by a state diagram in the form of a tree as shown in Figure 48. The corresponding transition table is shown in Figure 49*a*. It is easily optimized by merging. The resulting merged table is shown in Figure 49*b*. By successively partitioning this system (do it) we find that no additional optimizing is possible. The final optimal state diagram is shown in Figure 50.

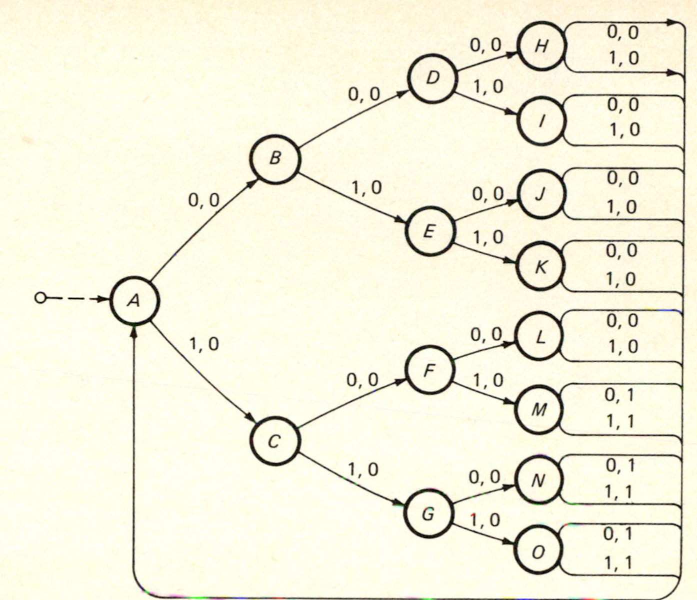

Figure 48

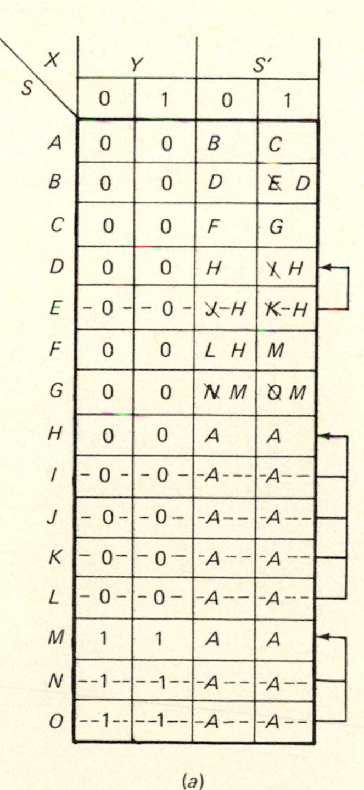

S \ X	Y 0	Y 1	S′ 0	S′ 1
A	0	0	B	C
B	0	0	D	~~E~~ D
C	0	0	F	G
D	0	0	H	~~I~~ H
E	~~0~~	~~0~~	~~J~~ H	~~K~~ H
F	0	0	~~L~~ H	M
G	0	0	~~N~~ M	~~O~~ M
H	0	0	A	A
I	~~0~~	~~0~~	~~A~~	~~A~~
J	~~0~~	~~0~~	~~A~~	~~A~~
K	~~0~~	~~0~~	~~A~~	~~A~~
L	~~0~~	~~0~~	~~A~~	~~A~~
M	1	1	A	A
N	~~1~~	~~1~~	~~A~~	~~A~~
O	~~1~~	~~1~~	~~A~~	~~A~~

(a)

S \ X	Y 0	Y 1	S′ 0	S′ 1
A	0	0	B	C
B	0	0	D	D
C	0	0	F	G
D	0	0	H	H
F	0	0	H	M
G	0	0	M	M
H	0	0	A	A
M	1	1	A	A

State Equivalences

$D = E$

$H = I = J = K = L$

$M = N = O$

(b)

Figure 49

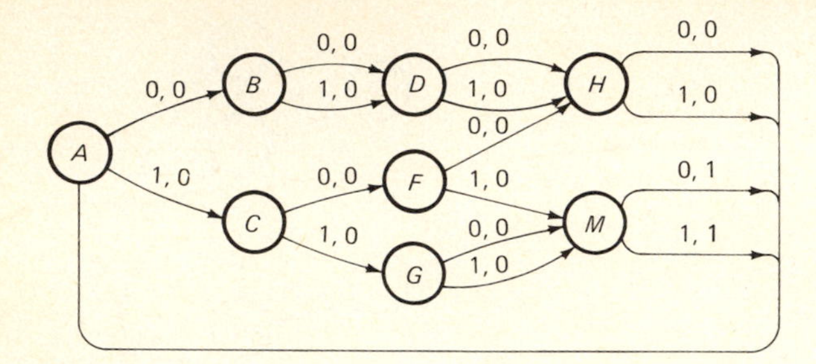

Figure 50

EXERCISE SET 6

1. Apply the partitioning procedure to the system of Figure 49 to show that it cannot be optimized further.
2. If the BCD error-detecting system of the example has the least significant digits applied first, construct the optimal state diagram. *Ans.*: It has only six states.
3. Apply the algorithm of Figure 43 to the system of Figure 42*a* and show the sequence of instantaneous descriptions for at least the first merge and first substitution.
4. Extend the algorithm of Figure 43 to print out the optimized transition table.
5. Optimize the sequential systems of Figure 51, and comment.

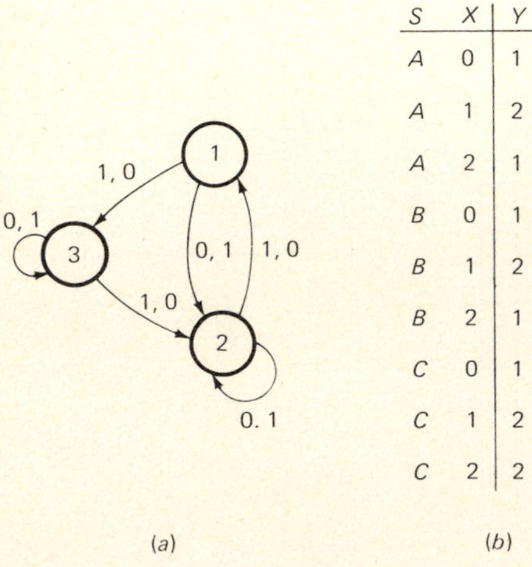

S	*X*	*Y*	*S′*
A	0	1	*B*
A	1	2	*A*
A	2	1	*A*
B	0	1	*B*
B	1	2	*A*
B	2	1	*A*
C	0	1	*B*
C	1	2	*A*
C	2	2	*C*

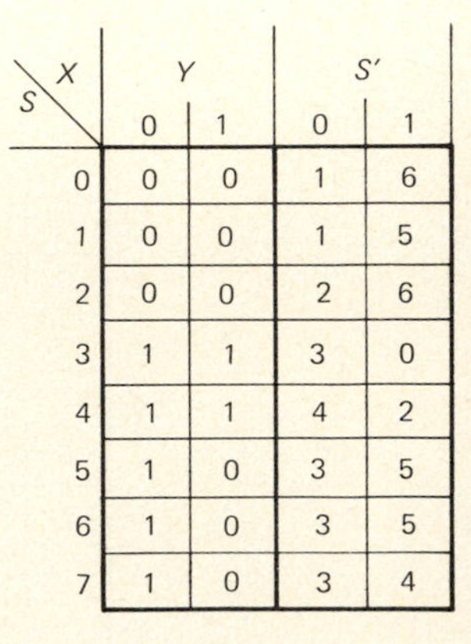

S \ *X*	*Y*		*S′*	
	0	1	0	1
0	0	0	1	6
1	0	0	1	5
2	0	0	2	6
3	1	1	3	0
4	1	1	4	2
5	1	0	3	5
6	1	0	3	5
7	1	0	3	4

(*a*) (*b*) (*c*)

Figure 51

5. IMPLEMENTATION OF SEQUENTIAL SYSTEMS

The basic components of large sequential systems are flip-flops, which are also called multivibrators or memory elements. The states of the flip-flops are usually also the outputs and are labeled by the symbol Q

$$Q = S = Y$$

Some Practical Components

The four most commonly used binary flip-flops are known as the set-reset flip-flop, the JK flip-flop, the toggle or T flip-flop, and the delay or D flip-flop. The electronic aspects of these flip-flops are considered in the Appendix; the behavioral aspects are considered next.

The set-reset flip-flop is described in Figure 52. Usually both inputs have value 0. If input S (set) then has value 1, the device will be set at state 1 (flip to ON state) and remain there with output 1, even when the set value is changed to 0. Now if the R (reset) has value 1, the state will change to 0 (flop to OFF state) and remain there when the input R is returned to value 0. The state will change again only when the set input S has value 1.

This flip-flop could serve as a simple store or memory device since the output indicates which of the two inputs S or R last had the value 1.

Usually a system is arranged so that both inputs R and S do not have the value 1 since physical devices may be indeterminate in these cases sometimes yielding an output of 0 and at other times yielding an output of 1. However, there is a *set-dominant set-reset flip-flop* which, in such a case, behaves as indicated on the dotted lines of Figure 52. Often the NOT of the output is conveniently available as a second separate terminal.

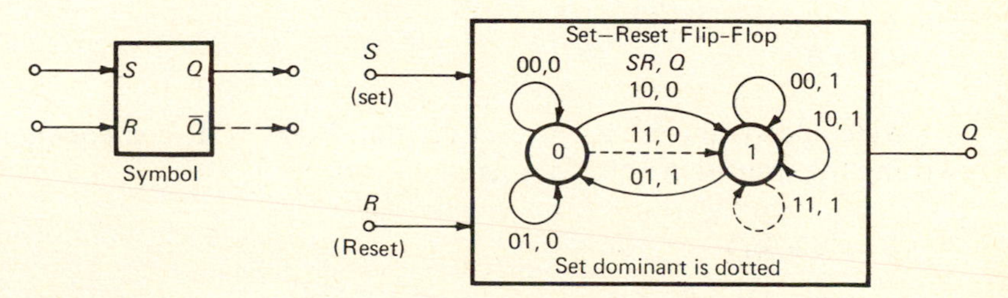

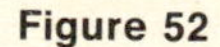
Figure 52

S	X	Y	S′
Q	SR	Q	Q′
0	00	0	0
0	01	0	0
0	10	0	1
0	11	0	?
1	00	1	1
1	01	1	0
1	10	1	1
1	11	1	?

(a)

Q \ SR	00	01	10	11
0	0	0	1	?
1	1	0	1	?

(Q′)

(b)

S R	Q′
0 0	Q
0 1	0
1 0	1
1 1	?

(c)

Figure 53

This set-reset flip-flop can also be described by a table-of-combinations, as in Figure 53*a*, a transition table, as in Figure 53*b*, or most conveniently by a brief modified transition table, as in Figure 53*c*. The brief table is called a *characteristic table*; it will be explained in detail later (for the JK flip-flop).

For the set-dominant set-reset flip-flop (where the undetermined condition "?" has value 1) the transition equation can be written from Figure 53*a* as

$$\begin{aligned} Q' &= \bar{Q}S\bar{R} \vee \bar{Q}SR \vee Q\bar{S}\bar{R} \vee QS\bar{R} \vee QSR \\ &= (\bar{Q}S\bar{R} \vee \bar{Q}SR) \vee (Q\bar{S}\bar{R} \vee QS\bar{R}) \vee (QS\bar{R} \vee QSR) \\ &= \bar{Q}S \vee Q\bar{R} \vee QS \\ &= (\bar{Q} \vee Q)S \vee \bar{R}Q \\ &= S \vee \bar{R}Q \end{aligned}$$

For the ordinary (non-set-dominant) set-reset flip-flop we must impose the condition that the set S and reset R not be applied at the same time; i.e.,

$$RS = 0$$

We could synthesize a set-reset flip-flop from NOR components by rewriting the transition equations as

$$\begin{aligned} Q' &= S \vee \bar{R}Q \\ &= S \vee \overline{(R \vee \bar{Q})} \end{aligned}$$

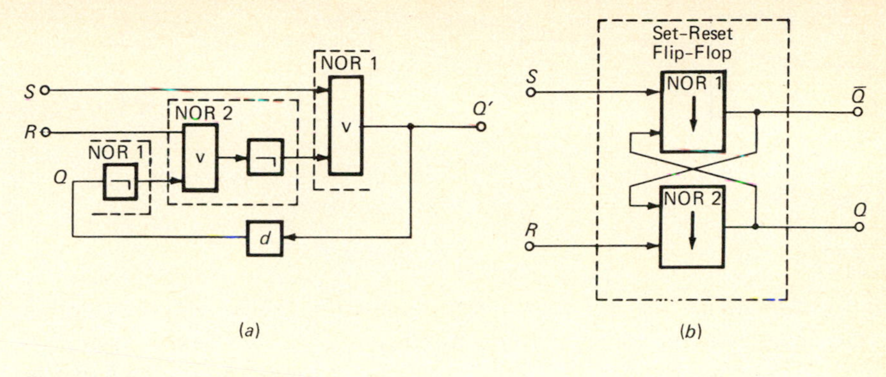

Figure 54

The corresponding diagram is shown in Figure 54*a*. The delay is assumed inherent in the components. By redrawing each combination of ORs and NOTs as a NOR we arrive at the symmetrical interconnections of NORs of Figure 54*b*.

The JK flip-flop shown in Figure 55 is similar to the *SR* flip-flop, except in the last case. When both J and K have value 1, the state changes to the opposite value. The characteristic table is interpreted below:

$JK = 00$, no inputs applied, no state change (stay)
$JK = 01$, input K only applied, state becomes 0 (reset)
$JK = 10$, input J only applied, state becomes 1 (set)
$JK = 11$, J and K applied, state reverses (toggle)

The next-state equation for the JK flip-flop can be determined to be

$$Q' = J\bar{Q} \vee \bar{K}Q$$

The toggle flip-flop shown in Figure 56*a* has a single input T. It changes state (or toggles) each time an input value of 1 is applied. The next-state function is given by the expression

$$Q' = T \veebar Q = T\bar{Q} \vee \bar{T}Q$$

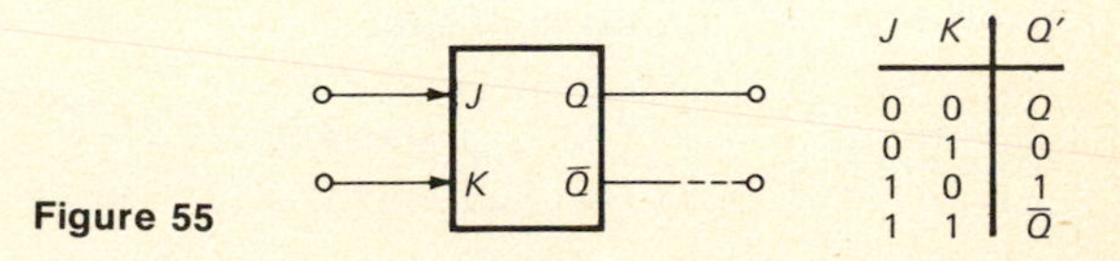

J	K	Q′
0	0	Q
0	1	0
1	0	1
1	1	$\bar{Q}$

Figure 55

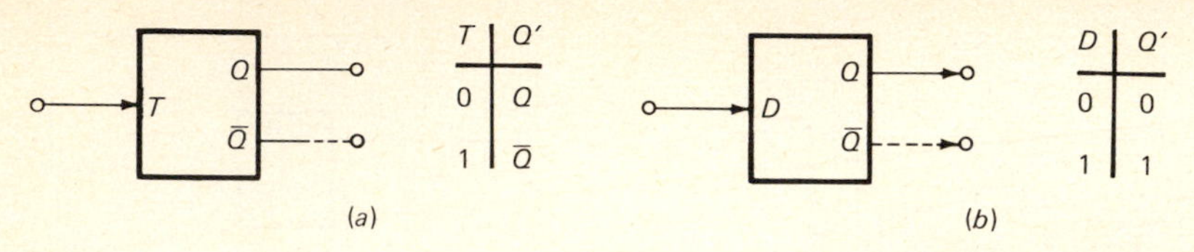

Figure 56

The delay flip-flop or D flip-flop shown in Figure 56*b* is essentially the delay element which we have already used to synthesize systems.

It is sometimes convenient to construct one type of flip-flop from another type. For example, a delay flip-flop can be constructed using a set-reset flip-flop as shown in Figure 57*a*. This can be verified by writing the corresponding switching algebraic equations:

$$\begin{aligned} Q' &= S \vee \bar{R}Q \\ &= X \vee \bar{\bar{X}}Q \\ &= X \end{aligned}$$

More explicitly in terms of time this becomes

$$Y(t+1) = X(t)$$

which does indeed describe a delay element.

The *clocked delay* is a particularly useful flip-flop that can be controlled by some clock signal. For example, a clocked delay element is shown in Figure 57*b*. In this case the corresponding equations are

$$\begin{aligned} Q' &= S \vee \bar{R}Q \\ &= XC \vee \overline{(\bar{X}C)}Q \\ &= XC \vee XQ \vee \bar{C}Q \\ &= XC \vee XQ(C \vee \bar{C}) \vee \bar{C}Q \\ &= XC \vee \bar{C}Q \end{aligned}$$

More explicitly in terms of time

$$Y(t+n) = \begin{cases} X(t) & \text{if clock } C(t) = 1 \\ Y(t) & \text{if clock } C(t) = 0 \end{cases}$$

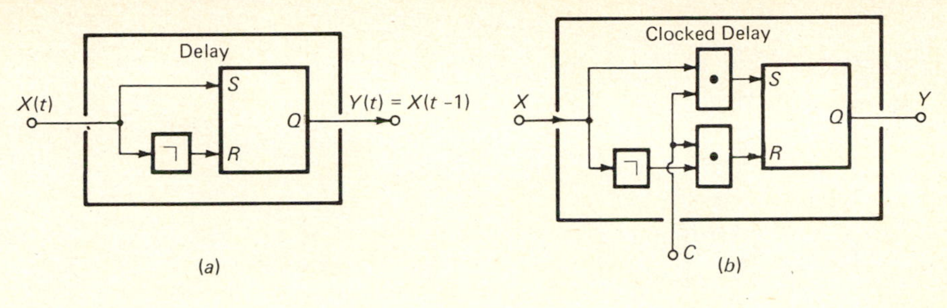

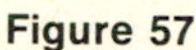

Figure 57

In such a clocked delay element, the delay is as long as the interval between clock pulses (and this may vary). This formal sampled delay is to be distinguished from the actual real-time delay. This difference is illustrated in Figure 58.

Notice that the real-time output is not simply a translation in time of the real-time input, but the form of the pulses is distorted. In this case, a short input pulse followed by a long one comes out as a long pulse followed by a short one. However, the "formal" time delay is precisely 1 unit of formal time.

EXERCISE SET 7

1. Construct a T flip-flop from a JK flip-flop.
2. Construct a D flip-flop from a JK flip-flop.
3. Construct a D flip-flop from a T flip-flop. *Hint*: $Q \veebar Q = 0$.
4. Construct a state diagram describing the system of Figure 59 if the flip-flop is a set-rest flip-flop of the set-dominant type.

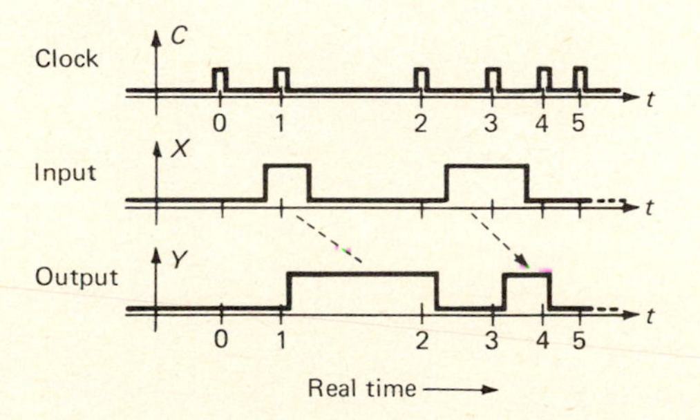

t	0	1	2	3	4	5	6
$X(t)$	0	1	0	1	0	0	
$Y(t)$		0	1	0	1	0	0

Formal time ⟶

Figure 58

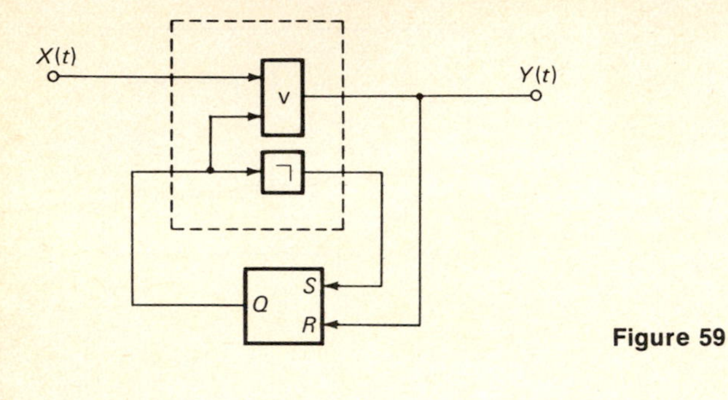

Figure 59

Implementation Method

Large sequential systems can be constructed from interconnections of instantaneous systems and smaller sequential systems. These smaller sequential systems need not be simply delay elements as previously but may be other elements, such as flip-flops. For such purposes a flip-flop is best represented by an *excitation table*, which indicates the inputs required for each state change. Figure 60 illustrates the excitation table for a toggle flip-flop as well as the brief transition table it was derived from. For example, the fourth row of the transition table corresponds to the third row of the excitation table, indicating that a state change from 1 to 0 requires in input value $T = 1$.

Let us implement the previous sequential adder (repeated in Figure 61*a*) using toggle flip-flops. The corresponding state changes are rewritten in Figure 61*b*, and the required inputs T to the flip-flop are determined from the excitation table of Figure 60. Then in Figure 61*c* the outputs S and T of the instantaneous network are shown as functions of the inputs A, B, and Q of this network. From this table we can write the expressions

$$S = A \veebar B \veebar C$$

$$T = \bar{A}\bar{B}Q \vee AB\bar{Q}$$

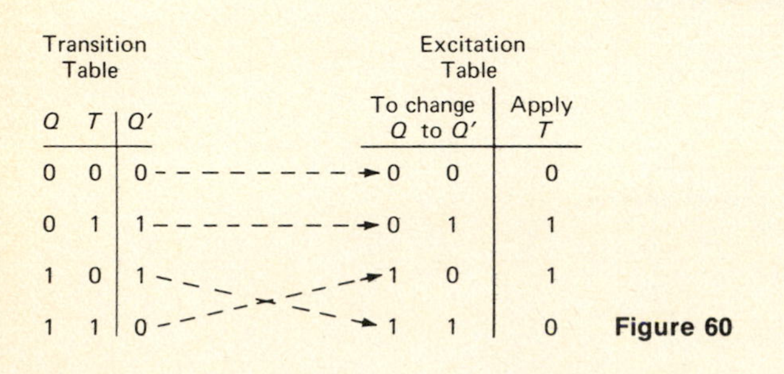

Transition Table

Q	T	Q'
0	0	0
0	1	1
1	0	1
1	1	0

Excitation Table

To change Q	to Q'	Apply T
0	0	0
0	1	1
1	0	1
1	1	0

Figure 60

A	B	Q	S	Q'
0	0	0	0	0
0	0	1	1	0
0	1	0	1	0
0	1	1	0	1
1	0	0	1	0
1	0	1	0	1
1	1	0	0	1
1	1	1	1	1

(*a*)

Change		Requires
Q	Q'	T
0	0	0
1	0	1
0	0	0
1	1	0
0	0	0
1	1	0
0	1	1
1	1	0

(*b*)

Inputs			Outputs	
A	B	Q	S	T
0	0	0	0	0
0	0	1	1	1
0	1	0	1	0
0	1	1	0	0
1	0	0	1	0
1	0	1	0	0
1	1	0	0	1
1	1	1	1	0

(*c*)

Figure 61

which is shown implemented in Figure 62. Notice that the complement $\bar{Q}$ of the flip-flop is also used as an input to the instantaneous network.

Other components can also be used to implement sequential systems. For example, a two-input sequential component such as a JK flip-flop can be used in a way similar to the one-input toggle flip-flop. The transition table and corresponding excitation table for the JK flip-flop are shown in Figure 63. The transition table indicates, for example, that to change state value 0 to value 1 requires either input $JK = 10$ or $JK = 11$; that is, $J = 1$ and $K = 0$ or 1. This is indicated as $JK = 1d$ (where d is a don't-care value) on the excitation table.

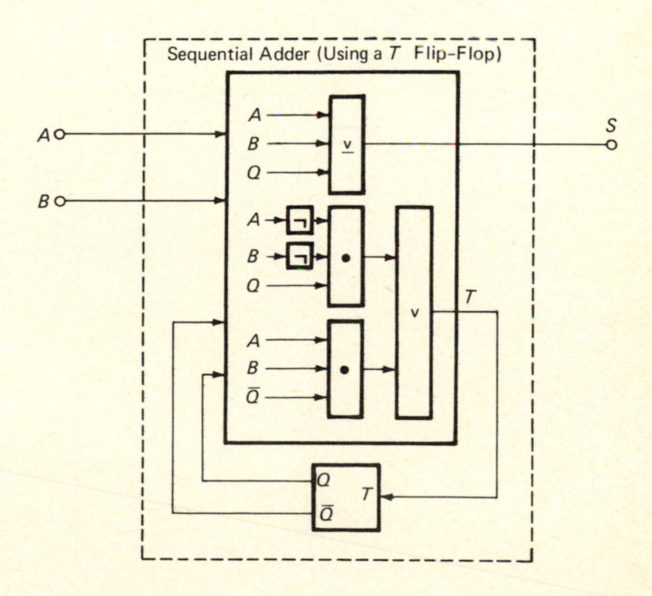

Figure 62

Q \ JK (Q')	00	01	10	11
0	0	0	1	1
1	1	0	1	0

Change Q to Q'		Requires J	K
0	0	0	*d*
0	1	1	*d*
1	0	*d*	1
1	1	*d*	0

Figure 63

This excitation table can now be applied to the sequential adder in the same way as the previous example, except that there are two flip-flop inputs J and K and therefore two outputs J and K of the instantaneous system. This is shown in Figure 64. A Karnaugh map can now be drawn for each output of the instantaneous network as shown in Figure 65. The final JK flip-flop implementation of the sequential adder is shown in Figure 66.

As another more complex example, let us attempt to implement the previous ternary counter using set-reset flip-flops. The transition table and excitation table for the set-reset flip-flop are given in Figure 67.

The excitation table is now applied to the ternary counter in the same way as in the previous examples, except that now two flip-flops are necessary to code the three states. This is shown in Figure 68.

Karnaugh maps can now be drawn for each input of each set-reset flip-flop. For the first flip-flop the maps are shown in Figure 69. For the

A	B	Q	S	Q'
0	0	0	0	0
0	0	1	1	0
0	1	0	1	0
0	1	1	0	1
1	0	0	1	0
1	0	1	0	1
1	1	0	0	1
1	1	1	1	1

Change Q	Q'	Requires J	K
0	0	0	*d*
1	0	*d*	1
0	0	0	*d*
1	1	*d*	0
0	0	0	*d*
1	1	*d*	0
0	1	1	*d*
1	1	*d*	0

Inputs A	B	Q	Outputs J	K	S
0	0	0	0	*d*	0
0	0	1	*d*	1	1
0	1	0	0	*d*	1
0	1	1	*d*	0	0
1	0	0	0	*d*	1
1	0	1	*d*	0	0
1	1	0	1	*d*	0
1	1	1	*d*	0	1

Figure 64

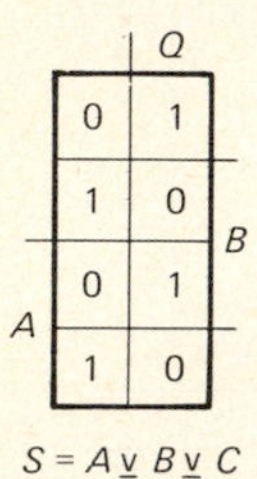

$S = A \veebar B \veebar C$

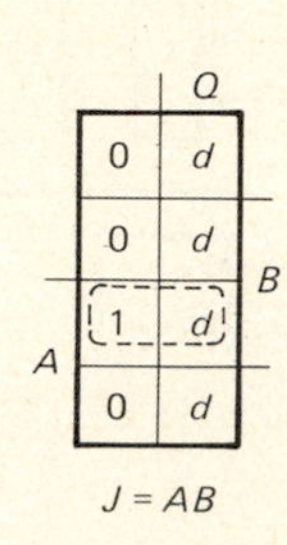

$J = AB$

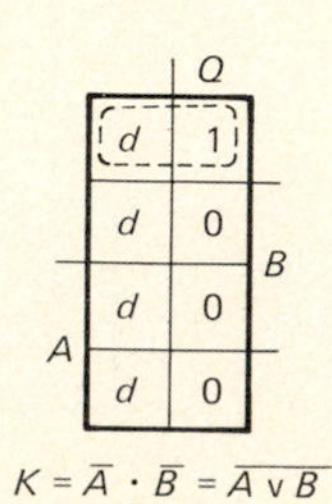

$K = \overline{A} \cdot \overline{B} = \overline{A \vee B}$

Figure 65

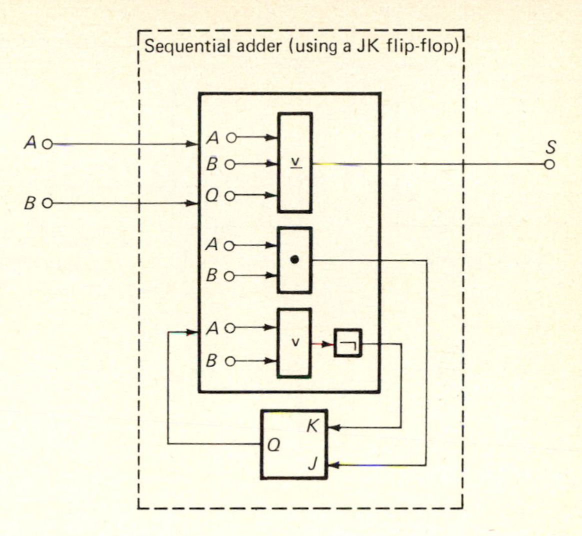

Figure 66

Q \ SR	00	01	10
0	0	0	1
1	1	0	1

To change Q	to Q′	Apply S	R
0	0	0	*d*
0	1	1	0
1	0	0	1
1	1	*d*	0
d	*d*	*d*	*d*

Figure 67

Q_1	Q_2	X	Y	Q'_1	Q'_2
0	0	0	0	0	0
0	0	1	0	0	1
0	1	0	0	0	1
0	1	1	0	1	0
1	0	0	0	1	0
1	0	1	1	0	0
1	1	0	*d*	*d*	*d*
1	1	1	*d*	*d*	*d*

Q_1	Q'_1	S_1	R_1
0	0	0	*d*
0	0	0	*d*
0	0	0	*d*
0	1	1	0
1	1	*d*	0
1	0	0	1
1	*d*	*d*	*d*
1	*d*	*d*	*d*

Q_2	Q'_2	S_2	R_2
0	0	0	*d*
0	1	1	0
1	1	*d*	0
1	0	0	1
0	0	0	*d*
0	0	0	*d*
1	*d*	*d*	*d*
1	*d*	*d*	*d*

Figure 68

second flip-flop the maps are shown in Figure 70. Notice that in both cases $S_i R_i = 0$. The final set-reset implementation of the ternary counter is shown in Figure 71. Note that the complements of all flip-flops are used.

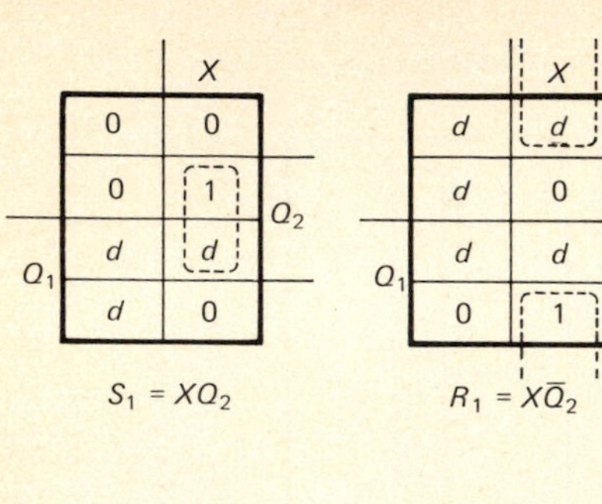

Figure 69

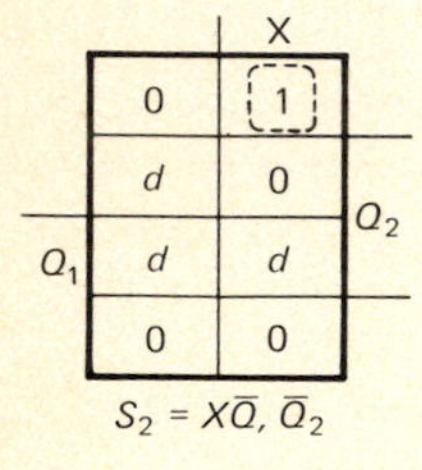

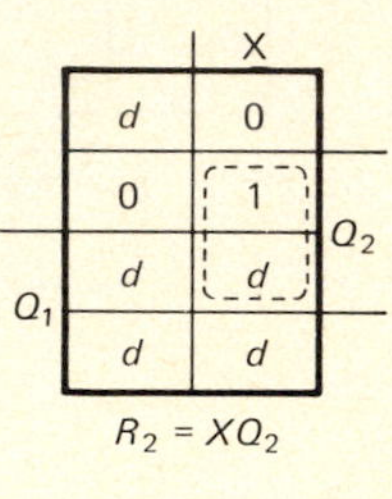

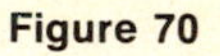
Figure 70

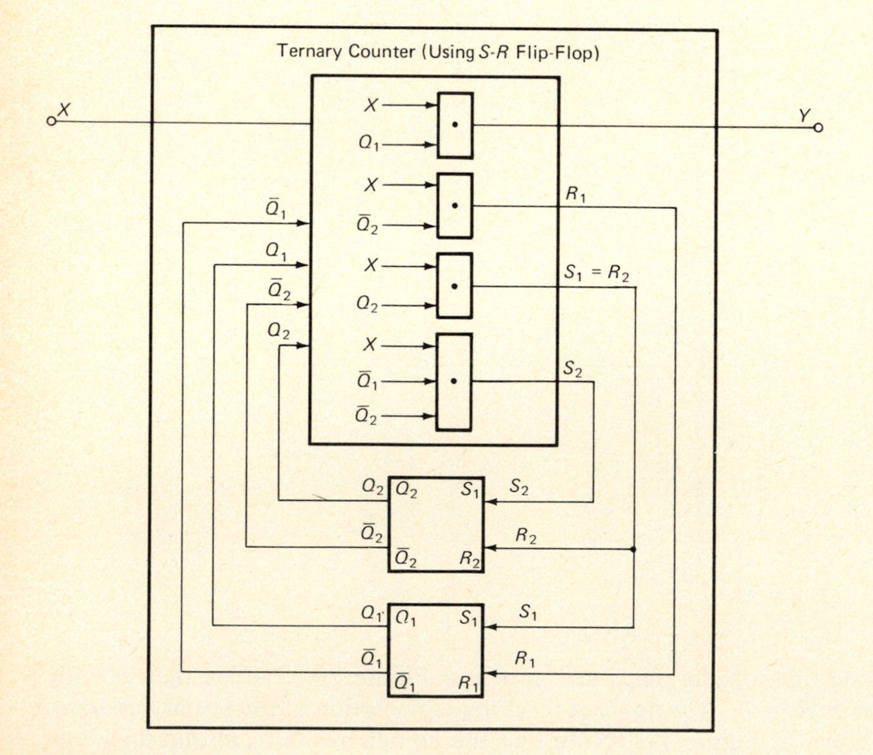

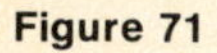
Figure 71

Other Possible Components

Although the four flip-flops considered here are the most common, there are many other possible binary flip-flops. For example, consider the general flip-flop of Figure 72.

The number of such flip-flops can be determined as follows. The first row, corresponding to $Q = 0$, could be filled in 2^4 or 16 ways. But two of these (having all 0s and all 1s) do not cause transitions (do not flip), and so only 14 are possible. Similarly there are only 14 ways of filling in the second row. By the product principal there are

$$14 * 14 = 196$$

different ways of filling in the table and therefore 196 possible binary flip-flops. We can construct sequential systems using any one of these 196 devices by first constructing the corresponding excitation table and then proceeding by the previous method.

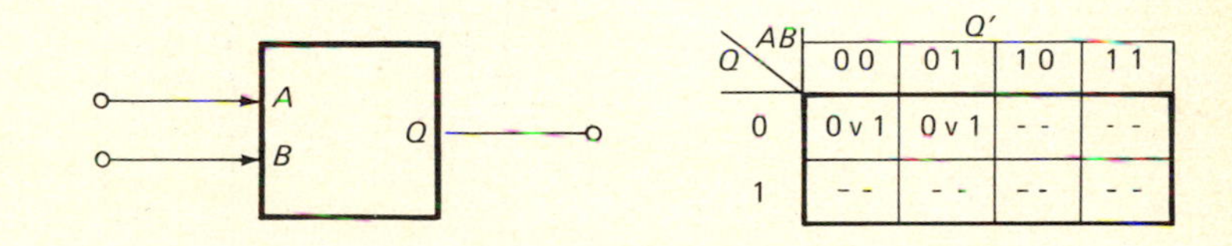

Figure 72

Example Using the flip-flop of Figure 73, which we will call an *MN flip-flop*, synthesize a binary sequential adder.

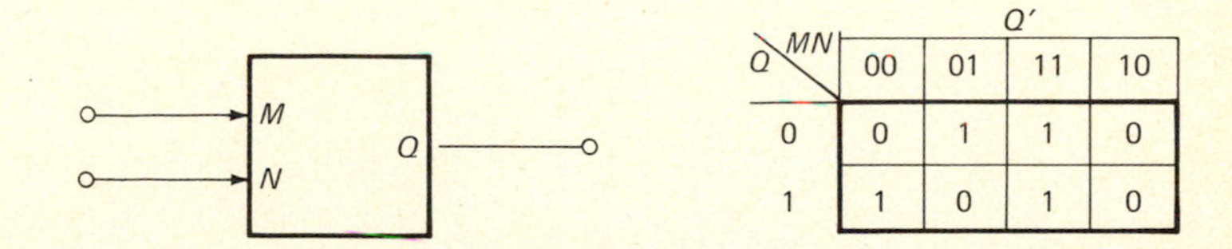

Figure 73

Solution The excitation table of this flip-flop is determined in Figure 74*a* or more briefly in Figure 74*b*. Notice that the *d* is a *free* don't-care value which may be assigned any value 0 or 1. However, the don't-care values labeled *a* and *b* are *bound*, so that value *a* may be chosen as 0 or 1 but it must have this same value wherever symbol *a* occurs.

Change Q	Q′	Requires MN
0	0	00 v 10
0	1	01 v 11
1	0	01 v 10
1	1	00 v 11

Q	Q′	M	N
0	0	*d*	0
0	1	*d*	1
1	0	*a*	$\bar{a}$
1	1	*b*	*b*

Figure 74

The required instantaneous network can be determined from the tables of Figure 75*a* and the corresponding Karnaugh maps of Figure 75*b*. From the Karnaugh maps we can choose many possible networks. A simple choice of don't-care values is given in Figure 75*c*, and the corresponding network is shown in Figure 76.

A	*B*	*Q*	*S*	*Q'*	*M*	*N*
0	0	0	0	0	*d*	0
0	0	1	1	0	*a*	$\bar{a}$
0	1	0	1	0	*d*	0
0	1	1	0	1	*b*	*b*
1	0	0	1	0	*d*	0
1	0	1	0	1	*c*	*c*
1	1	0	0	1	*d*	1
1	1	1	1	1	*e*	*e*

(*a*)

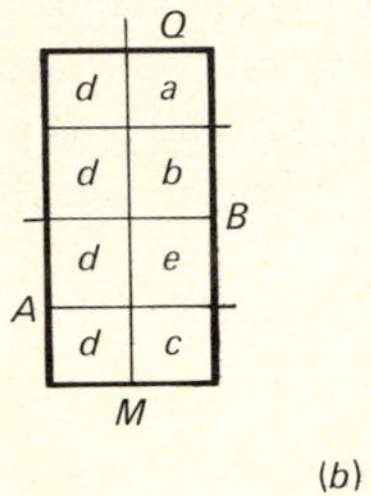

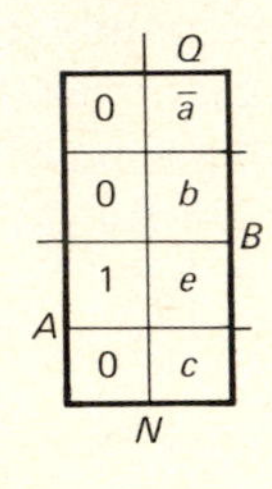

(*b*)

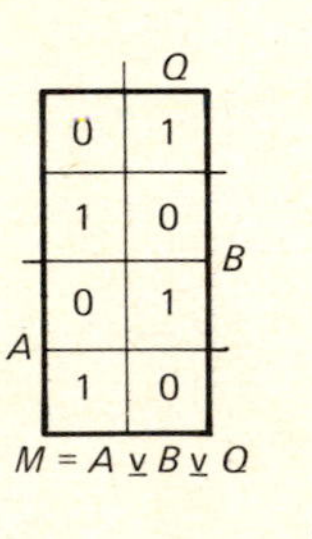

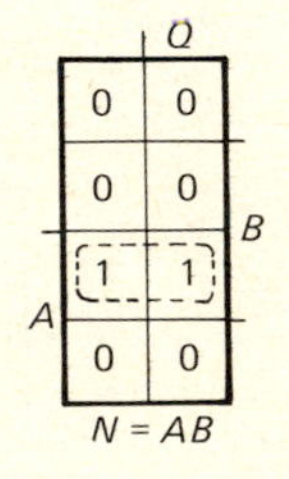

(*c*)

Figure 75

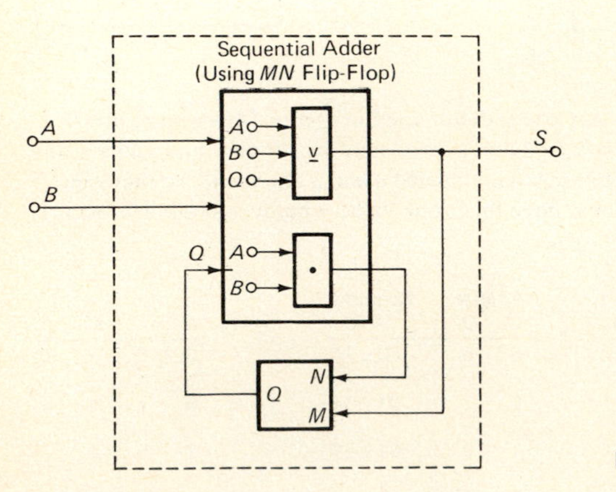

Figure 76

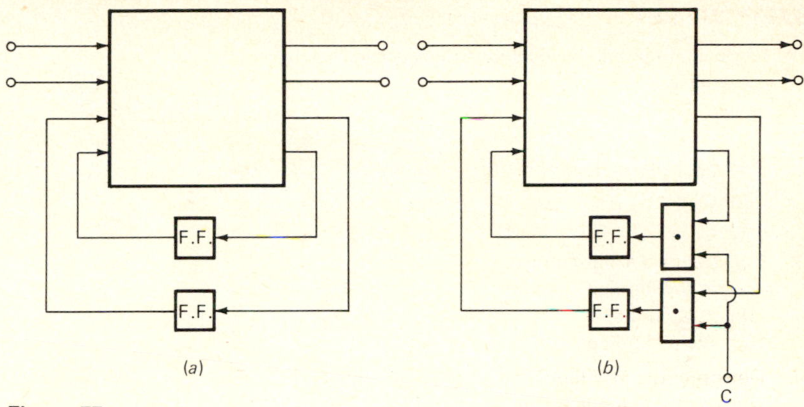

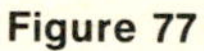
Figure 77

Timing Considerations

In the practical implementation of sequential systems there may be many difficulties involving timing. For example, it is extremely difficult to have two or more delays, as in Figure 77*a*, precisely equal and feeding back at precisely the proper times.

One way of eliminating most timing problems is to use a clocked or gated system, as in Figure 77*b*. A transition is possible only when a clock pulse is applied to the AND gates. This pulse must be long enough to cause the states to change, but it must be short enough to stop before any signal propagates through the network causing the inputs at the gates to change again. All the next-state values should also have settled down before the next clock pulse.

The relation between the pulse duration A, the operating time B (to propagate through the flip-flop and network), and the time between pulses C is shown in Figure 78. For proper operation these times must satisfy the condition

$$A < B < C$$

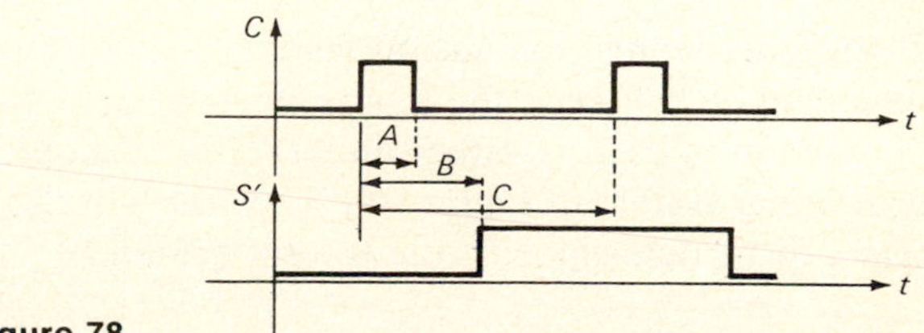

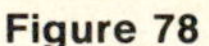
Figure 78

EXERCISE SET 8

1. Implement the binary sequential subtractor using a
   (*a*) T flip-flop
   (*b*) SR flip-flop
   (*c*) JK flip-flop
2. Construct the excitation tables for the following flip-flops:
   (*a*) $Q' = Q \veebar \bar{A}$ (*b*) $Q' = QA \vee B$
   (*c*) $Q' = \bar{A}Q \vee \bar{B}$ (*d*) $Q' = A \veebar B \veebar Q$
3. Construct a JK flip-flop from:
   (*a*) A SR flip-flop
   (*b*) A T flip-flop
   (*c*) A JK flip-flop
4. Construct the MN flip-flop of Figure 73 from a JK flip-flop in two different ways. Also construct a D flip-flop from an MN flip-flop in two different ways.
5. Given a flip-flop having two inputs A, B, and output C described by the equations

$$Q' = A \veebar B \veebar Q \quad \text{and} \quad C = Q$$

use it, along with any other two-valued components, to construct
   (*a*) A toggle flip-flop (by inspection) (*b*) A set-reset flip-flop
   (*c*) A sequential adder (*d*) A ternary counter
6. Construct an algorithm to generate systematically all 196 flip-flops and their corresponding excitation tables. Are all the 196 flip-flops essentially different from an engineering point of view? Why?

Ans.: No, since, for example, $Q' = Q \veebar A$ and $Q' = Q \veebar B$ are the same except for a renaming of inputs A and B

7. Implement the ternary counter using:
   (*a*) T flip-flops
   (*b*) JK flip-flops

6. LARGER INTERCONNECTIONS OF SEQUENTIAL SYSTEMS

Sequential systems may be interconnected in many ways to yield other larger sequential systems. In general, connecting a system having m states to one having n states yields a resulting system with as many as $m * n$ states.

Interconnections without Feedback

If the interconnection has no feedback loops (so that the outputs of one part cannot influence the inputs to that part) the method of finding the equivalent system is rather direct. For example, consider the cascaded sequential system of Figure 79. The states of the required equivalent system can be determined from the set product

$$\{A,B\} \times \{C,D\} = \{AC,AD,BC,BD\}$$

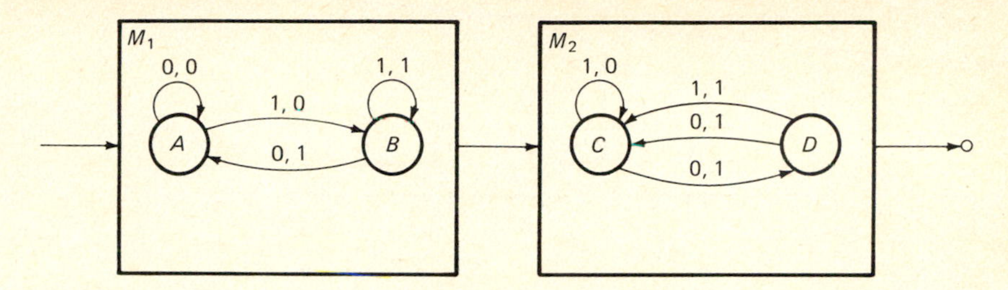

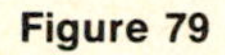
Figure 79

For each state and input we can determine the output and next state as illustrated in Figure 80. Continuing in this way for all possible transitions yields the transition table and state diagram of Figure 81.

Shift Registers. A particularly interesting type of interconnected system is the serial shift register. It consists of n cascaded delay elements (called *cells*), as shown in Figure 82. At each time interval (when the clock is applied) all the digits are shifted 1 unit to the right. The input digit occupies the leftmost cell, and the digit in the rightmost cell is shifted out and lost.

For example, in the given shift register the digits which were applied serially at the previous three time intervals are available at the output all at one time:

$$Y_1(t) = X(t-1) \qquad Y_2(t) = X(t-2) \qquad Y_3(t) = X(t-3)$$

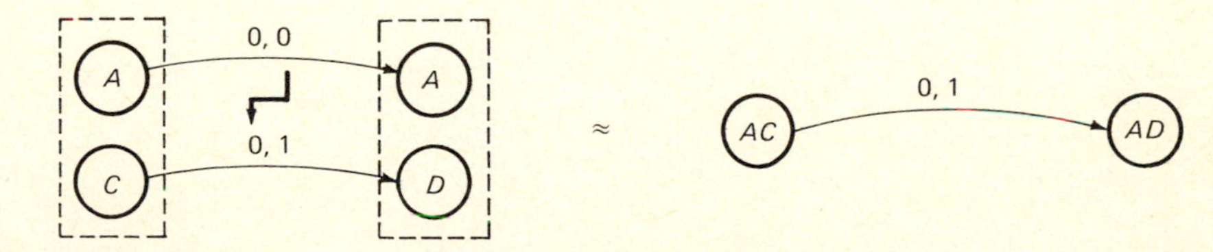

Figure 80

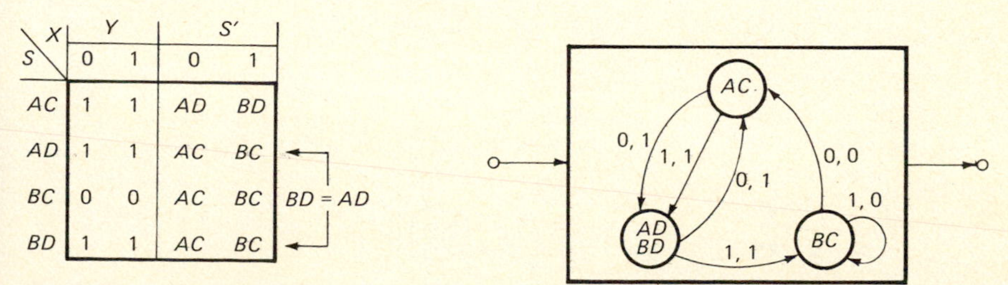

S \ X	Y: 0	Y: 1	S′: 0	S′: 1
AC	1	1	AD	BD
AD	1	1	AC	BC
BC	0	0	AC	BC
BD	1	1	AC	BC

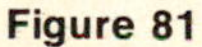
Figure 81

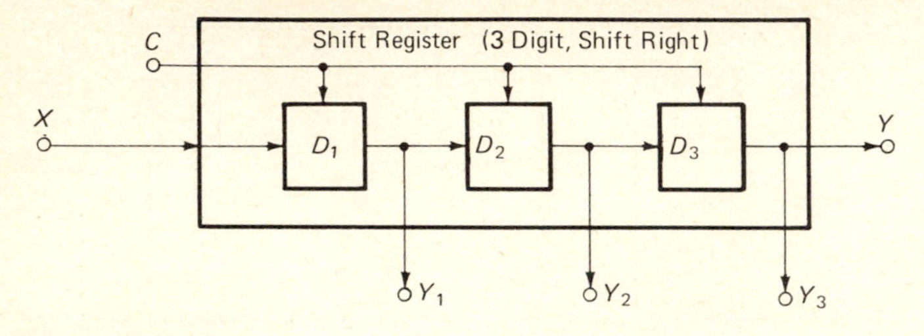

Figure 82

Each transition of this system has the form of Figure 83. The complete state diagram for this *three-cell binary shift-right register* is shown in Figure 84. Such shift registers are the basic components of larger systems such as digital computers and communication systems. In such applications they have many more cells, usually ranging from 8 to 64. In the next chapter shift registers will be used in constructing a computer, and in the Appendix shift registers will be described from an electronics point of view.

A Binary Coded Counter. Another interesting cascade interconnection consists of toggle flip-flops and AND components, as shown in Figure 85. The corresponding state diagram is shown in Figure 86. Notice that when the combination of digits of the state is viewed as a binary number, this number increases by 1 for each input pulse. This system therefore counts the number of pulses applied to it and indicates it as a binary code.

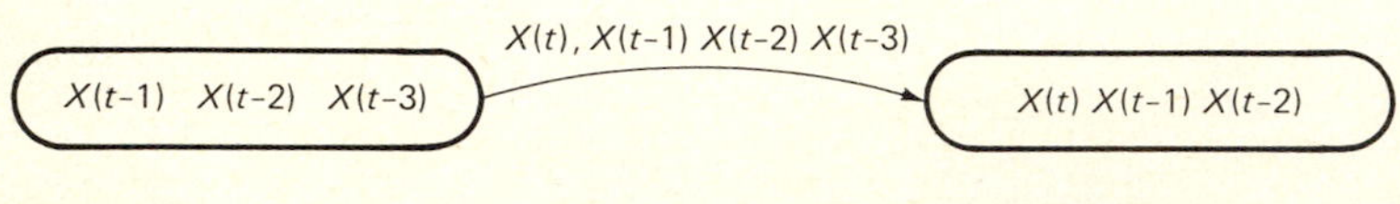

Figure 83

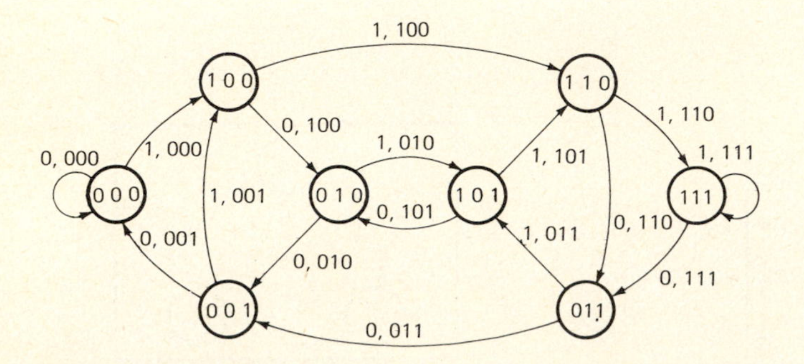

Figure 84

Similarly a three-stage binary counter counts from 0 to 7 as shown in Figure 87. Notice that the outputs are the states and so are not shown. In general a system with n such stages counts from 0 to $2^n - 1$ and then starts again at 0. Thus 20 such stages can count to over 1 million.

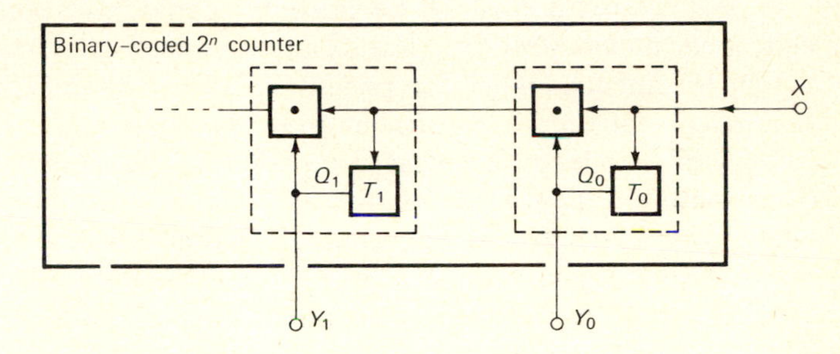

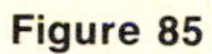
Figure 85

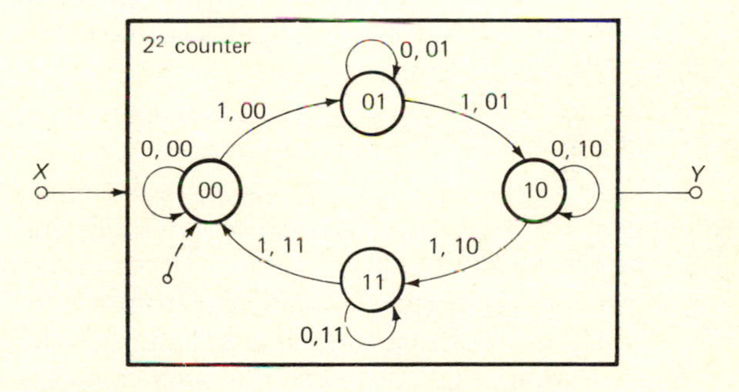

Figure 86

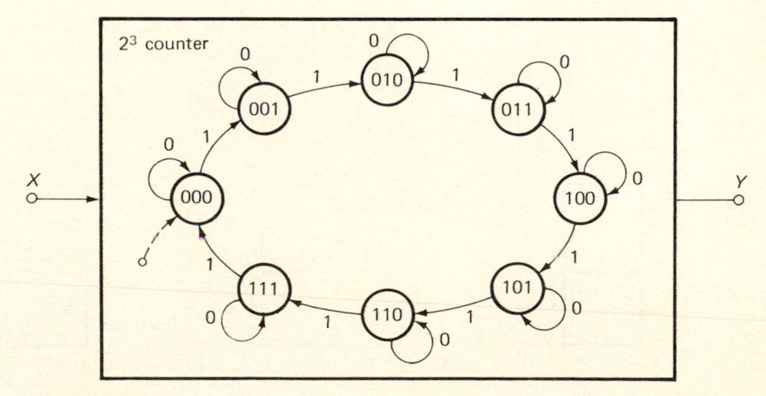

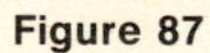
Figure 87

Interconnections with Feedback

Sequential systems may also be interconnected with feedback in a closed loop so that the output of one system can ultimately influence its own input. In such cases at least one of the systems in the loop must be *state-determined*; the output must be a function of the state only and not the input. This essentially introduces some delay in the loop.

If this condition is not satisified, we could have both mathematical and physical inconsistency. For example, consider the interconnection of two instantaneous binary components of Figure 88*a*. (This is essentially the network of Figure 20 without any delay elements.) If the input *A* is fixed at value 0, let us see what value *B* has. If *B* is assumed to have value 0, the output *C* is also 0. Then *C* is NOTted and fed back as value 1, which contradicts the original assumption that *B* was 0. Alternatively, in Figure 88*b*, if *B* is assumed to have value 1, the output *C* has value 1 and the NOT of *C* (which is *B*) has value 0, which is a contradiction again. The value of *B* is indeterminate; it cannot be determined to be 0 or 1.

This mathematical inconsistency also corresponds to a physical problem. For example, a mechanical implementation is given in Figure 88*c*. When the input *A* is released to value 0, the movable T-shaped part of the OR moves to the left whereas the input *B* moves to the right, thus the two parts bind somewhere in between value 0 and 1.

An example of a proper feedback interconnection is shown in Figure 89. In this case both systems are state-determined. The behavior of the combined system is given in the table-of-combinations of Figure 90. Notice the structure of the table. It is convenient for "bookkeeping" since we can proceed systematically from the left to the right. Try it for any row. The resulting state diagram and transition table is shown in Figure 91. Notice that this system is also state-determined.

EXERCISE SET 9

1. Determine the systems which are equivalent to the interconnections of Figure 92.
2. Which of the interconnections of Figure 93 are improper (inconsistent)?

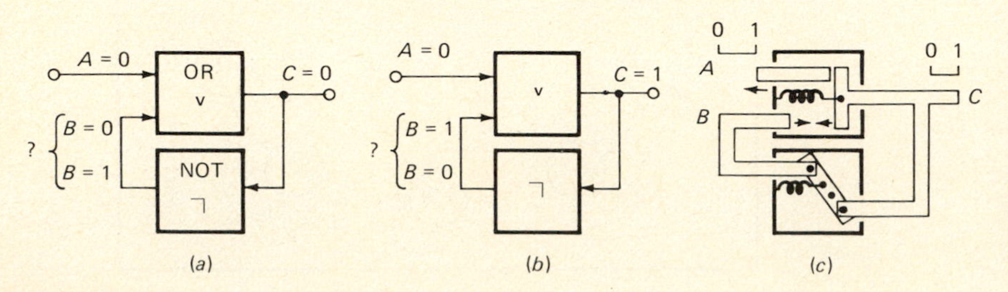

Figure 88

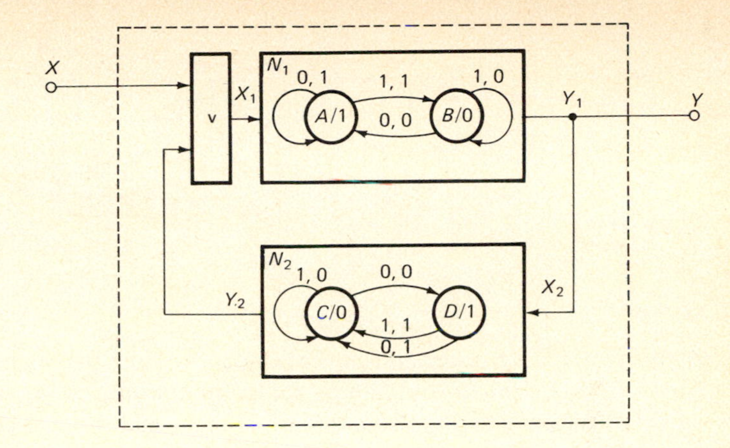

Figure 89

S		X	N_1			N_2			Y	S′	
S_1	S_2	X	X_1	Y_1	S'_1	X_2	Y_2	S'_2	Y	S'_1	S'_2
A	C	0	0	1	A	1	0	C	1	A	C
A	C	1	1	1	B	1	0	C	1	B	C
A	D	0	1	1	B	1	1	C	1	B	C
A	D	1	1	1	B	1	1	C	1	B	C
B	C	0	0	0	A	0	0	D	0	A	D
B	C	1	1	0	B	0	0	D	0	B	D
B	D	0	1	0	B	0	1	C	0	B	C
B	D	1	1	0	B	0	1	C	0	B	C

Figure 90

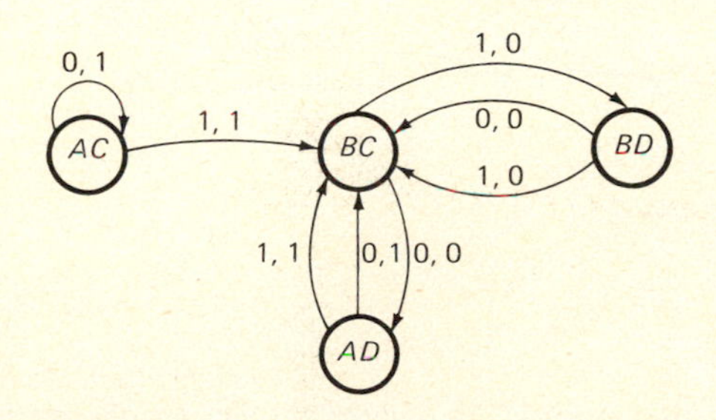

S \ X	Y: 0	Y: 1	S′: 0	S′: 1
AC	1	1	AC	BC
AD	1	1	BC	BC
BC	0	0	AD	BD
BD	0	0	BC	BC

Figure 91

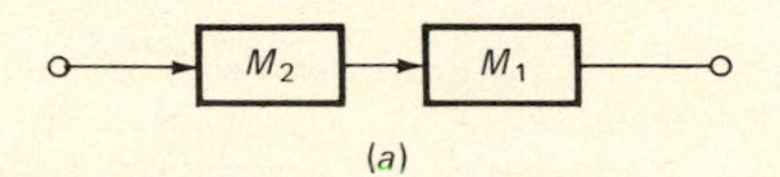

(a)

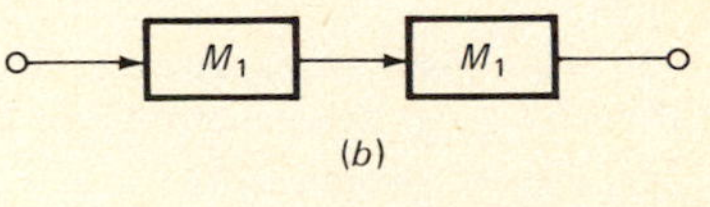

(b)

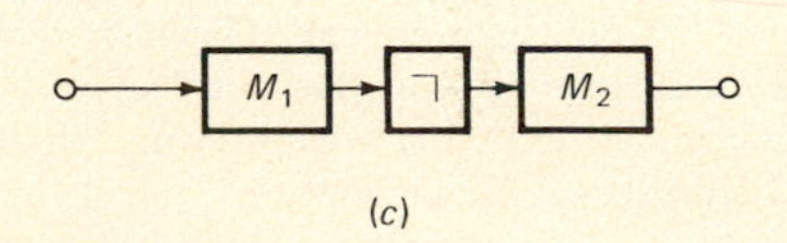

(c)

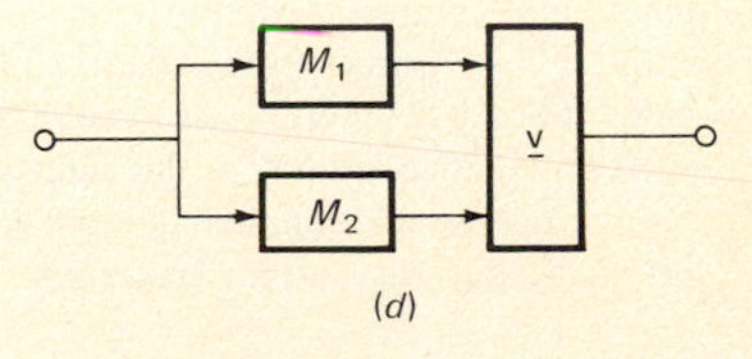

(d)

Figure 92

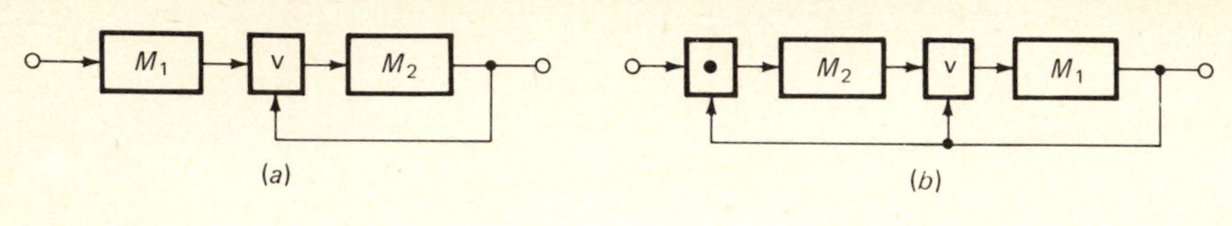

Figure 93

SUMMARY

In this chapter we encountered the concept of time in a digital or discrete way. It was applied to systems for computation, communication, and control. We analyzed the sequential behavior and the feedback structure of such systems. We synthesized such systems first using ideal delays, and then we implemented the systems using more practical flip-flops. But we avoided many timing problems (races, hazards) by concentrating on synchronous, clocked systems. Also we did not optimize the state assignments, but we did optimize the number of states.

PROBLEMS

1. Construct the state diagram of a system having output value 1 if the input sequence contains an odd number of 1s; otherwise the output is 0. How could such a system be useful?

2. Construct the state diagram of a sequential system which has an output value of 1 if the input consists of an odd number of 0s or an even number of 1s (or both). (Assume zero is an even number.)

3. Construct a state diagram for the given NOT element having unit delay.

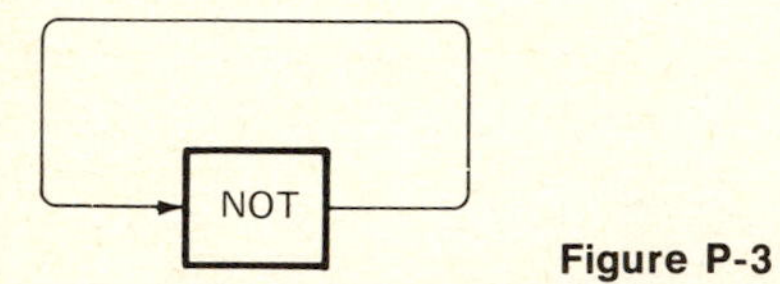

Figure P-3

4. How many sequential machines are possible if the number of states is p, the number of input values is q, and the number of output values is r? *Hint*: For $p = q = r = 7$, this number exceeds 10^{73}, the total number of atoms in all 100 billion galaxies of the visible universe!

5. How many states are necessary to describe a binary sequential system with two delay elements if one delay is two-thirds the length of the other delay?

6. *Fluidic* systems make use of jets of fluids flowing through paths and chambers cut into plastic and ceramic materials. In such systems the jets have a tendency to adhere to the walls. For example, in the figure suppose the jet from the source X is adhering to the left wall. Then it remains on this wall exiting at output C until a jet is applied at input A. This causes the jet to switch to the right wall, remaining there even when the input jet at A is removed. Now a jet applied at input B alone can cause a switch to the left wall again. Model this two-input–two-output sequential system and represent it in two ways.

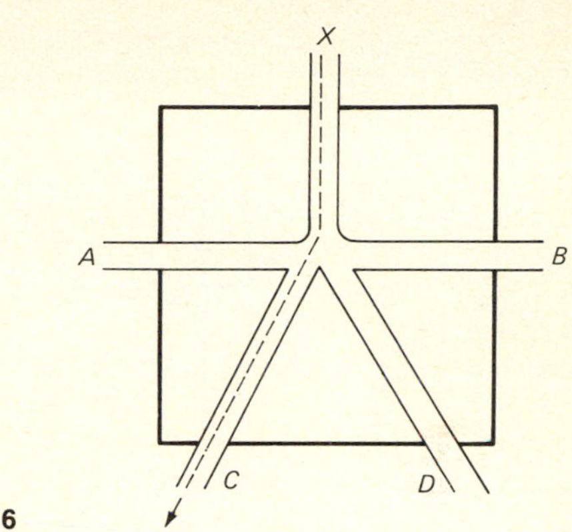

Figure P-6

7. Draw a state diagram describing the following system. A commutator samples the output and input of a system, alternating at each time interval. Draw also a state diagram describing the commutating sampler (or *multiplexer*) alone.

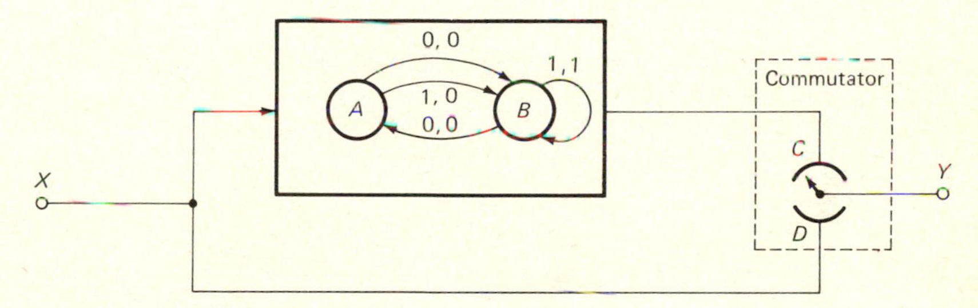

Figure P-7

8. A *neuron* is often modeled as shown. It has inputs which are called *excitatory* (indicated by arrows) and other inputs called *inhibitory* (indicated by bulbs). A neuron is binary, having values called active (firing, value 1) or inactive (value 0). A neuron fires at one time instant $t + 1$ if and only if at the previous instant t the number of active excitatory inputs equals or exceeds the threshold T and no inhibit input is active. Construct the simplest state diagram for the given interconnection of such neurons. (The answer has three states.)

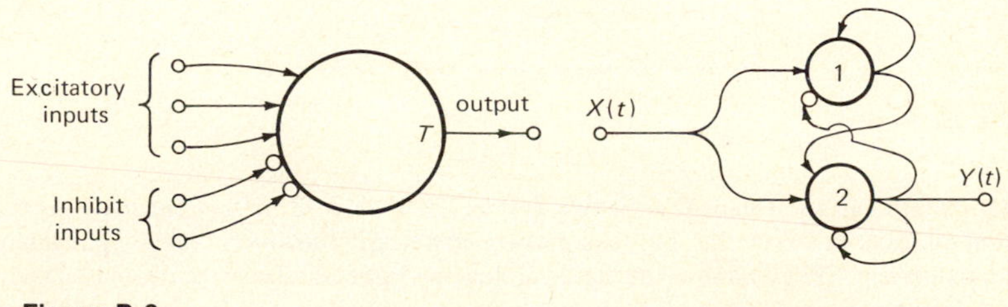

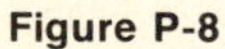
Figure P-8

9. Construct a state diagram corresponding to the given network. Does the network of Figure P-9 suggest any ideas which could be useful in design?

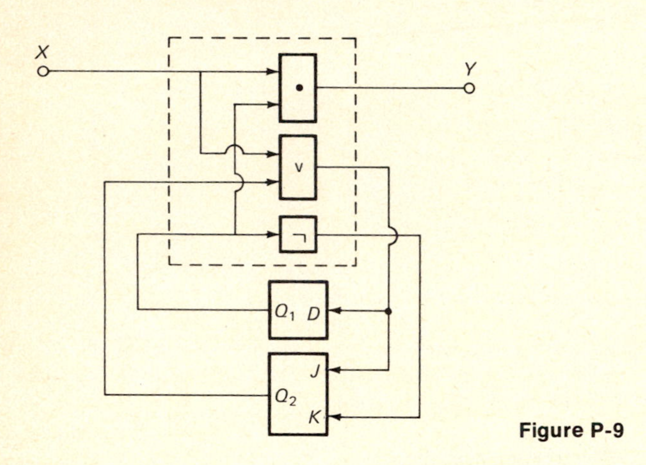

Figure P-9

10. Draw the state diagram of the equivalent sequential system corresponding to the given interconnection of systems.

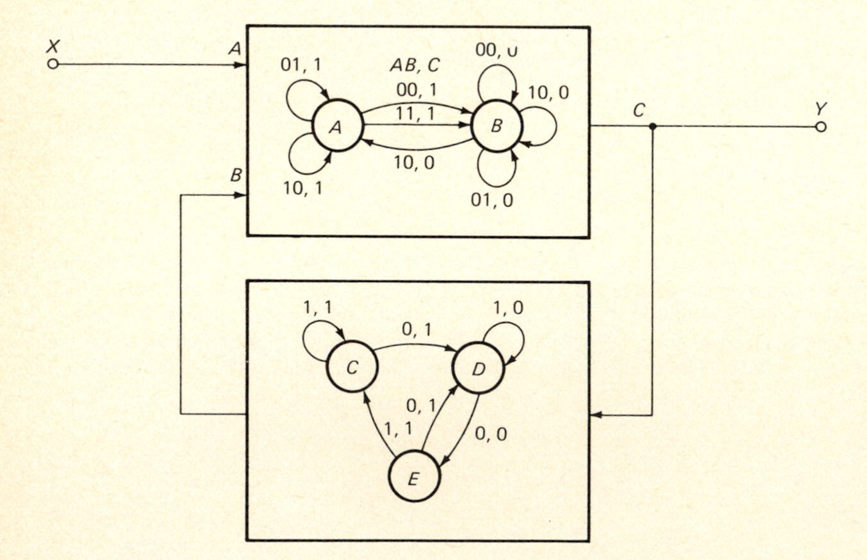

Figure P-10

11. Show the equivalence of the two *systems* of Figure 38 by first constructing the transition tables of both systems and then making up a larger transition table by placing one table on the other. Then optimize the states of this one large system by partitioning, and observe the final partitions.

12. Which of the four common flip-flops should be used to implement the next-state function

$$S' = \bar{S} \vee X$$

if only NORs can be used in the instantaneous network?

13. Using JK flip-flops construct a simple:

(*a*) Binary subtractor

(*b*) BCD error detector (of Figure 50)

(*c*) Control system of Problem 6 in Exercise Set 2

14. Construct and optimize a state diagram for a sequential system which has an output value of 1 for the input sequence 1111:

(*a*) Every time a consecutive four inputs have value 1 as in the example

X 101111111001

Y 000001111000

(*b*) If the sequence is broken up into blocks of four digits (called *bytes*) and the four 1s appear within one block as in the example

X 1011 1111 1001

Y 0000 0001 0000

15. Design a simple system to yield an output of 1 at each third time interval only if the input sequences 000, 011, 100, 110 occur in a block. Use only NORs and JK flip-flops.

16. In a communication system messages are sent in blocks of digits of constant length 3. The first two digits represent the "information," and the third digit is a parity check digit which is chosen so that an even number of 1s is always transmitted. Construct the state diagram of a receiver to decode such messages, indicating whether an error has occurred. Then synthesize such a decoder intuitively.

17. Design in detail the coin-operated dispenser considered in Section 2.

18. Prove that a sequential system (having a fixed finite number of states) cannot multiply any two arbitrarily long numbers. (*Hint*: Multiply 2^n by 2^n.)

19. *Project*. Construct an algorithm in flow-diagram form to optimize the number of states of any sequential system by the method of partitioning.

20. *Project*. Attempt to model some practical system (from areas such as medicine, biology, economics, sociology, or business) as a sequential system. For example, in geology, the agents of temperature, pressure, and weathering transform the state of rocks (igneous, metamorphic, sedimentary, etc). A state diagram could show that igneous rock under heat and pressure becomes metamorphic rock, and metamorphic rock under weathering becomes sediments, which under pressure becomes sedimentary rock.

Chapter 6

Stochastic Systems

Dynamic Nondeterministic Systems

Stochastic systems are probabilistic systems involving a variable which is usually interpreted as time. They can also be thought of as sequential systems involving probabilities. In either view they involve both probabilities and time. Stochastic systems are often called Markovian systems and also dynamic probabilistic systems.

1. MODELING, IDENTIFICATION, AND REPRESENTATION

In this chapter not only are the probabilistic systems discrete as previously, but time is also discrete. The systems are sampled at discrete stages, thus yielding sequences of random events. Many practical systems can be modeled in this way. As usual, we will start by modeling the relay.

Modeling a Relay with Error Bursts

A relay can be modeled as a stochastic system in a number of different ways. First we will model the error-burst behavior; other behaviors will be modeled later. In relays, as well as in many other systems, errors often tend to occur in bursts. Thus if there is an error in one time interval, the probability of error in the next time interval increases. A typical sequence from a system of this type could be

$$NNNN\underline{EEE}NNNNNNNNNNNN\underline{EEEE}NNNNNNN\underline{EEE}NNNNN\underline{E}$$
$$NN\underline{EEE}NNN\underline{EE}N\underline{E}\ N$$

where E denotes an error and N denotes no error. The bursts of error are underlined. These modes of behavior (such as E and N) are known as states.

Identification

The bursts of no error are longer than the bursts of error; this indicates that the probability of changing from the error state to the nonerror state is larger than the probability of changing from the nonerror state to the error state. These conditional probabilities of change, known as the state *transition probabilities,* can be computed from the data sequence as follows:

$$p[S(t+1)=N \mid S(t)=E] = \frac{\text{no. of times that } N \text{ follows } E}{\text{total no. of times } E \text{ occurs}} = \frac{n[EN]}{n[E]} = \frac{7}{17} \approx 0.4$$

$$p[S(t+1)=E \mid S(t)=N] = \frac{n[NE]}{n[N]} = \frac{7}{34} \approx 0.2$$

Note that the change to the best state (of no error) is twice as probable as the change to the worst state (of error).

In general the transition probabilities from state S_i to state S_j are described by any of the following abbreviations:

$$p[S(t+1) = S_j \mid S(t) = S_i] = p[S_j(t+1) \mid S_i(t)] = p[S_j' \mid S_i] = p_{ij}$$

It is important to realize that these probabilities are not probabilities of error but probabilities of *changing* between states of error and nonerror. The probability of error of this system could be computed from the sequences as

$$p[\text{Error}] = \frac{\text{no. of times error occurs}}{\text{total length of sequence}} = \frac{17}{51} = 0.33$$

Later we will see how the transition probabilities can also be used to find the probability of error.

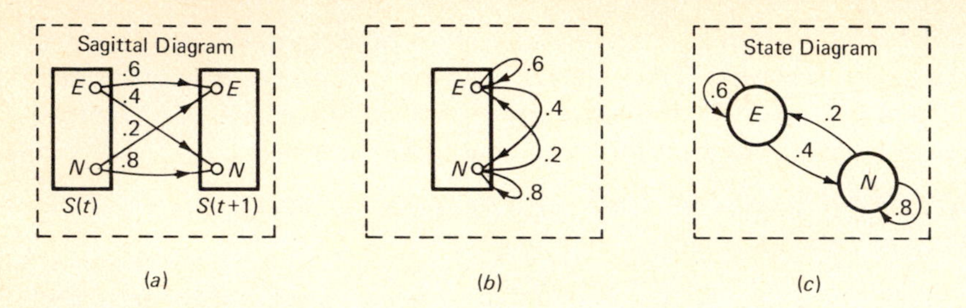

Figure 1

Representation of Stochastic Systems

The stochastic model can be represented by a *sagittal diagram* as shown in Figure 1*a*. However, the time behavior is intuitively more apparent from another state-diagrammatic representation of Figure 1*c*. The *state diagram* can be obtained from the sagittal diagram through the intermediate diagram of Figure 1*b* by recognizing that the input set and output set are always identical. Thus only one set is necessary, with the arrows drawn between states within the same set. The states in the state diagram are redrawn for convenience as circles rather than points.

A third representation is that of a *transition matrix* having as many rows and columns as there are states. The element p_{ij} in the ith row and jth column represents the probabilities of transition from state S_i to state S_j. The transition matrix for the error-burst relay is shown in Figure 2. Notice that the probabilities in each row sum to 1.

Examples of Stochastic Systems

Transportation System. Another example of a stochastic system is the taxicab system shown in Figure 3. The states represent three relevant areas of a city near transportation junctions:

S_1-area at or near the bus terminal, including the center of town
S_2-area at or near the train depot, including hotels nearby
S_3-area near the airport

$p[S_j(t+1) | S_i(t)]$

$i \backslash j$	E	N
E	.6	.4
N	.2	.8

Figure 2

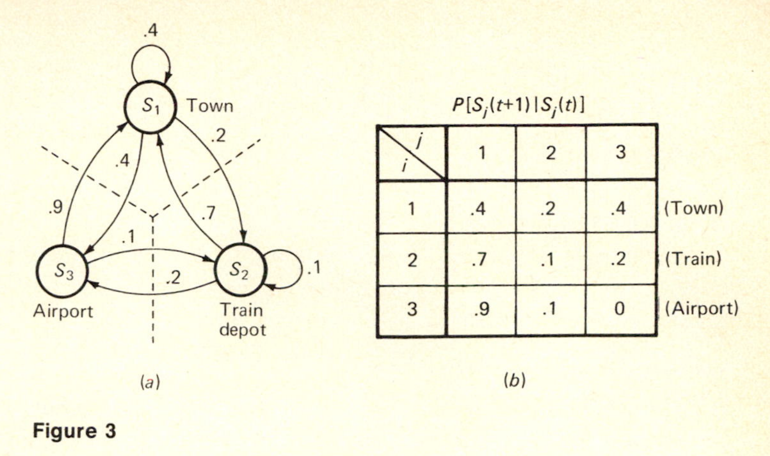

$P[S_j(t+1) \mid S_i(t)]$

i \ j	1	2	3	
1	.4	.2	.4	(Town)
2	.7	.1	.2	(Train)
3	.9	.1	0	(Airport)

(b)

Figure 3

The state diagram illustrates, for example, that if a cab is at the airport, its next transition is into town with probability 0.9 and to the train depot with probability 0.1.

Replacement System. The replacement of parts in a machine can also be modeled as a stochastic system. For example, consider a gear train, in which individual gears are often replaced. When a gear is replaced the process of breaking in tends to wear away the gears it meshes with, thus increasing their probability of being the next gears to be replaced. The probabilistic dependence could be found experimentally and described by a transition matrix, as in Figure 4. The probability p_{ij} is the probability of replacing gear G_j given that gear G_i was replaced last.

Queues. The waiting lines seen in barber shops, ticket offices, or breadlines can also be modeled as stochastic systems. The states indicate the number of individuals waiting. For example, the state diagram of Figure 5 could describe the queue in a barber shop in which there are only

Gear Train

G_1 G_2 G_3

(a)

i \ j	1	2	3
1	.1	.7	.2
2	.4	.2	.4
3	.3	.5	.2

(b)

Figure 4

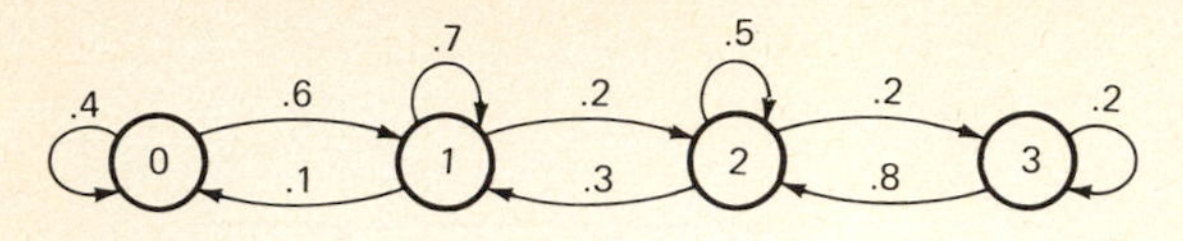

Figure 5

three chairs to wait in; the length of queue is limited to 3. Note also that the transitions are limited from one state to its next higher or lower state only. If the time interval between samples is increased, there could be a greater transition change.

Modeling of Stochastic Systems

All the above-mentioned systems, as well as many others, have some characteristics in common. We will now attempt to abstract some of these characteristics, which we will use later to arrive at a more general model.

1. The system is observed or sampled at discrete intervals of time called *stages*. The stages are labeled with integer values 1, 2, 3, ... to indicate the order of occurrence of events. The time intervals between stages need not be equal, although they usually are.
2. At each stage there is a finite number of possible outcomes, which we have called *states*. The system can be in one and only one state at a time. In other words, the states are collectively exhaustive and mutually exclusive (like basic events). Later we will have a more general concept of this extremely important idea of state.
3. The transition between states at two different stages is of the probabilistic form

$$p[S(t+1) \mid S(t), S(t-1), \ldots, S(1), S(0)] = p[S(t+1) \mid S(t)]$$

In other words, the state at a given time $S(t+1)$ is dependent only on the immediately preceding state $S(t)$; it is independent of any earlier states. Furthermore, this probabilistic dependence is the same at all stages; the transition probabilities are constant, independent of time.

Even though these characteristics seem quite general, they are in fact restrictive and will be extended later.

EXERCISE SET 1

1. *Weather model.* Draw the state diagram corresponding to the following description of the weather in some city. If on one day the weather is good, then on the next day the probability of good weather is 5 times greater than the probability of bad weather. If on one day the weather is bad, then on the next day the probability of bad weather is twice as large as the probability of good weather.

2. *Coin toss.* Make an intuitive guess which of the following two sequences more closely resembles the experiment of flipping a coin for a sequence of 50 trials.

(*a*) HHTTHHHTTHTHTHHTTTHTHTTHTHHTHTHTHHTTTTTHTHTH HTHTHT

(*b*) HTTTTHTTHTTHHHHTHHTTHHHHHHTTTHTHHTHTTTHHTHTH TTTTTT

Verify your intuitive guess by drawing and comparing state diagrams. What would be the state diagram for an ideal coin?

3. *Linguistic system.* A simple language consists of three letters A, B, C and a space. Letter C never follows itself, but it is followed by A with probability 0.7. Letter B is never followed by C but follows itself with probability 0.1. Letter A follows itself with probability 0.4 and is followed by all other symbols with equal probability. Any letter is followed by a space with probability 0.2; after a space the letter B is three times as probable as the other two letters. Construct the state diagram describing this language.

Another Example: Time-dependent Activity

Consider a system such as a telephone exchange, highway, or service depot whose activity varies with time; 100 samples are taken of this system, each sample lasting for four consecutive time intervals (stages). The results are tabulated on the table-of-combination and set diagram of Figure 6. There event $A_i = 1$ indicates that during the ith time interval the system is active, whereas $A_i = 0$ (or $\bar{A}_i = 1$) indicates that the system is inactive.

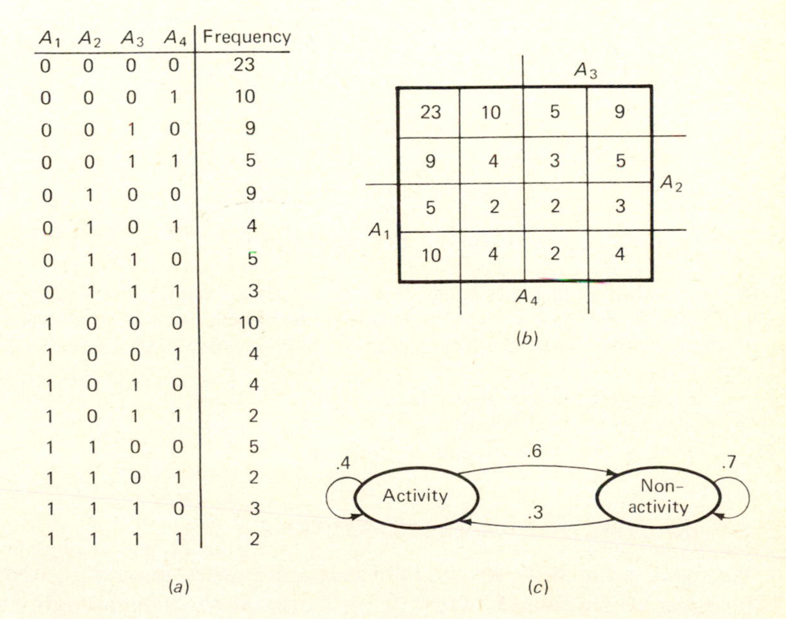

A_1	A_2	A_3	A_4	Frequency
0	0	0	0	23
0	0	0	1	10
0	0	1	0	9
0	0	1	1	5
0	1	0	0	9
0	1	0	1	4
0	1	1	0	5
0	1	1	1	3
1	0	0	0	10
1	0	0	1	4
1	0	1	0	4
1	0	1	1	2
1	1	0	0	5
1	1	0	1	2
1	1	1	0	3
1	1	1	1	2

(*a*)

Figure 6

The probability that the system is inactive during the interval $i + 1$, given that it is active in the ith interval, is

$$p[A_{i+1} = 0 \mid A_i = 1] = \frac{p[(A_{i+1} = 0) \text{ and } (A_i = 1)]}{p[A_i = 1]}$$

This could be evaluated for $i = 1$, 2, and 3 to obtain

$$p[A_2 = 0 \mid A_1 = 1] = \tfrac{20}{32} = 0.62$$
$$p[A_3 = 0 \mid A_2 = 1] = \tfrac{20}{33} = 0.60$$
$$p[A_4 = 0 \mid A_3 = 1] = \tfrac{21}{33} = 0.63$$

Notice that this probability of change from activity to inactivity is approximately constant at a value 0.6 for all intervals. Similarly the probability of change from inactivity to activity is constant at a value of 0.3. This system can therefore be modeled by the stochastic system of Figure 6*c*.

EXERCISE SET 2

1. Find the probability of activity for each time interval; i.e., compute $p[E_i = 1]$ for $i = 1$, 2, 3, 4.
2. Find the probability that the system is active for a total of x intervals of the 4 intervals, where $x = 0$, 1,2,3,4. Sketch the pdf.
3. Find the probability of activity in the ith interval, given that the system is active in interval $i + 1$; that is, compute $p[A_i = 1 \mid A_{i+1} = 1]$ for $i = 1$, 2,3. Comment on the physical as well as mathematical aspects of this problem.
4. Determine the probabilities below and comment:

$$p[(A_4 = 1) \mid (A_3 = 0)]$$
$$p[(A_4 = 1) \mid (A_3 = 0) \text{ and } (A_2 = 1)]$$
$$p[(A_4 = 1) \mid (A_3 = 0) \text{ and } (A_2 = 0)]$$
$$p[(A_4 = 1) \mid (A_3 = 0) \text{ and } (A_2 = 1) \text{ and } (A_1 = 1)]$$

5. The table of Figure 7 describes the smog on successive days, $S(t)$ and $S(t + 1)$, and the dependence on the temperature $T(t)$ (where the high level is indicated by 1 and a lower level by 0). Construct a state diagram relating the smog level on one day to the level on the succeeding day:
   (*a*) When the temperature is low.
   (*b*) When the temperature is high.
   (*c*) When the temperature is not specified.

2. ANALYSIS OF STOCHASTIC SYSTEMS

Analysis is the study of the behavior and properties of a system and can be done in many ways. We will use three approaches: an indirect method

$T(t)$	$S(t)$	$S(t+1)$	Probability
0	0	0	.10
0	0	1	.05
0	1	0	.05
0	1	1	.10
1	0	0	.15
1	0	1	.10
1	1	0	.10
1	1	1	.35

Figure 7

of iteration, a more direct method of inversion, and another method of simulation.

But before discussing these methods, we must derive a set of equations, called transition equations, to describe any stochastic system.

Transition Equations

Given a model of a stochastic system, it is quite easy to write the corresponding transition equations. The taxicab system will be used to illustrate this method. Suppose that at one instant, say t, there are n_1 cabs in state S_1, and n_2 cabs in state S_2, · · · , and in general n_i cabs in state S_i. Then at the next instant, $t + 1$, the number of cabs in state S_j can be determined by considering the incoming transitions to state S_j. For example, Figure 8 emphasizes a portion of Figure 3. It indicates that 40 percent of the cabs in state S_1 return to state S_1, 70 percent of those in S_2 go to S_1, and 90 percent of those in S_3 also transfer to S_1, yielding the equation

$$n_1(t + 1) = 0.4n_1(t) + 0.7n_2(t) + 0.9n_3(t)$$

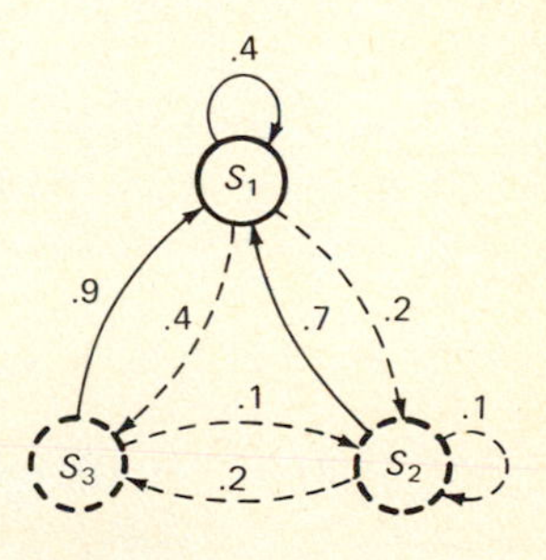

Figure 8

Doing this for all states yields the recurrence relations

$$n_1(t+1) = 0.4n_1(t) + 0.7n_2(t) + 0.9n_3(t)$$
$$n_2(t+1) = 0.2n_1(t) + 0.1n_2(t) + 0.1n_3(t)$$
$$n_3(t+1) = 0.4n_1(t) + 0.2n_2(t) + \quad 0n_3(t)$$

Such recurrence relations are called *transition equations* and have the form

$$n_j(t+1) = \sum_i p_{ij} n_i(t)$$

To make this result more general (and unfortunately more confusing) the number n_i of cabs in each state can be replaced by the probability P_i of a cab's being in that state. This leads to the general form of the transition equations

$$P_j(t+1) = \sum_i p_{ij} P_i(t)$$

Analysis by Iteration

The transition equations can be used to determine the distribution of cabs at any time if the distribution at the immediately preceding time is known. Thus if the initial distribution (at time $t = 0$) is given, these equations can be iterated, or used successively, to generate the distributions at all later times. For example, consider an initial distribution in which all 100 cabs start in state S_1 (town), i.e.,

$$n_1(0) = 100 \qquad n_2(0) = 0 \qquad n_3(0) = 0$$

The substitution of these values into the transition equations determines the distribution at the next interval, $t + 1$:

$$n_1(1) = 0.4n_1(0) + 0.7n_2(0) + 0.9n_3(0) = 0.4(100) = 40$$
$$n_2(1) = 0.2n_1(0) + 0.1n_2(0) + 0.1n_3(0) = 0.2(100) = 20$$
$$n_3(1) = 0.4n_1(0) + 0.2n_2(0) + \quad 0n_3(0) = 0.4(100) = 40$$

After another time interval has elapsed, the distribution is

$$n_1(2) = 0.4(40) + 0.7(20) + 0.9(40) = 66$$
$$n_2(2) = 0.2(40) + 0.1(20) + 0.1(40) = 14$$
$$n_3(2) = 0.4(40) + 0.2(20) + \quad 0(40) = 20$$

After still other time intervals $t = 3, 4,5$

$$n_1(3) = 54.2 \qquad n_1(4) = 59.6 \qquad n_1(5) = 57.1$$
$$n_2(3) = 16.6 \qquad n_2(4) = 15.4 \qquad n_2(5) = 16.0$$
$$n_3(3) = 29.2 \qquad n_3(4) = 25.0 \qquad n_3(5) = 26.9$$

Notice that the numbers are not integers, thus indicating that these numbers represent the average number of cabs in a state. The results of this continued iterative process can be plotted as shown in Figure 9.

Observation of this behavior indicates a highly significant phenomenon. After some time the average number of cabs in each state tends to remain constant. This phenomenon will be referred to as an *equilibrium condition,* a condition of no change. An interpretation of this phenomenon is that at any instant (sufficiently removed from the initial instant) an average of about 58 percent of the cabs are in state S_1 (town), 16 percent are in state S_2 (train depot), and 26 percent are in state S_3 (airport). In short, from the behavior of individual cabs we have determined the behavior of the entire system of cabs.

It is important to realize that neither the transition equations nor the computations from the equations indicated the presence of an equilibrium. The computations yielded a set of numbers, and it was observed that these numbers tended to an equilibrium. This illustrates again the significance of the process of observation or induction rather than deduction.

The equilibrium distributions when expressed as probabilities rather than numbers or percentages are known as *equilibrium probabilities.*

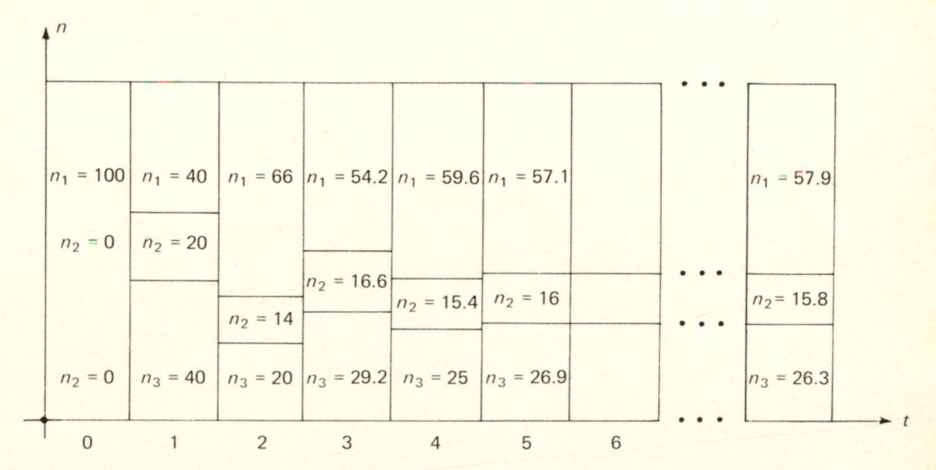

Figure 9

Thus with any stochastic system there are two types of probabilities associated with opposing tendencies: (1) transition probabilities describe change, and (2) equilibrium probabilities describe constancy.

Matrix Interpretation. This process of iteration may also be viewed as an operation on matrices by simply rewriting the transition equations in matrix form as

$$[n_1(t+1), n_2(t+1), n_3(t+1)] = [n_1(t), n_2(t), n_3(t)] \times \begin{bmatrix} 0.4 & 0.2 & 0.4 \\ 0.7 & 0.1 & 0.2 \\ 0.9 & 0.1 & 0.0 \end{bmatrix}$$

Also n iterations correspond to n multiplications of the transition matrix. Given the initial distribution $[p_1(0), p_2(0), \ldots, p_n(0)]$ and the transition matrix T, the final distribution after d stages can be determined from the matrix product

$$[p_1(d), p_2(d), \ldots, p_n(d)] = [p_1(0), p_2(0), \ldots, p_n(0)]T^d$$

EXERCISE SET 3

1. Find the equilibrium probability (of error) for the original error-burst relay using this method of iteration. Compare this to the value determined from the sequence of data.

2. Find, by observation alone, the equilibrium probabilities corresponding to the stochastic systems of Figure 10. All arrows shown have nonzero probabilities.

3. Construct a flow diagram describing the computation of the probability distribution after a duration of d stages of a stochastic system described by a given transition matrix, starting in a given initial distribution.

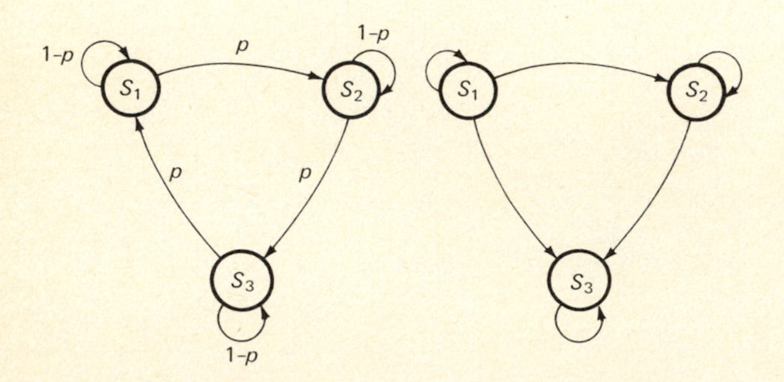

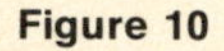

Figure 10

Analysis by Inversion

The previous method of iteration is an indirect method since it takes some time to converge. A more direct method proceeds by recognizing that at equilibrium there is no change in distribution between two adjacent time intervals. For example, in the taxicab system we can write

$$n_1(t+1) = n_1(t)$$

$$n_2(t+1) = n_2(t)$$

$$n_3(t+1) = n_3(t)$$

Substituting these equations into the previous transition equations and eliminating any reference to time yields the simpler equations

$$n_1 = 0.4n_1 + 0.7n_2 + 0.9n_3$$

$$n_2 = 0.2n_1 + 0.1n_2 + 0.1n_3$$

$$n_3 = 0.4n_1 + 0.2n_2 + \quad 0n_3$$

We could attempt to solve these three equations in many ways. However, all methods will fail to give a solution.

The reason for the difficulty is essentially that more information is required (or more formally, the equations are linearly dependent). After some searching, it will be observed that nowhere in the equations is there mention of the total number of cabs in the system. This is an omission which certainly should affect the solution. The additional equation

$$n_1 + n_2 + n_3 = 100$$

along with any two of the previous equations can now be solved by any method (such as substitution, etc.).

After some computation, the solution for the equilibrium distribution is

$$n_1 = 57.9$$

$$n_2 = 15.8$$

$$n_3 = 26.3$$

This agrees with the equilibrium distribution obtained by the previous iterative computation. Notice that no use was made of any initial distribution, thus suggesting that the equilibrium distribution is independent of the initial distribution.

In the general case the transition equations have equilibrium probabilities P_i rather than numbers n_i, and the required additional equation is simply the property of normality

$$P_1 + P_2 + P_3 \cdots + P_n = 1$$

Comparison of Iteration and Inversion Methods Although these methods of iteration and inversion provide the same values for the equilibrium probabilities, they differ in many ways. The method of iteration is concerned primarily with the initial step-by-step behavior and so would be most useful in indicating such things as the time required to achieve equilibrium. The method of inversion, on the other hand, while computationally more direct, is concerned with only the final behavior. We will see that each method has its proper place.

Example For a two-state stochastic system as given in Figure 11 determine the equilibrium probabilities P_i and P_j.

Solution For the state S_i we can write

$$P_i(t+1) = [1 - p_{ij}]P_i(t) + p_{ji}P_j(t)$$

which at equilibrium can be written in abbreviated form as

$$P_i = [1 - p_{ij}]P_i + p_{ji}P_j$$

The equation at state S_j is written similarly, but it is exactly the same as the above equation (try it). Finally, realizing that the equilibrium probabilities sum to 1, we obtain the second equation

$$P_i + P_j = 1$$

Substituting for P_j from the last equation into the first equation leads to

$$P_i = [1 - p_{ij}]P_i + p_{ji}[1 - P_i]$$

which can be solved for the equilibrium probabilities

$$P_i = \frac{P_{ij}}{P_{ij} + P_{ji}} \qquad P_j = \frac{P_{ij}}{P_{ij} + P_{ji}}$$

Notice the convenience of this computation; we simply take a ratio of the two probabilities between the two states. Unfortunately this is not a general result but holds only for systems having two states.

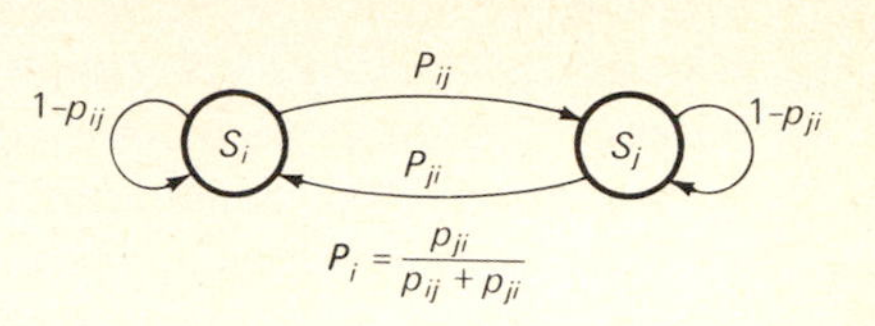

Figure 11

EXERCISE SET 4

1. Use the result of Figure 11 to check the equilibrium probability of
   (*a*) The error-burst relay of Figure 1
   (*b*) The time-dependent activity of Figure 6*c*
   (*c*) The weather system of Problem 1, Exercise Set 1

2. For a two-state system described by

$$p_{01} = 1 \qquad \text{and} \qquad p_{10} = 1 \qquad \text{and} \qquad p_1(0) = 0.8$$

find the equilibrium probabilities by inversion and iteration methods. Comment!

3. The gear-train system of Figure 4 is to be used on a manned space ship. If only 50 gears can be taken aboard, how many of each type of gear should be taken?

4. In the queueing problem of Figure 5, find the average number of people waiting.

Ans.: $\frac{17}{21} \approx 1.4$

5. It is often computationally convenient to set one of the equilibrium probabilities at a constant value, say 1, and then to solve for the other equilibrium probabilities. Then all these probabilities are normalized to get their sum equal to 1. Try this method on the error-burst relay, the taxicab system, the queueing system, and the gear train.

Analysis by Simulation (Monte Carlo Method)

Another method of studying a system is to *simulate* it by constructing a copy, or prototype, which behaves like the system. It is one of the simplest methods of analysis and often works even when most other methods fail because of mathematical difficulties.

Random Numbers. Simulation is usually accomplished with the aid of a table of random numbers. This table is simply an array of the decimal digits 0, 1, 2, 3, ..., 9 each digit occurring equiprobably and independent of any nearby digits. A short table of 250 random digits follows in Figure 12. The spaces in the table are for convenience of application. You may think of these numbers as obtained by selecting balls, with replacement,

Table of Random Numbers

89731	85785	36220	52879	04604	89918	90438	99982	58187	51539
72618	80754	21291	96433	09214	24449	03948	03856	81590	24202
54724	70817	94617	57487	76728	00628	92814	83811	54995	19334
11192	77618	58308	52561	96864	81530	13638	79720	52571	40507
08901	04808	25832	51404	68377	41388	94935	40894	29304	87665

Figure 12

from an urn containing 10 physically identical balls, one corresponding to each digit. The digits themselves are not significant; only their distribution is important. We could equally well use 10 letters, colors, or other symbols, but numbers are convenient.

Such a table can be used to simulate the flipping of a coin by taking a sequence of these digits and transforming it by the following rule:

If the random number is 0, 1, 2, 3, or 4, replace it with H (head).
If the random number is 5, 6, 7, 8, or 9, replace it with T (tail).

A sample coin toss corresponding to the first row of random numbers is

8 9 7 3 1	8 5 7 8 5	3 6 2 2 0	5 2 8 7 9	0 4 6 0 4
TTTHH	TTTTT	HTHHH	THTTT	HHTHH

Of course a longer sequence is necessary to ensure that the relative frequency has reached a stable value. Notice that a head could correspond to any other five digits (such as 0, 2, 4, 6, 7) since they will also come up half the time.

The simulation of a stochastic system is similar but slightly more complex because there is a set of rules for each state. For example, the error-burst relay of Figure 13*a* could be simulated using the following rules:

If state is *E* and random number is 0, 1, 2, 3, or 4, the next state is *N*.
If state is *E* and random number is 5, 6, 7, 8, or 9, the next state is *E*.
If state is *N* and random number is 0 or 1, the next state is *E*.
If state is *N* and random number is 2, 3, ..., 9, the next state is *N*.

These rules are summarized on the diagram of Fig. 13*b*.

We could start the random numbers where we previously left off (in the middle of the first row) to obtain the sequence:

8 9 9 1 8	9 0 4 3 8	9 9 9 8 2	5 8 1 8 7	5 1 5 3 9
NNNNE	*EENNN*	*NNNNN*	*NNNEE*	*EENNN*
7 2 6 1 8	8 0 7 5 4	2 1 2 9 1	9 6 4 3 3	0 9 2 1 4
NNNNE	*EENNN*	*NNENN*	*EEENN*	*NEENEN*

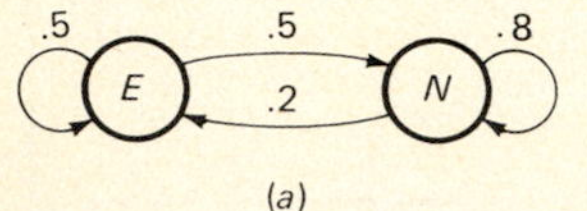

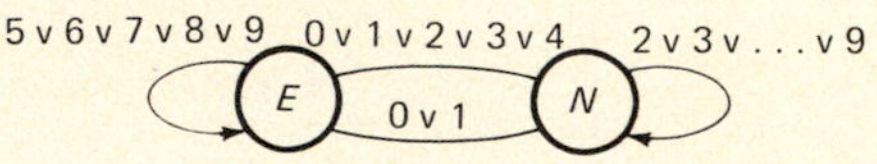

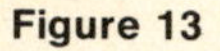
Figure 13

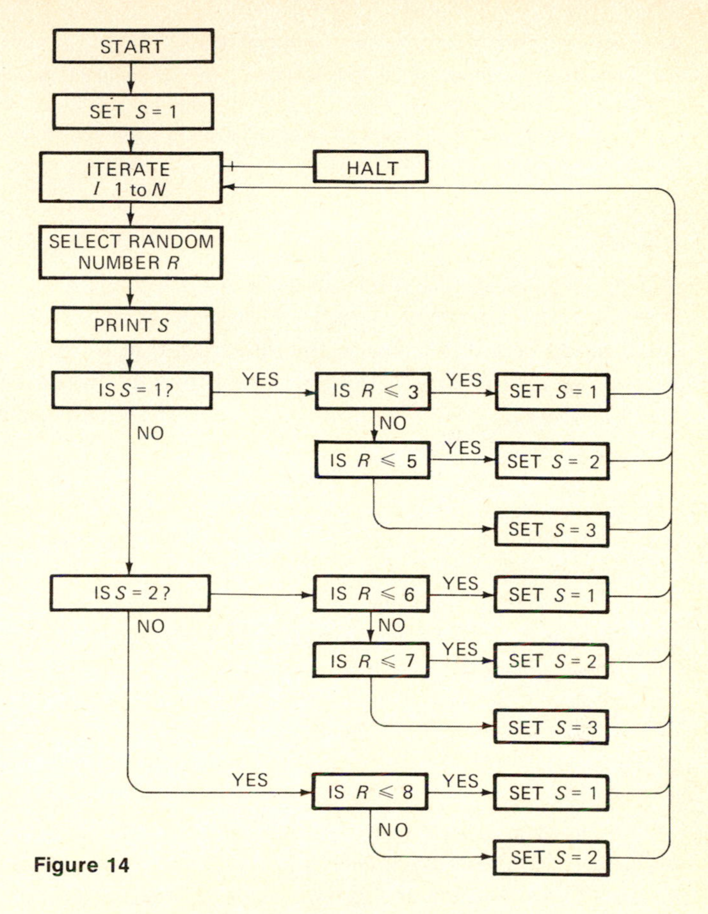

Figure 14

Notice that this is the sequence we started with in this chapter. Notice also that from this sample sequence we modeled the relay in Figure 1, which is not identical to Figure 13. This is a result of the short sequence of 50 observations. For long sequences digital computers are convenient. An algorithm for simulating the taxicab system is shown in Figure 14.

EXERCISE SET 5

1. Simulate a thick coin having probabilities

$$p\,[\text{Edge}] = 0.10 \qquad p\,[\text{Head}] = 0.44 \qquad p\,[\text{Tail}] = 0.46$$

by considering the random digits two at a time making up the numbers 00 to 99.

2. How could you simulate the throwing of one die in a simple way?

3. Simulate in two ways the throwing of two dice.

4. Construct an algorithm in flow-diagram form to simulate the queueing system of Figure 5.

5. Using the algorithm of Figure 14, with initial state 1, simulate five taxicabs, each for 20 stages or transitions. Compare.

Ergodicity

It is apparent that various methods of determining an equilibrium distribution may yield the same result. However, it may not be apparent that there could be a difference in the interpretation of these results. One interpretation is that after a long period of time the portion P_i of this time is spent in state S_i. Another interpretation is that at any instant a portion P_i of cabs is in state S_i.

These interpretations correspond to two ways of considering complex systems: (1) *microscopically,* concentrating on individual elements, or (2) *macroscopically*, concentrating on ensembles of elements. The first interpretation (microscopic) corresponds to observing one typical cab for a long time and determining that it spends a portion P_i of its time in state S_i. The second interpretation (macroscopic) corresponds to observing (monitoring on radio perhaps) all the cabs at one instant and determining that portion P_i of all the cabs are in state S_i. The resulting values of P_i need not be the same for both cases. The fact that the values for these two interpretations are the same is a consequence of the property of ergodicity.

A system is said to be *ergodic* (or statistically homogeneous) if the behavior of one sequence over a long period of time is the same as the behavior of an ensemble of sequences at one instant of time. In other words, for an ergodic system any sample sequence is typical of the entire system. A more quantitative concept of ergodic behavior will follow later.

Nonergodicity

The concept of ergodicity is so natural that it may be difficult to imagine a nonergodic system. Therefore two examples of nonergodic systems follow.

Maintenance System. The state diagram of Figure 15 represents a taxicab system as described by someone concerned with maintenance. The states represent the physical condition of the cabs

State S_1 indicates first-class condition.
State S_2 indicates second-class condition.
State S_3 indicates a condition beyond repair.

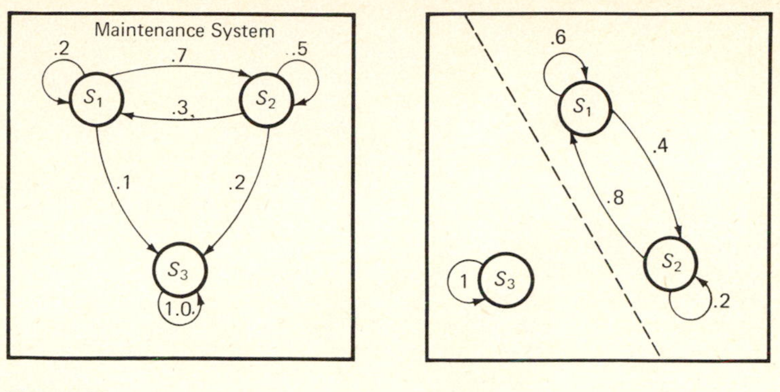

Figure 15

Figure 16

Each transition represents a 1-month sampling interval.

Notice that all cabs will eventually end up in state S_3, which is appropriately called an *absorbing state* or *sink state*. Since some cars may reach this state immediately whereas others may take years to reach it, observing any one car would not be typical of all the others. Thus this system is not ergodic.

Separate Ergodic Systems. The state diagram of Figure 16 describes a taxicab system in which the states are again areas of a city. However, in this case one part of the system is separated from the rest (by flooded roads, private agreement, or legal reasons) so that there are no transitions between the two separate systems.

There are no typical cabs in this system since the behavior of any cab depends on which of the two parts it is in. Thus the entire system is not ergodic, although each of the two parts is ergodic. Notice also that the equilibrium distribution in this case is dependent on the initial distribution.

EXERCISE SET 6

1. Determine the equilibrium distribution of the maintenance system by inspection.

2. For the system of Figure 16, determine the equilibrium distribution if the initial distribution is

(*a*) $p_3 = 0.8$ (*b*) $p_3 = 0.4$

3. Determine the pdf of the lifetime of a cab (described by Figure 15) using the method of iteration. Find the average lifetime by iteration and simulation. Try also some direct analytic method.

4. The *half-life* of a system is defined as the time after which the probability of being absorbed is greater than 0.5. Determine the half-life of the cabs of Figure 15.

5. Construct two other examples of nonergodic systems.

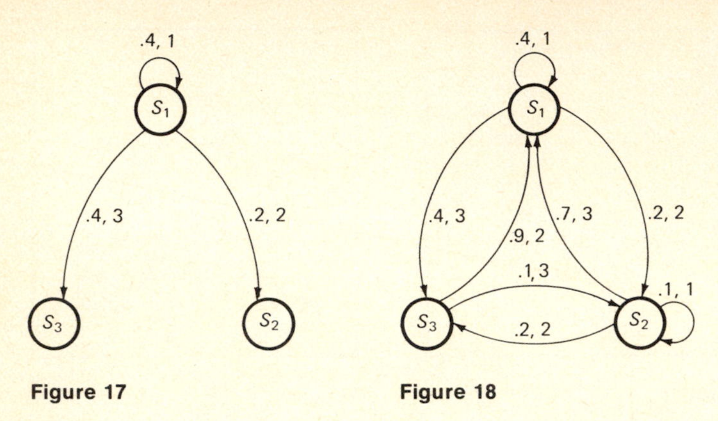

Figure 17

Figure 18

Systems with Output

The purpose of a model is to describe the behavior of systems. The previous model of a taxi system is not yet a very accurate description since it does not involve the cab fare or profit. In general there is usually such a numerical quantity or measure associated with all systems. This measure will be referred to as the *output* and will be denoted by the symbol Y.

In the taxi system, the output could be the fare or profit, which is clearly a function of the transition. It is therefore indicated along with the probability of transition. The partial state diagram of Figure 17 indicates the probabilities and outputs associated with transitions from S_1. The output is seen to be simply a random variable defined on the transitions.

The expected value of output associated with state S_1 can be determined by averaging as

$$\begin{aligned} Y_1 &= (0.4)(1) + (0.2)(2) + (0.4)(3) \\ &= 2 \text{ units per transition} \end{aligned}$$

An output could be associated with all the transitions on a state diagram, as indicated in Figure 18, and the average output Y_i associated with state S_i computed from

$$Y_i = E[Y(S_i)] = \sum_j p[S_j | S_i] Y_{ij} = \sum_j p_{ij} Y_{ij}$$

For example, the average output associated with state S_2 is

$$\begin{aligned} Y_2 = \sum_j p_{2j} Y_{2j} &= (0.7)(3) + (0.1)(1) + (0.2)(2) \\ &= 2.6 \text{ units per transition} \end{aligned}$$

Similarly for state S_3

$$Y_3 = (0.9)(2) + (0.1)(3) = 2.1 \text{ units per transition}$$

Knowing this average output Y_i associated with each state S_i and knowing the probability P_i of occurrence of each state (at equilibrium), it is possible to average over all the n states to obtain the overall average output Y of the entire system:

$$\begin{aligned} Y = \sum_i P_i Y_i = \sum_i P_i \sum_j p_{ij} Y_{ij} \\ &= (0.579)(2) + (0.158)(2.6) + (0.263)(2.1) \\ &= 2.12 \text{ units per transition} \end{aligned}$$

The average output could also be determined by simulation. The table of Figure 19 is a simulation showing the outputs of 10 cabs for one day of 20 transitions. The output of cab j at transition t is given in the table as $Y(t,j)$.

Notice that there are two averages indicated; a time average $A[Y(j)]$ shown at the bottom, and a probabilistic average $E[Y(t)]$ shown at the right.

t \ j	$Y(t,j)$ 1	2	3	4	5	6	7	8	9	10	Pdf of Y: $Y=1$	$Y=2$	$Y=3$	Average $E[Y(t)]$
1	3	2	2	1	2	2	1	3	1	2	.3	.5	.2	1.90
2	2	2	3	3	1	2	1	2	2	3	.2	.5	.3	2.10
3	2	3	2	2	1	3	3	2	3	2	.1	.5	.4	2.30
4	3	3	2	3	3	2	2	3	3	1	.1	.3	.6	2.50
5	1	2	2	2	2	1	1	2	2	3	.3	.6	.1	1.80
6	3	2	1	2	2	1	3	1	1	2	.4	.4	.2	1.80
7	2	3	1	3	2	2	2	2	3	2	.1	.6	.3	2.20
8	1	1	1	1	2	3	3	2	2	2	.4	.4	.2	1.80
9	2	3	3	1	3	1	2	3	3	1	.3	.2	.5	2.20
10	2	2	2	2	2	1	1	2	2	2	.2	.8	.0	1.80
11	2	1	3	3	2	1	2	2	3	3	.2	.4	.4	2.20
12	3	3	2	1	3	3	3	1	2	1	.3	.2	.5	2.20
13	2	2	1	3	2	3	3	2	1	1	.3	.4	.3	2.00
14	2	3	3	2	3	3	2	2	3	1	.1	.4	.5	2.40
15	3	2	2	2	3	3	3	2	2	3	.0	.5	.5	2.50
16	1	1	1	3	3	2	2	3	1	2	.4	.3	.3	1.90
17	3	2	3	3	3	3	3	1	2	3	.1	.2	.7	2.60
18	2	3	2	2	1	2	3	3	2	2	.1	.6	.3	2.20
19	1	3	3	1	1	1	2	2	3	1	.5	.2	.3	1.80
20	3	2	2	3	2	1	2	1	2	1	.3	.5	.2	1.90
$A[Y(j)] =$	2.15	2.25	2.05	2.15	2.15	2.05	2.20	2.05	2.10	2.00				

Figure 19

The *time average* $A[Y(j)]$ corresponds to observing one cab labeled j for T transitions, summing all the individual outputs, and then dividing by T to get the average output per transition from

$$A[Y(j)] = \frac{1}{T} \sum_t Y(t,j)$$

For example, consider cab 5

$$A[Y(5)] = \tfrac{1}{20}[2 + 1 + 1 + 3 + \cdots + 1 + 2]$$
$$= 2.15 \text{ units per transition}$$

The *probabilistic average* $E[Y(t)]$ corresponds to observing all the cabs for one instant t, determining the pdf of the output at this instant, and computing the expected value of the output form

$$E[Y(t)] = \sum_y Y(t)p[Y(t)]$$

For example, at time interval $t = 2$, the pdf of the output is

$$p[Y = 1] = 0.2 \qquad p[Y = 2] = 0.5 \qquad p[Y = 3] = 0.3$$

Then the average output is

$$E[Y(2)] = (1)(0.2) + (2)(0.5) + (3)(0.3)$$
$$= 2.10 \text{ units per transition}$$

Notice that for this system all the time and probabilistic averages essentially agree at a value of about 2.15, which was also the theoretical value computed before this simulation. This equivalence of time average and probabilistic average is often given as a quantitative definition of the property of ergodicity. Thus if a system has this property, we can choose to compute whichever average is most convenient. Usually it is easier to perform a *simple experiment* on a single object for a long period of time. However, if a system is nonergodic, it is necessary to perform *multiple experiments* on many objects at one time.

Stationarity. The simulation of Figure 19 illustrates other concepts. Notice that the time averages tend to vary less than the probabilistic averages. This is essentially because there are 20 samples of time, as opposed to only 10 samples of cabs.

Also in this case, the pdf of the output is essentially the same at each interval of time. If more cabs were sampled, this constancy would be more evident. Such a system is said to have the property of stationarity, or time invariance; i.e., a system is *stationary* if the pdf is independent of time.

Athough all ergodic systems are stationary, not all stationary systems are ergodic. For example (an extreme one), the system of Figure 20 has exactly the same pdf at every instant of time $p_0 = 0.4$ and $p_1 = 0.6$. However, any one sequence is certainly not typical of all others, therefore making the system nonergodic. The "separated system" of Figure 16 is a less extreme example of a stationary system which is not ergodic.

EXERCISE SET 7

1. Is the maintenance scheme of Figure 15 stationary?
2. Extend the algorithm of Figure 14 to describe this system with output. Extend it to compute all averages also.

*Extending Memory Length

In this chapter we have interpreted the concept of state as that entity which determines (probabilistically) the next event. Until now we have had no reason to differentiate between events and states, since the next event depended only on the immediately previous event. However, there are systems in which the next event is dependent on the preceding two events, and we say they have a memory length of 2 units. In such a case our interpretation of state still holds if the state is assigned to be the combination of the two previous events:

$$S(t) = E(t)E(t-1)$$

For example, the next day's weather may depend not only on today's weather but on yesterday's weather also. If the weather is classified as either B (bad) or G (good), there are four states corresponding to the preceding four combinations of weather:

$BB \quad BG \quad GB \quad GG$

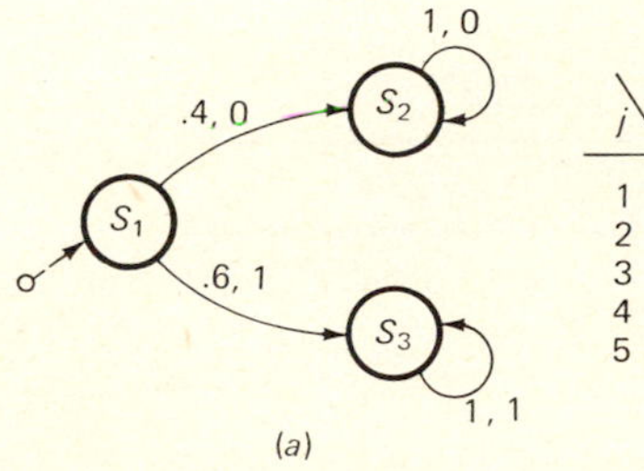

(a)

$j \backslash t$	1 2 3 4 5 6 7 . . .
1	0 0 0 0 0 0 0 0 0 0 . . . 0 0 0 . . .
2	1 1 1 1 1 1 1 1 1 1 1 . . 1 1 1 . . .
3	1 1 1 1 1 1 . . . 1 1 1 . . .
4	0 0 0 0 0 . . . . 0 0 0 .
5	1 1 1 1 1 1 1 1 . . . . 1 1 1 1 . . .

(b)

Figure 20

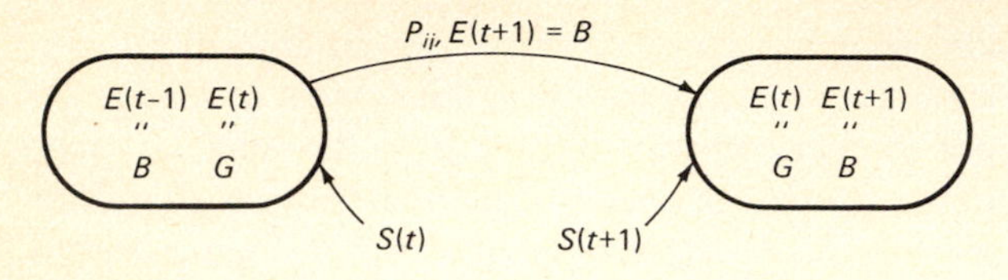

Figure 21

The 3-day sequence *BGB* indicates a transition from state *BG* to state *GB* with output *B* as shown in Figure 21. The probabilities p_{ij} can be determined from a sample sequence such as

GGGGBBBGGGGGGGGGGGGGGBBGBBGGGGGGGGGG
GGGGGGBGGGGGGGGBB

The state transitions could be counted conveniently by applying a mask or template along the sequence, as in Figure 22. This yields the matrix of joint frequencies of Figure 23*a*, which is converted to a matrix of transition probabilities in Figure 23*b*. The corresponding state diagram is shown in Figure 24. Notice the symmetry of the transitions. Notice also that direct transitions are not possible between all states; i.e., no lines connect states *GG* and *BB*.

In general a system is defined to have memory of length m if an event is dependent only on the previous m events and independent of all earlier events. The above methods (for $m = 2$) extend readily to any value of m.

EXERCISE SET 8

1. Observe the matrix of joint frequencies in Figure 23*a* and comment on its structure.
2. From the given sample sequence of weather, attempt to model the weather as a stochastic system of memory length 3. Comment.

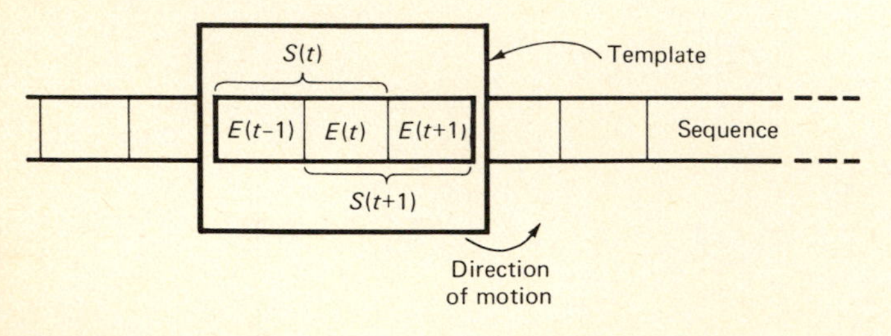

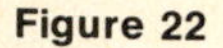

Figure 22

$i \backslash j$	$n[S_i(t) \cdot S_j(t+1)]$ BB	BG	GB	GG
BB	2	3	0	0
BG	0	0	1	3
GB	4	1	0	0
GG	0	0	4	32

(a)

$i \backslash j$	$p[S_j(t+1) \mid S_i(t)]$ BB	BG	GB	GG
BB	.4	.6	0	0
BG	0	0	.25	.75
GB	.8	.2	0	0
GG	0	0	.1	.9

(b)

Figure 23

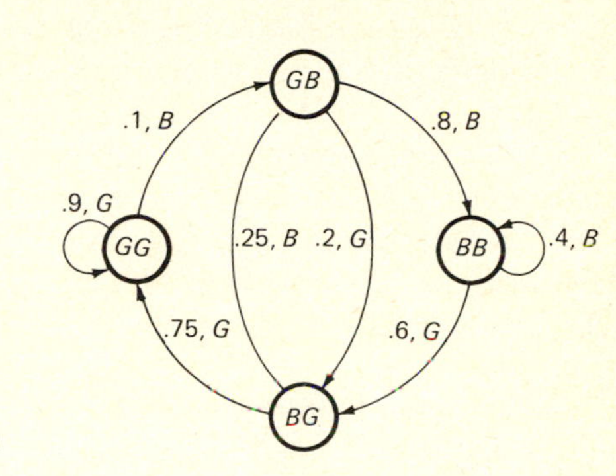

Figure 24

3. For the gear train of Figure 4*a*, the probability of replacing a gear depends on the previous two replaced, as indicated by the table of Figure 25. Each entry p_{ijk} indicates the probability of replacing gear g_k at time $t+1$ given that gear g_j was replaced at time t and gear g_i was replaced at time $t-1$.

(*a*) Construct the transition matrix for this system.

(*b*) Draw the state diagram without labeling the transitions. Notice its general form.

(*c*) Determine the probability of replacing gear g_2 then check this result by some other method.

i	j	p_{ijk} $k=1$	2	3
1	1	.5	.4	.1
1	2	.2	.4	.4
1	3	.1	.8	.1
2	1	.3	.3	.4
2	2	.3	.3	.4
2	3	.5	.4	.1
3	1	.0	.9	.1
3	2	.4	.3	.3
3	3	.4	.6	.0

Figure 25

3. OPTIMIZATION OF STOCHASTIC SYSTEMS

Stochastic Systems with Inputs

A manufacturing system which produces defective items in bursts can be modeled by a stochastic system as in Figure 26*a*. The outputs represent the gain (or profit) associated with the manufactured product.

In such systems there may be some change in probabilities or outputs depending on such factors or parameters as change of weather, time of day, condition of tools, change of operator, or difference in quality of materials. For example, changing to a poorer quality material (say grade *B*) may increase the profit per item, but it may also increase the probability of defect. This situation could be described by the model of Figure 26*b*.

The above two models of Figure 26 may be combined into the single model of Figure 27, in which the parameter causing the change in behavior is considered as the input. In this case the input values would be the two grades of raw materials, *A* and *B*. The labels on each transition arrow indicate in order:

1. The input X having value k
2. The transition probability p_{ij}^k
3. The output Y_i^k

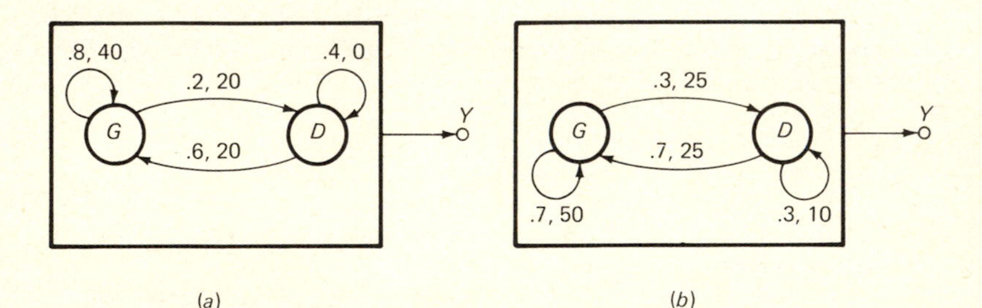

Figure 26

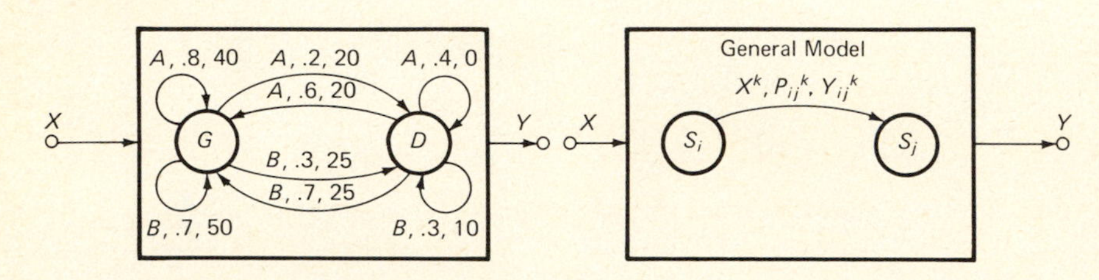

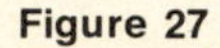

Figure 27

The state equilibrium probabilities P_i^k can be determined separately for each input k by the result (associated with Figure 11)

$$P_i^k = \frac{p_{ji}^k}{p_{ij}^k + p_{ji}^k}$$

For input A the equilibrium probabilities are

$$P_G^A = \frac{0.6}{0.2 + 0.6} = 0.75 \qquad \text{and} \qquad P_D^A = 1 - 0.75 = 0.25$$

Similarly for input B the equilibrium probabilities are

$$P_G^B = \frac{0.7}{0.3 + 0.7} = 0.70 \qquad \text{and} \qquad P_D^B = 1 - 0.7 = 0.30$$

Notice that the probability of defect P_D is smaller for input A, but this may not be the best input since the average output may be larger for the other input B.

This average output could be computed from

$$Y_i^k = \sum_j p_{ij}^k Y_{ij}^k$$

For input value A the average outputs associated with each state are

$$Y_G^A = p_{GG}^A Y_{GG}^A + p_{GD}^A Y_{GD}^A = (0.8)(40) + (0.2)(20) = 36$$

and

$$Y_D^A = (0.6)(20) + (0.4)(0) = 12$$

These outputs can now be averaged with the state equilibrium probabilities to determine the average output associated with each input. For example,

$$\begin{aligned} Y^A &= p_G^A Y_G^A + P_D^A Y_D^A \\ &= (0.75)(36) + (0.25)(12) = 30 \text{ units per item} \end{aligned}$$

This same process may be done in one step, by substitution:

$$Y^k = \sum_i P_i^k Y_i^k = \sum_i P_i^k \sum_j p_{ij}^k Y_{ij}^k$$

The average output associated with the second input B is

$$Y^B = P_G^B\ [p_{GG}^B Y_{GG}^B + p_{GD}^B Y_{GD}^B] + P_D^B [p_{DG}^B Y_{DG}^B + p_{DD}^B Y_{DD}^B]$$

$$= 0.7[(0.7)(50) + (0.3)(25)] + 0.3[(0.7)(25) + (0.3)(10)]$$

$$= 35.9 \text{ units per item}$$

Decision making in Stochastic Systems

Inputs are of two types, those which can be controlled, such as the quality of input material or the condition of tools, and those which cannot be controlled, such as the change of weather or time of day.

If inputs can be controlled, it is preferable to select and apply that input which optimizes some output or gain or profit. In general the optimal output is

$$Y^* = \max_k\ [Y^k]$$

In the previous system, where the inputs were the grades A and B (describing the quality of input materials), the optimal output is

$$Y^* = \max\ [Y^A, Y^B] = \max\ [30, 35.9]$$
$$= 35.9 \text{ units per item}$$

which results when the optimal input B is applied.

Of course we could hide the basic simplicity of this idea by performing various substitutions to get the hideous expression

$$Y^* = \max_k\ [\sum_i P_i^k \sum_j p_{ij}^k\ Y_{ij}^k]$$

However, if the input cannot be controlled but it is known that input X^k occurs with probability $p[X^k]$, about all that can be done is to estimate the overall average output as

$$Y = \sum_k Y^k p[X^k]$$

For example, if the inputs A and B of the previous system are equiprobable, the overall average output is

$$Y = Y^A p[X^A] + Y^B p[X^B]$$
$$= (0.5)(30) + (0.5)(35.9)$$
$$= 33.5 \text{ units per item}$$

Of course this overall average output can be generalized to

$$Y = \sum_k p[X^k] \sum_i P_i^k \sum_j p_{ij}^k Y_{ij}^k$$

EXERCISE SET 9

1. The stochastic system of Figure 28 represents a critical gear train on a spaceship. The table indicates some relevant random variables. During any journey there will be much wear and thus many replacements necessary. We are to plan for 25 replacements per journey. Since not all gears will fail with the same frequency (they are of different sizes and materials and in different positions), it would be unwise to carry an equal number of all three gears.

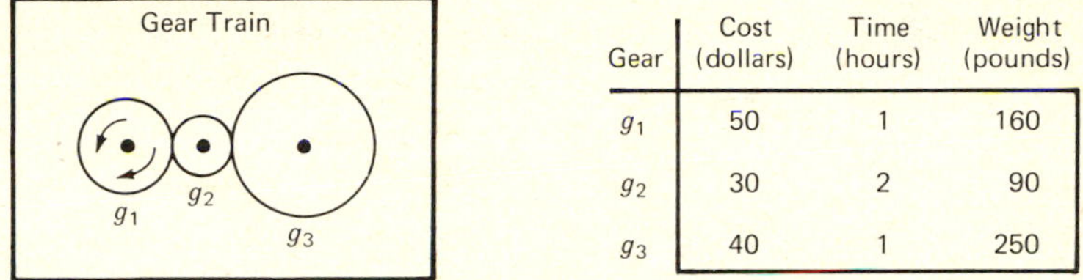

Gear	Cost (dollars)	Time (hours)	Weight (pounds)
g_1	50	1	160
g_2	30	2	90
g_3	40	1	250

Figure 28

The problem is to determine the optimal number of each gear that should be carried on a journey, the expected weight and cost of the gears, and also the average time to replace a gear. The driving gear g_1 rotates clockwise 70 percent of the time and counterclockwise 30 percent of the time. For each of these two directions the transition probabilities p_{ij} of Figure 29 indicate the probability of replacing gear g_j given that gear g_i was replaced last.

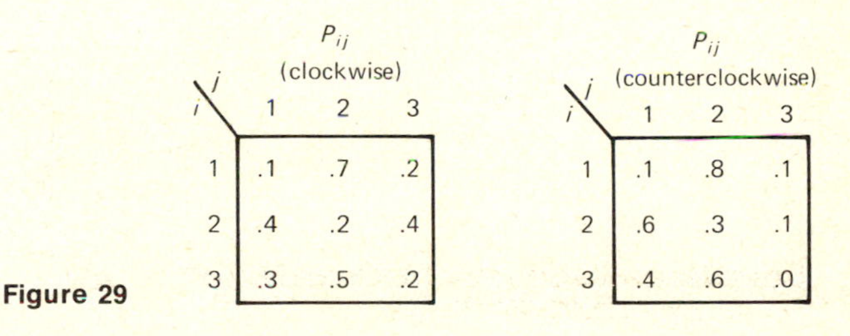

P_{ij} (clockwise)

i \ j	1	2	3
1	.1	.7	.2
2	.4	.2	.4
3	.3	.5	.2

P_{ij} (counterclockwise)

i \ j	1	2	3
1	.1	.8	.1
2	.6	.3	.1
3	.4	.6	.0

Figure 29

Control of Multi-stage Stochastic Systems

Consider a manufacturing system as in Figure 30 in which every item goes through three stages of treatment (such as painting). At each stage a controller examines the state of each item and decides which of two possible inputs to apply.

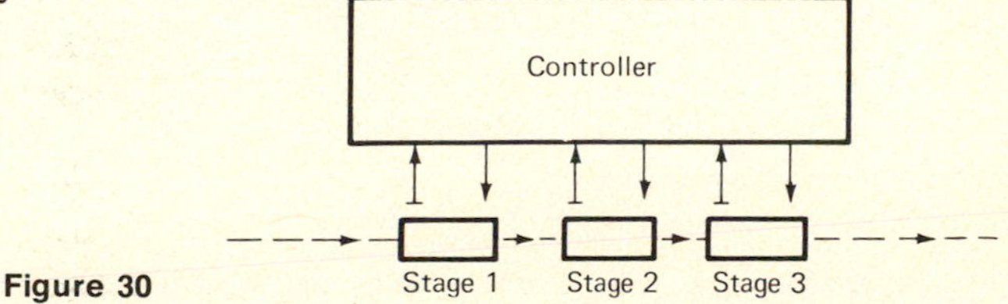

Figure 30

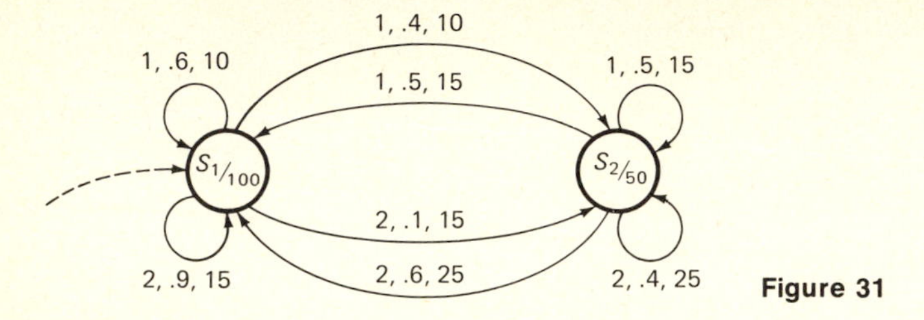

Figure 31

The first input (treatment) is inexpensive but results in a large probability (0.4) of having a second-class product. If at one stage the product is second-class, there is a probability that another treatment can convert it to a first-class state again. However, this second input is more expensive. All the probabilities and costs are shown in Figure 31.

The item is originally in the first-class state. If after the three treatments the item is still first-class, it sells for 100 units, whereas if it is second-class, it sells for only 50 units.

The goal is to optimize the expected gain (average profit), which is the selling price less the accumulated cost of inputs.

Dynamic Programming

One rather significant way of approaching such a multi-stage problem is to start at the last stage and determine the optimal inputs at this last step. This procedure is based on the fact that the last part of an optimal sequence is optimal. For example, consider the diagram of Figure 32, where the next to last state is S_1. If the input is $X = 1$, from Figure 32*a* the average gain is

$$\begin{aligned} E[G] &= (0.6)(100 - 10) + (0.4)(50 - 10) \\ &= 70 \text{ units} \end{aligned}$$

But if the input is $X = 2$, from Figure 32*b* the average gain is

$$\begin{aligned} E[G] &= (0.9)(100 - 15) + (0.1)(50 - 15) \\ &= 80 \text{ units} \end{aligned}$$

So the optimal gain is 80 units with optimal input $X^* = 2$.

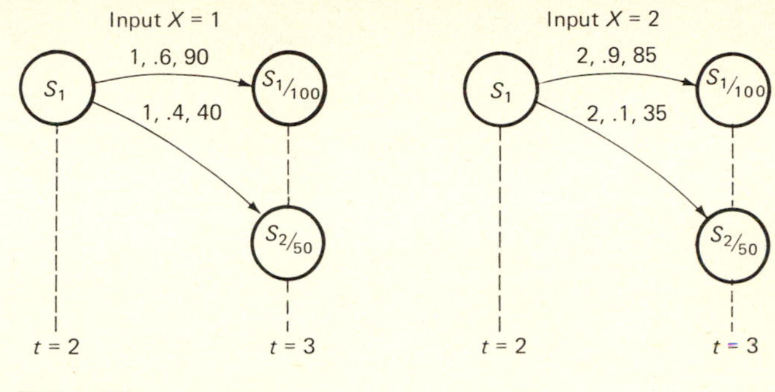

Figure 32

Similarly, if the state at this last stage is S_2, the average gain for input $X = 1$ is

$$\begin{aligned} E[G] &= (0.5)(100 - 15) + (0.5)(50 - 15) \\ &= 60 \text{ units} \end{aligned}$$

and the average gain for input $X = 2$ is

$$\begin{aligned} E[G] &= (0.6)(100 - 25) + (0.4)(50 - 25) \\ &= 55 \text{ units} \end{aligned}$$

The optimal gain is 60 units with optimal input $X^* = 1$ (if the system is in stage S_2 at the last stage).

Now that we know the optimal inputs and gains one stage from the end, we can retreat another stage from the end and optimize again. This second stage is shown in Figure 33.

If the state is S_1, the optimal gain is

$$\begin{aligned} G_1^* &= \max\ [(0.6)(70) + (0.4)(50), (0.9)(65) + (0.1)(45)] \\ &= \max [62,63] \\ &= 63, \text{ with optimal input } X^* = 2 \end{aligned}$$

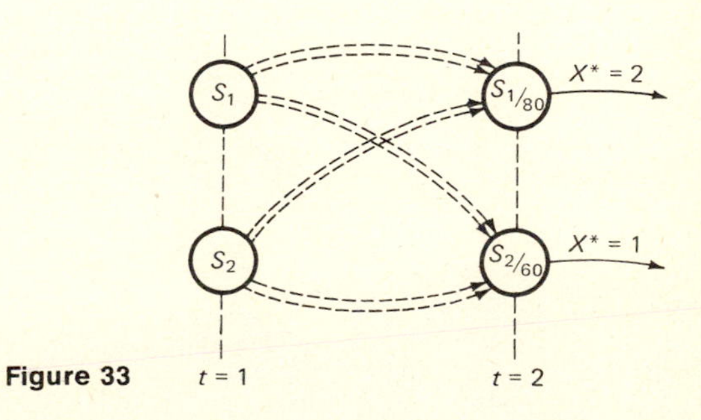

Figure 33

If the state is S_2, the optimal gain is

$$\begin{aligned} G_2^* &= \max[(0.5)(65) + (0.5)(45), (0.6)(55) + (0.4)(35)] \\ &= \max[55,47] \\ &= 55, \text{ with optimal input } X^* = 1 \end{aligned}$$

Again we retreat another step from the end, and in general the optimal gain $G^*(i,t)$ associated with state i at time t can be computed as

$$G^*(i,t) = \max_i [\sum_j p[S_j(t+1) \mid S_i(t)] \, G_{ij}(t+1)]$$

which is more conveniently expressed as

$$G^* = \max_i \; [\sum_j p_{ij} G_{ij}]$$

This process of retreating from the final stage is repeated until the first stage is reached. In this case, the first stage is shown in Figure 34. When in state S_1 the optimal gain is

$$\begin{aligned} G^* &= \max_i \; [\sum_j p_{ij} G_{ij}] \\ &= \max[(0.6)(53) + (0.4)(45), (0.9)(58) + (0.1)(40)] \\ &= \max[50,47.2] \\ &= 50 \text{ for optimal input } X^* = 1 \end{aligned}$$

Since the initial state is given as S_1, we need not consider the initial state of S_2.

The complete optimal controller can be described by the diagram of Figure 35. The policy can also be summarized verbally as follows:

At the first stage always apply input 1.
At the remaining two stages, if in the first state, apply input 2.
If in the second state, apply input 1.

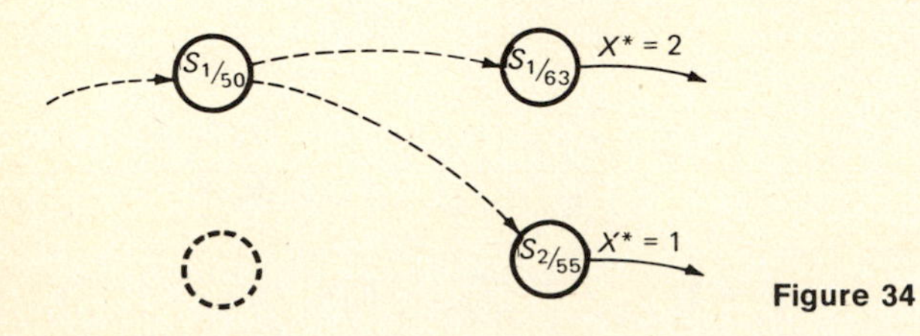

Figure 34

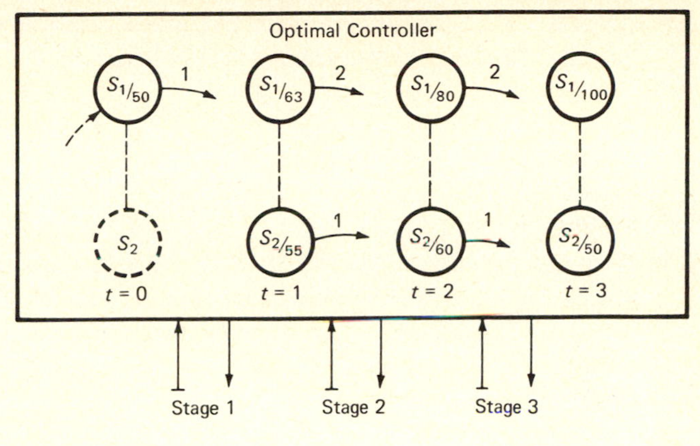

Figure 35

This procedure of finding the optimal controller is known as *dynamic programming*. It is based on the *principle of optimality*, developed by Richard Bellman, which states that "An optimal policy has the property that whatever the initial state and initial decision are, the remaining decisions must constitute an optimal policy with regard to the state resulting from the first decision." In other words, the second part of an optimal policy is also optimal. This view is often computationally convenient and philosophically interesting. It can also be extended to other types of systems such as continuous systems.

EXERCISE SET 10

1. Continue the given example for two more stages; i.e, assume it is a five-stage system.
2. In the given example the gain may be separated into two components, one associated with the next state and one associated with the present input. Show how this could be computationally convenient.
3. Construct an algorithm in flow-diagram form to describe this procedure for optimizing multi-stage systems.

4. CONCLUSION

General Model of Stochastic Systems

A discrete-time, finite-state, stochastic system consists of:

1. $S = \{S_1, S_2, ..., S_n\}$, a finite set of states
2. $X = \{X_1, X_2, ..., X_p\}$, a finite set of inputs
3. $Y = \{Y_1, Y_2, ..., Y_q\}$, a finite set of outputs

4. $R: S \times X \to S \times Y$, an input-output state relation which is decomposed into the two functions f and g, where

$$p[Y(t)] = f[S(t), X(t)]$$
$$p[S(t+1)] = g[S(t), X(t)]$$

The Relay Revisited

A relay in some cases can be modeled as a stochastic system as in Figure 36. For example, if a relay requires 2 amperes to operate and the input voltage ($X = 1$) results in a current near this critical value of 2 amperes, any slight fluctuation (of voltage, vibration, temperature, etc.) may cause or prevent a transition.

Interpretations of some transitions follow. If a relay has no current flowing in it (unenergized state $S = 0$) and an input voltage is applied ($X = 1$), the contact will be in its open position ($Y = 0$) and the current will steadily increase in the coil. If the current reaches the critical value, say 80 percent of the time, it causes the transition to state $S = 1$ (operate) with probability 0.8. Similarly when the relay is energized and the input voltage remains applied, 30 percent of the time this voltage is insufficient to maintain this state, causing a transition again. Another transition indicates sticking with probability 0.1.

If the probabilities have only the extreme values of 0 or 1, as in Figure 37, the system becomes a deterministic system involving time. Such dynamic deterministic systems are modeled as sequential systems, which are studied in another chapter.

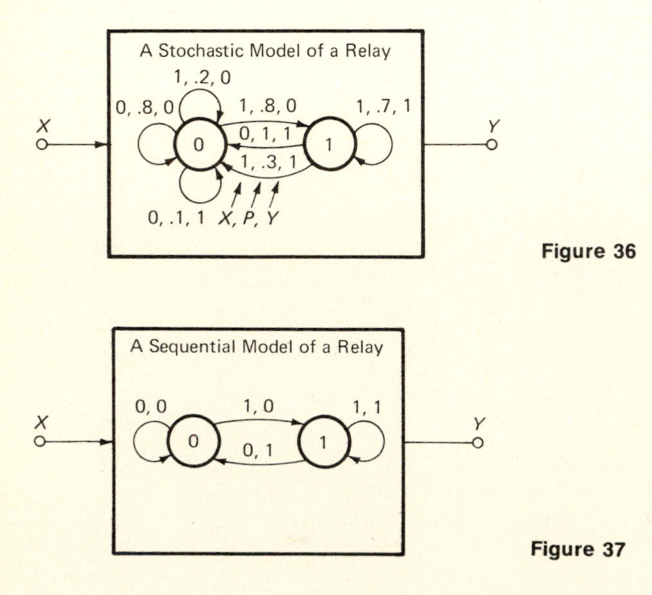

Figure 36

Figure 37

SUMMARY

Stochastic systems involve both probability and time. Thus they combine both uncertainties and change. They have been represented in many ways (sagittal diagram, state diagram, transition matrix) and analyzed in many ways (iteration, inversion, simulation). There were also various interpretations (the macroscopic and microscopic). Finally there were rather unusual ways of optimizing such systems.

PROBLEMS

1. Suppose that only two brands of some product are available, brand X and Y. When a person is using brand X, his next purchase will invariably be brand Y. If a person is using brand Y, half the time his next choice is brand X. How much more of brand Y is sold than brand X?

2. For the given communication system, with stochastic source:

(*a*) Determine the probability of error.

(*b*) Would an encoder in the form of a NOT decrease the probability of error?

(*c*) Determine the probability that the output of the channel has value 1.

(*d*) Can you construct a stochastic system equivalent to this combination?

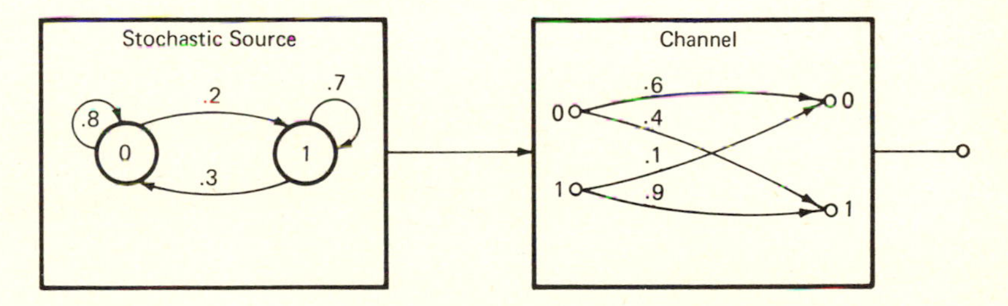

Figure P-2

3. Find the equilibrium probabilites for the given system if the initial distribution is given by

$$P[S_1(0)] = 0.2 \qquad P[S_2(0)] = 0.3 \quad P[S_3(0)] = 0.4 \qquad P[S_4(0)] = 0.1$$

$p[S_j(t+1) \mid S_i(t)]$

$i \backslash j$	1	2	3	4
1	.1	0	0	.9
2	.5	.5	0	0
3	0	0	1	0
4	.6	0	0	.4

Figure P-3

4. Find the equilibrium probabilities for the given system.

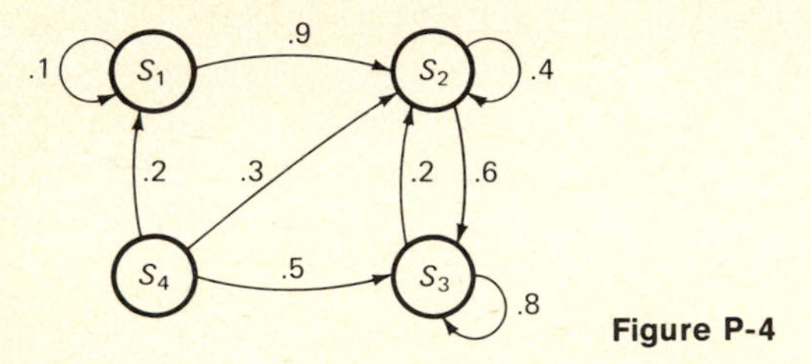

Figure P-4

5. Determine the probability distribution of the given system after eight stages by performing only three matrix products. Do also for five, six, and seven stages and comment.

	1	2	3	4
1	0	1	0	0
2	$1/3$	0	$2/3$	0
3	0	$2/3$	0	$1/3$
4	0	0	1	0

Figure P-5

6. The given system is a stochastic noise source having three states corresponding to the polarities of positive, negative, and zero voltage.

(*a*) By inspection of the symmetry, what can you conclude about some of the equilibrium probabilities?

(*b*) Using the above observation, determine the equilibrium probabilities for all the states.

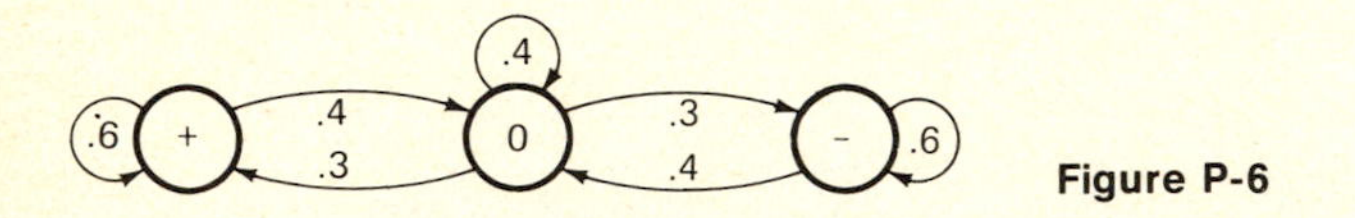

Figure P-6

7. A small four passenger airplane is booked and canceled according to the stochastic system in Figure P-7, where the state represents the total number of seats sold.

(*a*) Determine the equilibrium probabilities.

(*b*) What meaning do the equilibrium probabilities have?

(*c*) Determine also the average number of passengers on the plane.

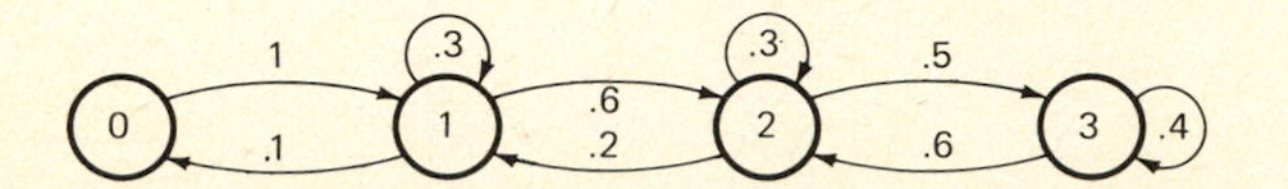

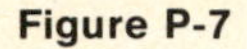

Figure P-7

8. The dice game called *craps* can be modeled as a stochastic system. First, two dice are thrown. If 7 or 11 shows up, you win; if a 2, 3, or 12 shows up you lose. Otherwise, you remember which number came up and keep throwing the two dice repeatedly until either this remembered number comes up (in which case you win) or until a 7 comes up (in which case you lose).

(*a*) Model this system, representing it as a state diagram or matrix (array). Indicate transition probabilities.

(*b*) Determine the probability of winning.

(*c*) Determine the expected length of a game.

9. A player has $x = 1$ dollar; his opponent (the house) has $y = 3$ dollars, the total amount of money involved being $n = x + y = 4$ dollars. At each stage of a game the player bets 1 dollar, winning with probability $p = 0.6$, and losing with probability $1 - p = 0.4$. The game ends when the player loses all his money, or gains the total n dollars.

(*a*) Draw the state diagram which describes this stochastic system.

(*b*) Find the probability that the game lasts for one stage. Find the probability that it lasts for two stages, then three and four stages also.

(*c*) Find the average amount of money the player has after three stages.

(*d*) How would you determine the average amount of winning?

(*e*) If the game continues until the player or opponent is "ruined," find the probability that the player is ruined (despite his probability of 0.6).

(*f*) The general problem of this type is called *random walk with absorbing barriers*. Find a more practical engineering application of this type of stochastic structure.

10. Construct a flow diagram to simulate the *gambler's-ruin* system of Problem 9 for N ruins, and compute the average length of the game before ruin.

11. A *birth and death* stochastic system is defined by the following limitation on the transition probabilities of Fig. P-11.

$$p_{ij} = 0 \qquad \text{if } j \neq i \text{ and } j \neq i + 1 \text{ and } j \neq i - 1$$

where the states are designated by the integers 0, 1, 2, ..., n.

(*a*) How can you recognize by inspection the state diagram of such a system?

(*b*) Prove that for such systems the equilibrium probabilities can be determined by the convenient result

$$P_{i+1} = \frac{P_i b_i}{C_{i+1}}$$

(*c*) Use the above computationally convenient result to find the equilibrium probabilities of:

1. The stochastic source of Problem 7
2. The airplane booking of Problem 8
3. The queueing system of Figure 5

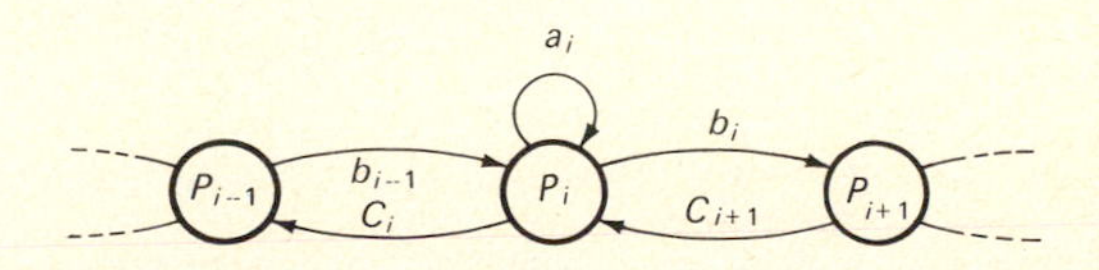

Figure P-11

12. A stochastic source transmits symbols 0 and 1 as described by the given state diagram. Occasionally a symbol from this source is erased or lost, as indicated in the sequence by the symbol?

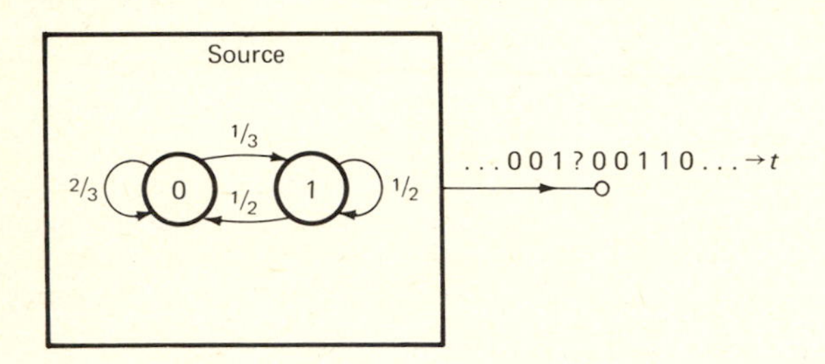

Figure P-12

(*a*) Find the probability that this symbol ? is actually a 0.
(*b*) If the preceding symbol was 1, what is the probability that this symbol ? is actually a 0?
(*c*) If the succeeding symbol is a 0, what is the probability that this symbol ? is actually a 0?
(*d*) If it is known that the preceding symbol was 1 and the succeeding symbol was a 0, what is the probability that this symbol ? was actually a 0?

13. The given set diagram describes the smog situation $S(t)$, $S(t+1)$ on two successive days and the dependence on the temperature $T(t)$ and humidity $H(t)$. For example, the temperature and humidity are both high with probability 0.35.

(*a*) Determine the probability of smog on any day.
(*b*) Determine the probability of smog on one day:
1. If the previous day was humid
2. If it was not humid
3. If it was hot
4. If it was not hot

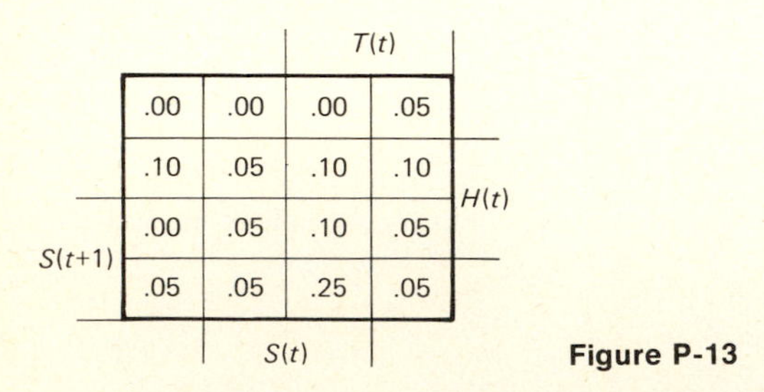

.00	.00	.00	.05
.10	.05	.10	.10
.00	.05	.10	.05
.05	.05	.25	.05

Figure P-13

(*c*) Which of these four situations is the next day's smog most independent of?
(*d*) Determine the combination of temperature and humidity which maximizes the probability of smog on the next day.

14. For the set diagram P-13:
(*a*) Construct the state diagram relating the smog on one day to the smog condition on the next day.
(*b*) Compute the equilibrium probability of smog, using two different methods.
(*c*) Construct the state diagram, with input H, having values of 0 (low humidity) and 1 (high humidity).

(*d*) Construct the state diagram, with input T, having values of 0 and 1.

(*e*) Construct the state diagram with inputs which are combinations of H and T.

15. Consider a system of taxicabs which travel between a town T and a suburb S. The driver knows that he can influence his transition probabilities p_{ij} and gains G_{ij} by taking any of three actions: A stopping and waiting for a radio dispatch; B going to the nearest taxi stand; or C driving around. The corresponding actions, probabilities, and gains are given in Figure P-15. What is the optimal action in the town? In the suburb? Construct the optimal policy to maximize the total accumulated gain if he wishes to be in town after three transitions (since then he stands to gain tips of 15 units whereas otherwise he stands to gain only 5 units).

	Action A		Action B		Action C	
i j	P_{ij}	G_{ij}	P_{ij}	G_{ij}	P_{ij}	G_{ij}
SS	.4	3	.3	3	.6	4
ST	.6	6	.7	4	.4	5
TS	.7	4	.8	5	.2	5
TT	.3	2	.2	4	.8	4

Figure P-15

16. The given map represents a number of towns and the roads between them, with town A having the only hospital. By extending the concept of dynamic programming, find the shortest distance from any town to town A. Find also the shortest distance from town A to all other towns (for dispatching an ambulance possibly).

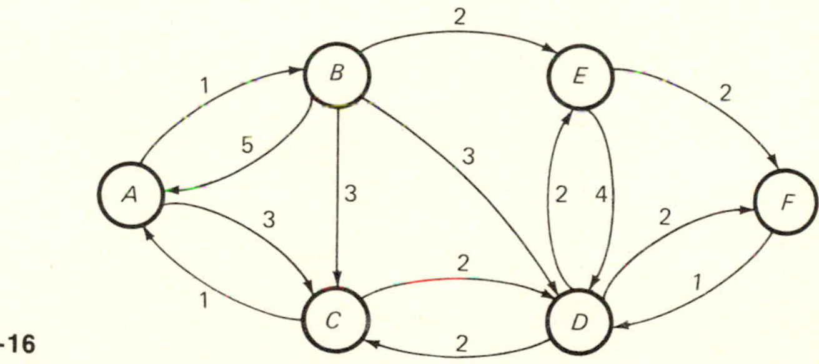

Figure P-16

17. *Project.* Investigate some practical systems (related to medicine, ecology, business, etc.) which behave in a stochastic way; then use the results of the investigation to make decisions and recommendations or take actions.

Chapter 7

Large Digital Systems

Digital Computation, Communication, and Control

In this chapter we consider large systems for computation, communication, and control. These large systems are made from interconnections of the smaller systems discussed in the previous chapters. Here we will emphasize computing systems, but the concepts encountered also apply to digital communication and control systems. As an example of a large system, we shall construct a simple digital computer. This particular computer is chosen not for great speed or minimum components but for its conceptual simplicity.

For computation we are usually interested in numbers (such as 120, −75, 3.14); for communication we are interested in characters (such as A, B, C, . . . , Y, Z, , ;, .), and for control we are interested in operations (such as shift, add, compare). All these entities (numbers, characters, and operations) can be coded as sequences of digits, and these sequences can be manipulated for computing, communicating, and controlling. The sequences and the devices for storing and manipulating them are the basic entities of this chapter.

1. REGISTERS

The basic components of large digital systems are registers. We will consistently use serial right-shift registers, which are simple cascade combinations of n memory elements (such as D flip-flops) as shown in Figure 1 (for $n = 4$). Such serial shift registers are considered theoretically in Chapter 5 and physically in the Appendix.

The state of the shift register is the combination of values at the output of the flip-flops. When a gate pulse is applied ($G = 1$), it applies n clock pulses, causing n shifts to the right. Thus the n digits of the state are shifted out sequentially and replaced by whatever sequence of n digits was applied at the input. For example, Figure 2 illustrates both physically and state-diagrammatically that a register in state 0111 when applied input sequence 1100 produces the output sequence 1110 and that the next state becomes 0011 (which is the input sequence but in reverse order). We may think of the input and output sequences as digits marching in a single file.

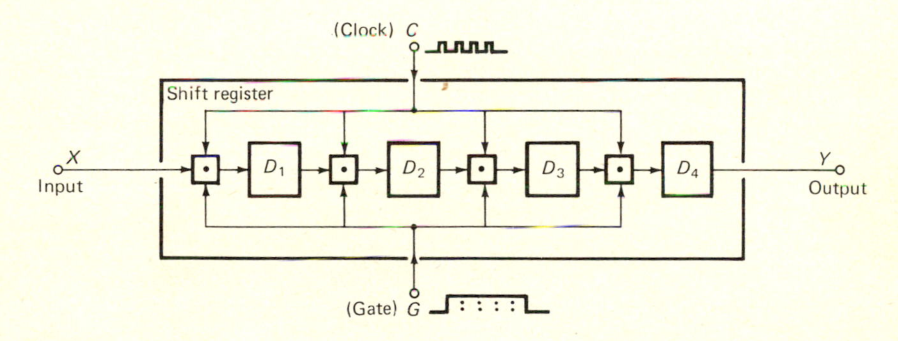

Figure 1

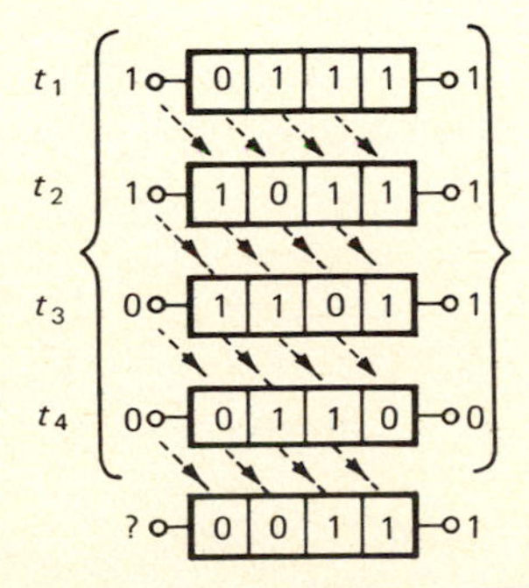

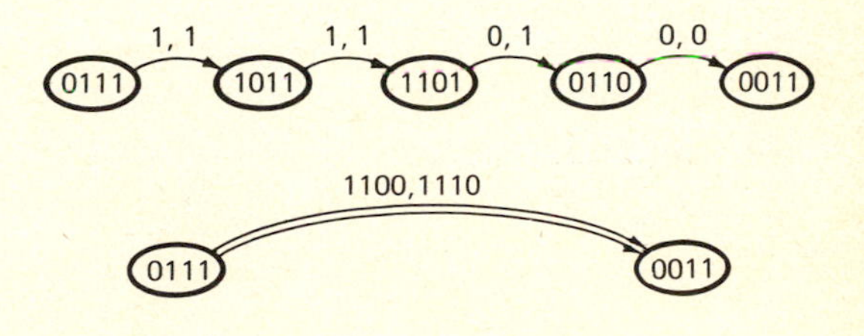

Figure 2

In our system we will use a register of eight binary digits (called *binits* or *bits*) as shown in Figure 3. More practical systems use registers having from 12 to 64 bits. However our simple registers illustrate all the basic concepts without introducing great complexity.

Any register can have associated with it a name, a location, a content, and a meaning. These will be illustrated below for the register of Figure 3.

The *name* of a register is any label or symbol such as

R or REG or RESULT

The *location* or address of a register is a method of referring to a register by some number such as

11001 or the equivalent decimal 25

The *content* or word is the ordered arrangement of the digits within the register. The content of a register R is usually denoted within parentheses as

$$(R) = 00100110$$

The *meaning* or interpretation indicates how the contents are coded as numbers, letters, operators, or any combinations of these. This coding will be considered in detail later.

Manipulation of Registers

Readout. To read out the contents of a shift register it is necessary only to apply the gate pulse. This then generates a sequence of binary digits, and it also destroys the content of the register by replacing it with the input sequence to this register. But if the output of this register is fed back into the input of the same register, as in Figure 4, its original content is replaced (recirculated) as it is read out. Thus the content is not destroyed by this method (known as *nondestructive readout*). So we have two possibilities of readout (destructive or nondestructive), either of which can be used, depending on the applications.

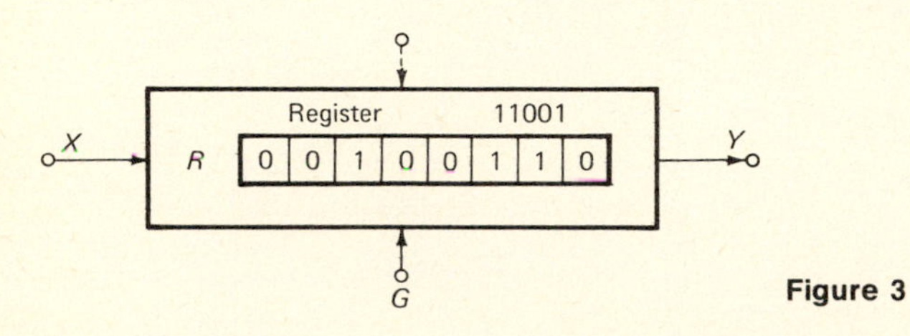

Figure 3

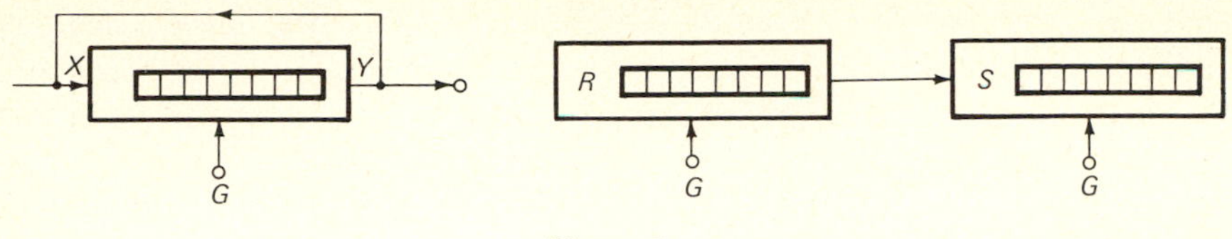

Figure 4

Figure 5

Transfer. Two registers can be interconnected as in Figure 5 to transfer the contents of one into the other. The transfer of the contents of register R into register S upon operation of gate G is denoted symbolically as

$$G{:}(R) \rightarrow S$$

In general the gate G is given as a switching algebraic expression.

Transforming. In addition to transferring sequences from one register into another, we are often interested in transforming or performing operations on the digits as they are transferred. The operations may be arithmetic (such as add or subtract), or they may be logical (such as AND, OR, NOT, EXCLUSIVE-OR) or they may be other (such as shift, compare, test for zero).

The network of Figure 6 shows how a variety of operations may be performed on the contents of registers A and B as they are transferred into register C. For example, when a value of 1 is applied to terminal f_0 (as well as to the register gates), the contents of A and B are added together

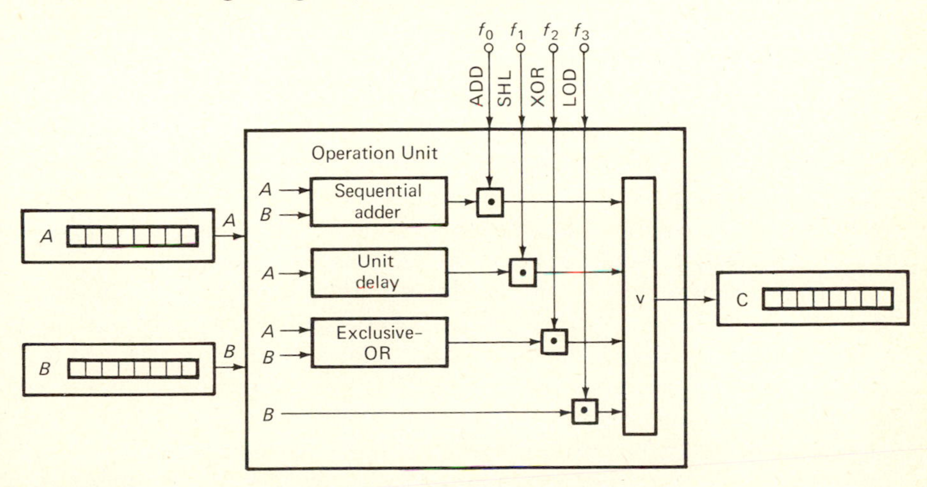

Figure 6

and the resulting sum is gated by the AND into register C. This is symbolically indicated as

$$f_0:(A) + (B) \rightarrow C$$

Similarly f_1 controls the transfer of the contents of A into C through a delay element. The contents of C become the contents of A but shifted by one digit to the left. Symbolically this left shift can be written

$$f_1:\ SL(A) \rightarrow C$$

or in greater detail

$$f_1:\ [a_7 a_6 a_5 a_4 a_3 a_2 a_1 a_0] \rightarrow [a_6 a_5 a_4 a_3 a_2 a_1 a_0 0]$$

Notice that this left-shift operation essentially performs the arithmetic operation of multiplying a binary number by its base 2. This is similar to the ordinary decimal system, where multiplication by the base 10 corresponds to shifting the decimal point one place to the right.

The application of f_2 alone simply performs the EXCLUSIVE-OR operation on the contents of A and B. Notice that if the contents of A and B are identical, C has value 0. This can be useful for comparing sequences. Finally the last operation, f_3, simply transfers the contents of register B into C, an operation known as *loading C*.

Of course other operations are possible; they include subtracting, incrementing (adding a constant), right shifting, and parity checking.

2. DIGITAL COMPUTERS

A digital computer can be thought of as consisting of registers arranged within three basic functional units as shown in Figure 7. The memory unit stores data and instructions, the operation unit performs arithmetic and logical operations, and the control unit coordinates, sequentially, the behavior of all the units. This "decomposition" is for conceptual convenience; physically the units may not be separate.

It is also necessary to have an input-output unit (or units) for communication with the external environment. However, since we are primarily concerned with the internal behavior and structure, we will ignore the input-output unit now.

The operation unit and the memory unit behave similarly in most digital computers, whereas the control units differ considerably depending on how they are used. For this reason we will consider the operation and memory units first, and then later we will consider various control units.

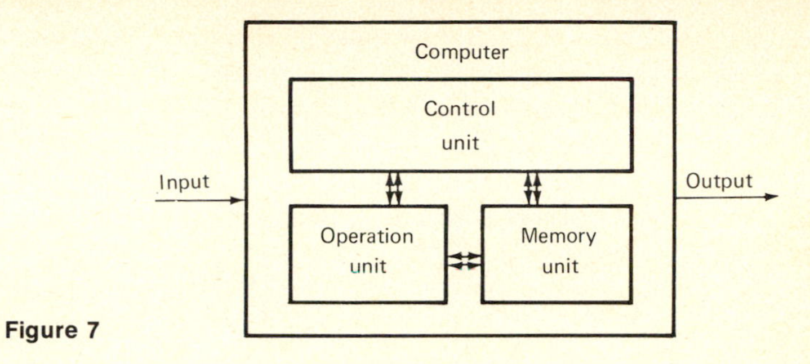

Figure 7

The Operation Unit

An *operation unit* (commonly called an *arithmetic unit*) is shown in Figure 8. Notice that the lower part, called the *operation network*, is similar to the network of Figure 6. The contents of registers A and B are fed into this network, exactly one of the eight operations f_1 is indicated, and the resulting sequence is fed back into register A. The A register associated with the operation unit is commonly known as the *accumulator*.

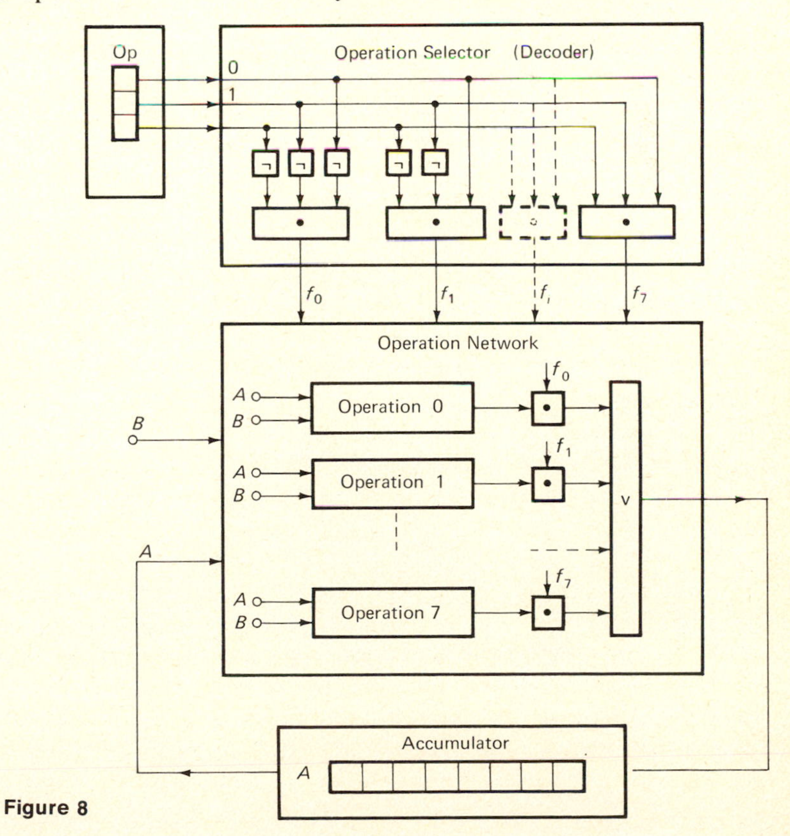

Figure 8

The *operation selector* (also called the operation decoder) is an array of gates which transform any n digits in a register marked Op to operate exactly one of the 2^n operations. For example the operation of addition could be coded as 000, and when this combination is in the operation register, only terminal f_0 has value 1, all others have value 0. Similarly subtraction could be coded as 001, and associated with f_1.

The Memory Unit

A memory unit (also called a storage unit) is shown in Figure 9. Notice that its structure is similar to the operation unit. It consists of 32 registers each with a five-digit address. It also has a register B, known as the *buffer*, which is an intermediate store to compensate for the difference in speeds between the memory (which is usually magnetic) and the rest of the system.

When an address is put into the select register S, the select network operates precisely one of the leads M_j, which in turn operates the corresponding memory register and output gate.

To write a word into the address indicated by S, it must first be placed in the buffer B and the write terminal W activated. Symbolically

$$W\colon (B) \to M\langle S\rangle$$

To read out of the memory, the appropriate address is indicated in S; then the read terminal R is activated. This causes the transfer of the word in the location $M\langle S\rangle$ indicated by S into the buffer B. It also recirculates the content back into the original memory register so as not to destroy it. Symbolically,

$$R\colon (M\langle S\rangle) \to B \qquad \text{and} \qquad R\colon (M\langle S\rangle) \to M\langle S\rangle$$

Single-address Organization

All digital computers have operation units and memory units which are similar in behavior to the units just described. Of course the structure may differ somewhat for different computers. Also all digital computers have in common a stored program in which the memory stores both the

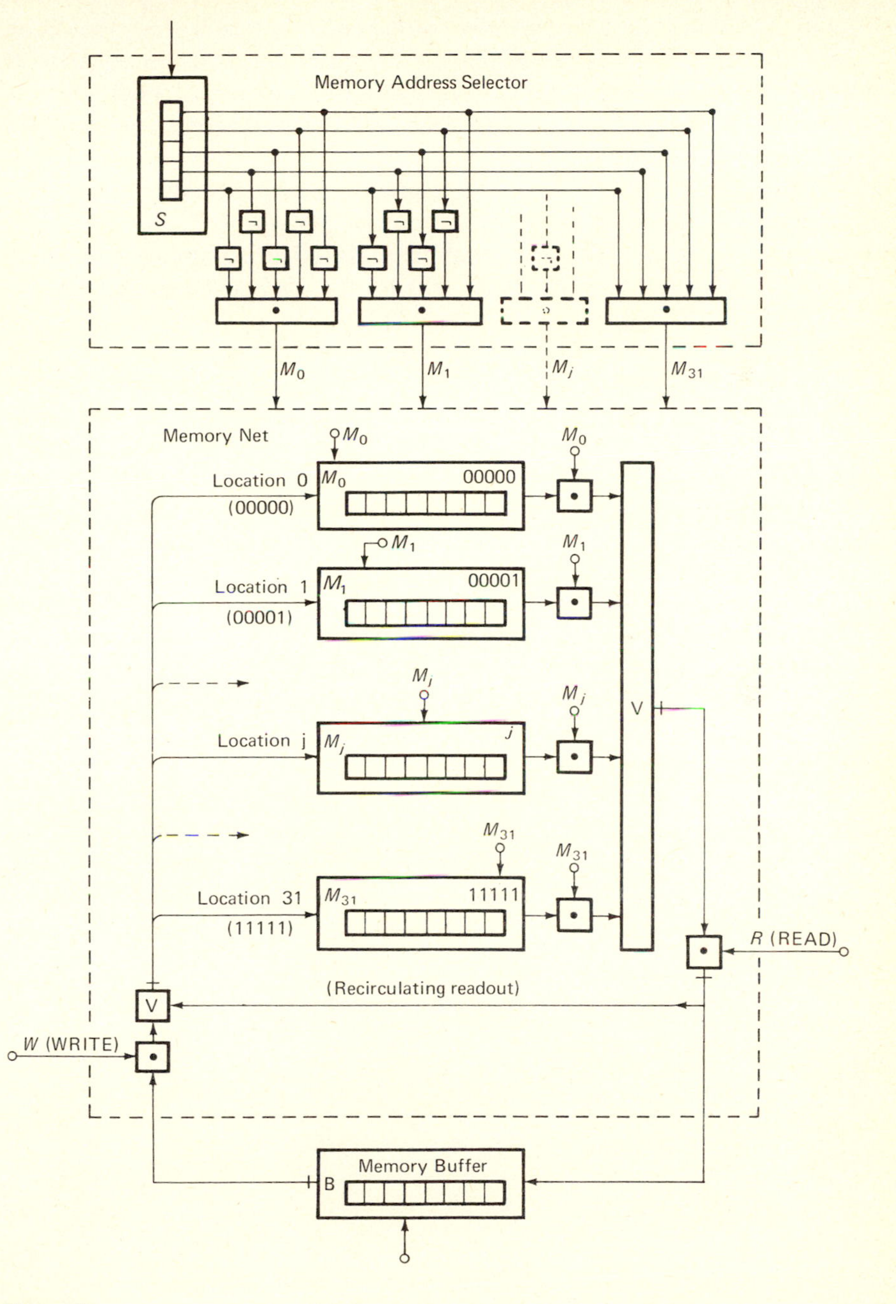
Memory Address Selector
S
M_0
M_1
M_j
M_{31}
Memory Net
Location 0
(00000)
00000
Location 1
(00001)
00001
Location j
j
Location 31
(11111)
11111
V
R (READ)
(Recirculating readout)
W (WRITE)
Memory Buffer
B

Figure 9

numbers and the algorithm for computing with these numbers. However, there are many ways of organizing the computations and therefore many different types of control units. We will first consider in detail a single-address type of organization; then later we will indicate more briefly other types of organizations.

In a single-address system the content (or word) of any memory register has two possible forms or meanings, as shown in Figure 10.

The word may represent a number in its positional notation, with the first digit indicating the sign (0 for positive, 1 for negative), as shown in the data word of Figure 10*a*. For example the word 00000110 could indicate the number 6, whereas 10000110 could indicate the negative number -6. With our 8-digit word we can represent 2^8 numbers ranging from -127 to $+127$, with a positive and a negative zero.

A word may also be coded as an instruction. An instruction word in a single-address system is shown in Figure 10*b*. It consists of an operation followed by a single address of the operand which is to be operated on.

Operations may be coded arbitrarily as sequences of digits. In this case we have 3 digits and thus $2^3 = 8$ possible operations. We will use the operation codes (called op-codes) given in Figure 11. There they are also represented by an abbreviated English symbol and a brief explanation. In a more practical computer there may be dozens or even hundreds of possible operations.

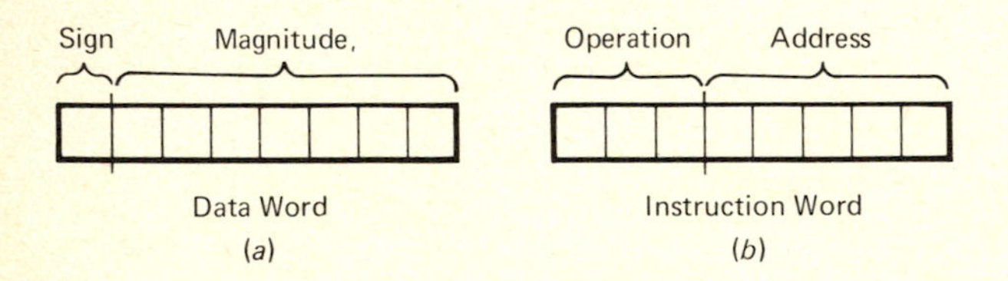

Figure 10

Op Code	Symbol	Operation
000	ADD	Add
001	SUB	Subtract
010	SHR	Shift right
011	LOD	Load
100	STO	Store
101	BRU	Branch unconditionally
110	BRP	Branch on positive
111	HLT	Halt

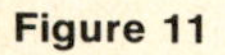

Figure 11

The second part of an instruction word specifies the address of the number to be operated on. In this case there are five digits and therefore only $2^5 = 32$ possible addresses we can refer to. Notice that by simply tripling the size, from 5 digits, we can address over 32,000 words of memory, which is much more practical.

The combination of an operation code and an address indicates what operation is performed and on what quantity. If an operation (such as ADD) requires two quantities, one quantity is specified by its address and the other is understood to be in the accumulator. For example, the word 00100110 has an operation part 001 (indicating subtraction) and an address part 00110 (indicating address 6). This instruction indicates that the contents of address 6 are to be subtracted from the contents of the accumulator and the result stored in the accumulator. Notice that this does not mean that the number 6 is to be subtracted from the accumulator.

It is also important to realize that any word, such as the previous 00100110, could have two meanings. It could represent a number (+38) or an instruction (of subtracting the content of address 6 from the accumulator). The proper meaning is determined through a program which is considered in the next section.

The Control Unit

A control unit for a single-address system consists of an instruction counter C, an instruction register I, and some timing networks. The instruction counter C serves essentially as a pointer indicating which word in memory is currently being addressed. Its contents are incremented (usually) so that it points at sequentially numbered memory registers. Thus it can cause instructions to be taken from the memory in sequential order. The instructions are then transferred to the instruction register I, which dissects the instruction word into its operation part (denoted Op $[I]$), and its address part (denoted Ad $[I]$).

We have now considered briefly the three main units of a computer having a single-address organization. The important registers of each unit are shown in Figure 12. Before starting a detailed synthesis of the computer let us first see how such an organization can be used to solve problems.

3. PROGRAMMING THE COMPUTER

Let us consider the problem of determining the amount of balloon payment on a loan. It will be a modification of a problem in Chapter 3 (Figure 31) for the convenience of our simple computer.

The problem is to find the balloon payment for a loan of $100 bor-

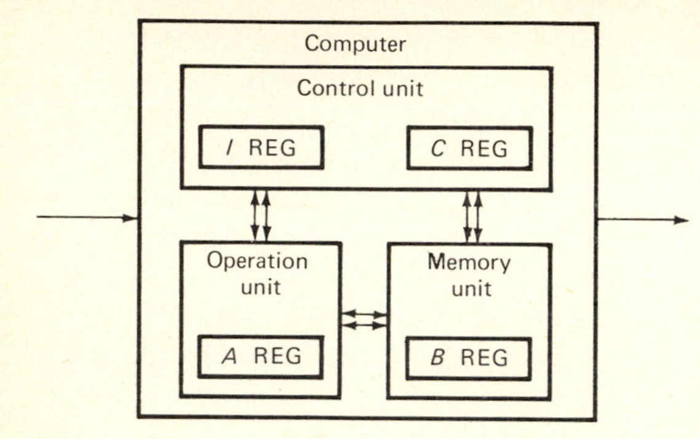

Figure 12

rowed for 5 years with payments of \$20 per year and a yearly interest rate of 12.5 percent on the unpaid balance (rounded off to the next lowest dollar).

Of course we could do this problem with five simple iterations:

$$\begin{aligned}
100 + \tfrac{100}{8} - 20 &= 100 + 12 - 20 = 92\\
92 + \tfrac{92}{8} - 20 &= 92 + 11 - 20 = 83\\
83 + \tfrac{83}{8} - 20 &= 83 + 10 - 20 = 73\\
73 + \tfrac{73}{8} - 20 &= 73 + 9 - 20 = 62\\
62 + \tfrac{62}{8} - 20 &= 62 + 7 - 20 = 49
\end{aligned}$$

The total amount of the balloon payment is \$49!

This computation can also be done by our simple computer. (In fact the interest rate of 12.5 percent was especially chosen since it corresponds to the ratio 1/8, and dividing by 8 in turn corresponds to three right shifts.)

The program for this balloon payment problem is shown in Figure 13. It consists of 16 words; the first 12 words are instructions, and the last 4 are numbers.

The first program (Figure 13*a*) indicates the precise words in each memory location; a formidable description. The second program (Figure 13*b*) is a symbolic description of the same program, which is easier to read. The English description (of Figure 13*c*) is a brief verbal representation of the program. Other higher levels of language (such as Basic, Algol or Fortran) are more convenient for programming, but they will not be considered here. These higher-level programming languages must ultimately be translated into a machine language of zeros and ones.

Machine Language			Symbolic Language		English Language
Location	Word		Location	Word	Comment
00000	01101100	Instruction words	ZRO	LOD SUM	Load Sum into Accum
00001	01000000		1	SHR	Shift three times
00010	01000000		2	SHR	(Divide by 8)
00011	01000000		3	SHR	to obtain interest
00100	00001100		4	ADD SUM	Add sum to interest
00101	00101110		5	SUB PYT	Subtract payment
00110	10001100		6	STO SUM	Store in SUM register
00111	01101101		7	LOD DRN	Load the Duration
01000	00101111		8	SUB INC	Subtract increment
01001	10001101		9	STO DRN	Store the new duration
01010	11000000		10	BRP ZRO	If DRN is + branch to 0
01011	11100000		11	HLT	Halt!
01100	01100100	Data words	SUM	100	The sum is 100 initially
01101	00000100		DRN	4	The duration is 4+1 = 5
01110	00010100		PYT	20	The payment is 20
01111	00000001		INC	1	The increment is 1
(*a*)			(*b*)		(*c*)

Figure 13

4. BEHAVIOR OF THE COMPUTER

After a program is stored in the memory and the start initiated, it behaves as indicated by the diagram of Figure 14. For example, let us follow the detailed computation of the program of Figure 13. First the instruction counter C is set to zero. It then is incremented sequentially as follows:

0. The sum is loaded from location 01100 (or 12) into the accumulator.

1,2,3. This sum is shifted three times to obtain the interest, which is left in the accumulator.

4. This interest in the accumulator is added to the sum from location 12, leaving the sum plus interest in the accumulator.
5. The payment (from location 14) is subtracted from the sum in the accumulator, leaving the new balance in the accumulator.
6. The sum from the accumulator is stored in location 12.
7. The duration (from location 13) is loaded into the accumulator.
8. The increment (a constant 1 from location 15) is subtracted from the accumulator.
9. The new duration is stored in location 13.
10. If the duration (in the accumulator) is positive, branch to location 0.
11. Halt! (when the duration becomes negative).

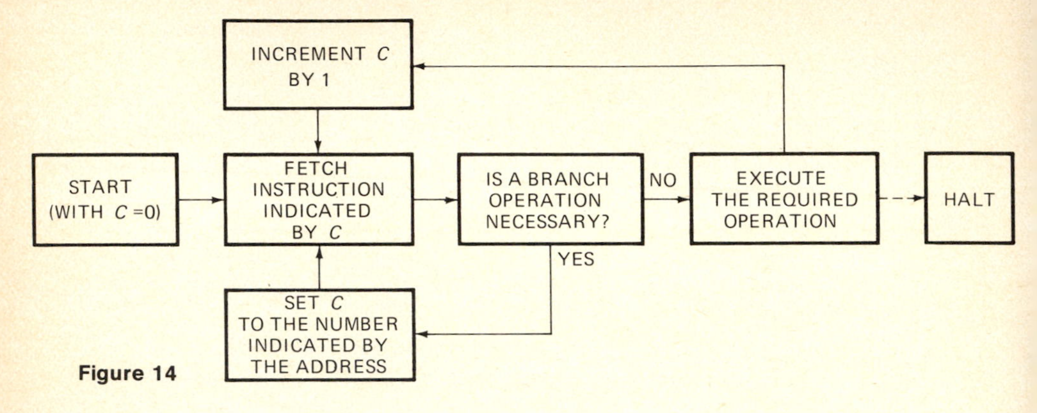

Figure 14

The entire detailed computation is shown in Figure 15. It shows the contents of the accumulator, the sum register, and the duration register at each interval (in decimal equivalent code for convenience).

Although this instantaneous description is extremely detailed for people, it is still not sufficiently detailed for a computer. We must extend the diagram of Figure 14 into more detailed transfer between registers. This will be done next.

The Timing Network

An important part of the control unit is the timing network, shown in Figure 16. The clock generates a sequence of pulses which go to the C gate of every register. These pulses also go into a divide network, which

A	S	D	A	S	D	A	S	D	A	S	D	A	S	D
???	100	4 (Initial state)												
100	100	4	92	92	3	83	83	2	73	73	1	62	62	0
50	100	4	46	92	3	41	83	2	36	73	1	30	62	0
25	100	4	23	92	3	20	83	2	18	73	1	15	62	0
12	100	4	11	92	3	10	83	2	9	73	1	7	62	0
112	100	4	103	92	3	93	83	2	81	73	1	69	62	0
92	100	4	83	92	3	73	83	2	62	73	1	49	62	0
92	92	4	83	83	3	73	73	2	62	62	1	49	49	0
4	92	4	3	83	3	2	73	2	1	62	1	0	49	0
3	92	4	2	83	3	1	73	2	0	62	1	-1	49	0
3	92	3	2	83	2	1	73	1	0	62	0	-1	49	-1
												-1	49	-1 (Final state)

Figure 15

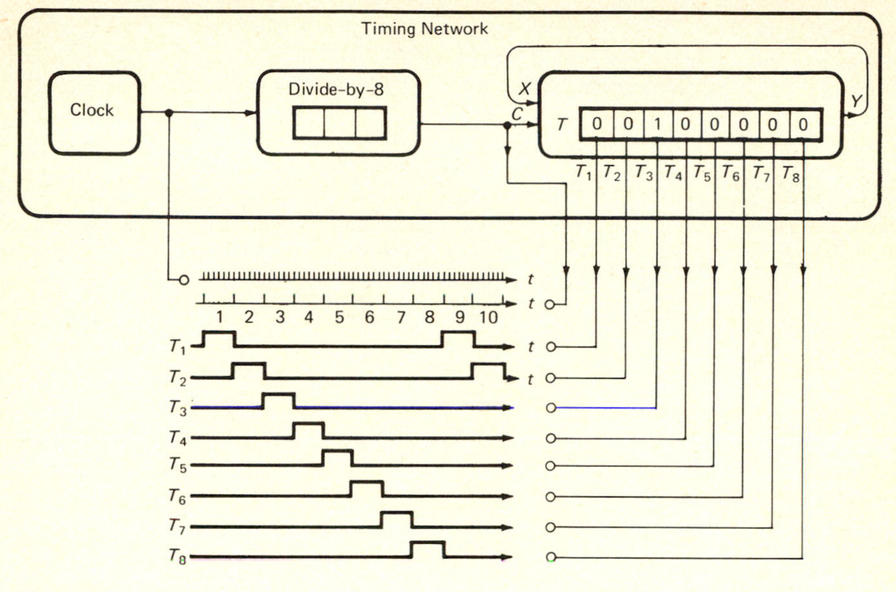

Figure 16

produces one output pulse for every eight input pulses. It is essentially the divide-by-8 network of Figure 87 in Chapter 5. This lower-frequency pulse then feeds the C input of a timing register T. The T register is a recirculating register (called a *ring counter*) with only one position having value 1. As this single bit moves from position to position, it sequentially applies a value of 1 first to T_1, then to T_2, etc., up to T_8, and then starts again at T_1. These terminals T_i will be used for sequencing operations between registers.

Sequencing the Register Transfers

We are now ready to specify the detailed sequencing between registers. This is summarized in Figure 17. Notice that it is simply a detailed elaboration of Figure 14.

Every cycle begins by fetching an instruction, which entails the following three transfers. During the first time interval T_1 the contents of the instruction counter C are transferred into the memory select register S

$$T_1: (C) \rightarrow S$$

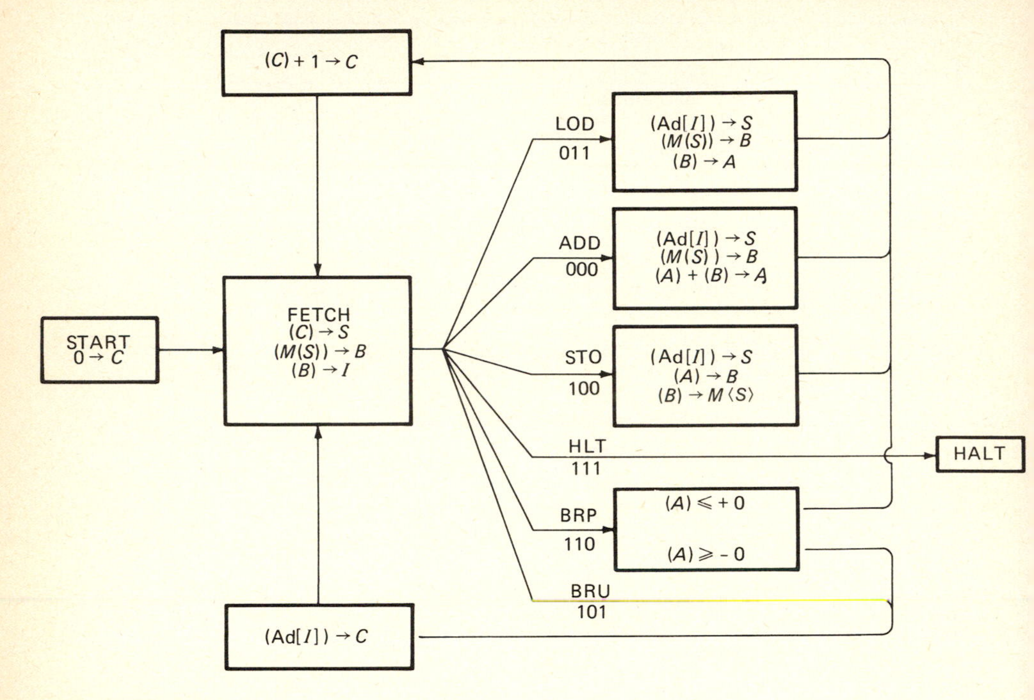

Figure 17

At the next time interval T_2, the contents of the memory register indicated by S are transferred into the memory buffer B

$$T_2\colon (M\langle S\rangle) \rightarrow B$$

At the third time interval T_3, the contents of the buffer are transferred into the instruction register I

$$T_3\colon (B) \rightarrow I$$

This concludes the *fetch part* of the cycle; the remainder is called the *execute* part of the cycle. The operation to be performed (executed) is determined by which operation f_i has been decoded by the operation-selection network. For example, if the operation is to load (indicated by the op-code 011 which operates f_3) the following sequence of transfers is performed. First, the contents of the address part of the instruction register are transferred to the select registers:

$$f_3T_4\colon (\mathrm{Ad}[I]) \rightarrow S$$

Then the contents of the memory register addressed by S are transferred to the buffer:

$$f_3 T_5: (M\langle S \rangle) \rightarrow B$$

Finally, the contents of the buffer are transferred into the accumulator.

$$f_3 T_6: (B) \rightarrow A$$

These three register transfers are sometimes called *micro operations,* and the sequence of them may be called a *micro program.*

Similarily, the operation of addition (op-code 000 and function f_0) consists of the three transfers (or micro operations)

$$f_0 T_4: (\text{Ad}[I]) \rightarrow S$$

$$f_0 T_5: (M\langle S \rangle) \rightarrow B$$

$$f_0 T_6: (A) + (B) \rightarrow A$$

The operation of storing (op-code 100, f_4) is slightly more unusual. It could consist of the three transfers

$$f_4 T_4: (\text{Ad}[I]) \rightarrow S$$

$$f_4 T_5: (A) \rightarrow B$$

$$f_4 T_6: (B) \rightarrow M\langle S \rangle$$

Since the first two transfers involve different registers, they can be performed at the same time, thus speeding up this cycle.

$$f_4 T_4: (\text{Ad}[I]) \rightarrow S, (A) \rightarrow B$$

$$f_4 T_5: (B) \rightarrow M\langle S \rangle$$

However, since our main goal is conceptual simplicity rather than speed, we will use the first set of transfer operations.

Following all the above operations the instruction counter C is incremented, and the cycle starts again.

$$T_7: (C) + 1 \rightarrow C$$

The branching operations differ considerably from all the above operations. For example, the unconditional branch (BRU, op-code 101, f_5) simply sets the instruction counter to the address indicated.

$$f_5 T_4: (\text{Ad}[I]) \rightarrow C$$

Then the cycle starts again.

The conditional branch (BRP, op-code 110, indicated by f_6) is performed only if the contents of the accumulator are positive (nonnegative actually). This is accomplished by using a small network $\bar{N}$ connected to the sign part of the accumulator, so that when the number is positive, the sign digit N has value 0 and the network $\bar{N}$ has value 1. The corresponding transfer operation is

$$f_6 T_4 \bar{N}\text{: } (\text{Ad}[I]) \rightarrow C$$

If the number in the accumulator is negative, there is no branch and the cycle starts again with the transfer

$$f_6 T_7 N\text{: } (C) + 1 \rightarrow C$$

All the operations are summarized in terms of their corresponding register transfers in Figure 18. This will be used later for designing.

The instantaneous description of the previous balloon-payment computation is shown in Figure 19. Only the first two instructions are illustrated, with the changing contents underlined.

5. STRUCTURE OF THE COMPUTER

The table of Figure 18 can be used to specify the detailed gating for various units of the computer. Consider, for example, the counting network of Figure 20.

	ADD	SUB	SHR	LOD	STO	BRU	BRP		HLT
	f_0	f_1	f_2	f_3	f_4	f_5	f_6		f_7
T_1	$(C) \rightarrow S$	$(C) \rightarrow S$	$(C) \rightarrow S$	$(C) \rightarrow S$	$(C) \rightarrow S$	$(C) \rightarrow S$	$(C) \rightarrow S$		$(C) \rightarrow S$
T_2	$(M(S)) \rightarrow B$	$(M(S)) \rightarrow B$	$(M(S)) \rightarrow B$	$(M(S)) \rightarrow B$	$(M(S)) \rightarrow B$	$(M(S)) \rightarrow B$	$(M(S)) \rightarrow B$		$(M(S)) \rightarrow B$
T_3	$(B) \rightarrow I$	$(B) \rightarrow I$	$(B) \rightarrow I$	$(B) \rightarrow I$	$(B) \rightarrow I$	$(B) \rightarrow I$	$(B) \rightarrow I$		$(B) \rightarrow I$
T_4	$(\text{Ad}[I]) \rightarrow S$	$(\text{Ad}[I]) \rightarrow S$	$\text{SR}(A) \rightarrow A$	$(\text{Ad}[I]) \rightarrow S$	$(\text{Ad}[I]) \rightarrow S$	$(\text{Ad}[I]) \rightarrow C$	$A \geqslant 0$	$A \leqslant 0$ $(\text{Ad}[I]) \rightarrow C$	$(\text{Ad}[I]) \rightarrow C$
T_5	$(M(S)) \rightarrow B$	$(M(S)) \rightarrow B$		$(M(S)) \rightarrow B$	$(A) \rightarrow B$				
T_6	$(A)+(B) \rightarrow A$	$(A) - (B) \rightarrow A$		$(B) \rightarrow A$	$(B) \rightarrow M(S)$				
T_7	$(C) + 1 \rightarrow C$	$(C) + 1 \rightarrow C$	$(C) + 1 \rightarrow C$	$(C) + 1 \rightarrow C$	$(C) + 1 \rightarrow C$		$(C) + 1 \rightarrow C$		
T_8									

Figure 18

	Time	Instruction Counter	Select Register	Instruction Register	Buffer Register	Accumulator	Micro Operations
	T_i	C	S	I	B	A	
LOAD SUM	T_0	*00000*	?	?	?	?	Initial state
	T_1	00000	*00000*				$(C) \to S$
	T_2	00000	00000		*01101100*		$(M(S)) \to B$
	T_3	00000	00000	*01101100*	01101100		$(B) \to I$
	T_4	00000	*01100*	01101100	01101100		$(\text{Ad},[I]) \to S$
	T_5	00000	01100	01101100	*01100100*		$(M(S)) \to B$
	T_6	00000	01100	01101100	01100100	*01100100*	$(B) \to A$
	T_7	*00001*	01100	01101100	01100100	01100100	$(C) + 1 \to C$
SHIFT RIGHT	T_1	00001	*00001*	01101100	01100100	01100100	$(C) \to S$
	T_2	00001	00001	01101100	*01000000*	01100100	$(M(S)) \to B$
	T_3	00001	00001	*01000000*	01000000	01100100	$(B) \to I$
	T_4	00001	00001	01000000	01000000	*00110010*	$\text{SR}(A) \to A$
	T_5	00001	00001	01000000	01000000	00110010	
	T_6	00001	00001	01000000	01000000	00110010	
	T_7	*00010*	00001	01000000	01000000	00110010	$(C) + 1 \to C$

Figure 19

The first row of Figure 18 indicates that the contents of the counter C are transferred into the select register at time T_1, and so the corresponding gate G_1 is simply described by

$$G_1 = T_1$$

Gate G_2 controls the transfer from the instruction register to the select register. This operation

$$(\text{Ad}[I]) \to S$$

appears only four times in the table of Figure 18, occurring for the combinations

$$f_0 T_4 \quad \text{or} \quad f_1 T_4 \quad \text{or} \quad f_3 T_4 \quad \text{or} \quad f_4 T_4$$

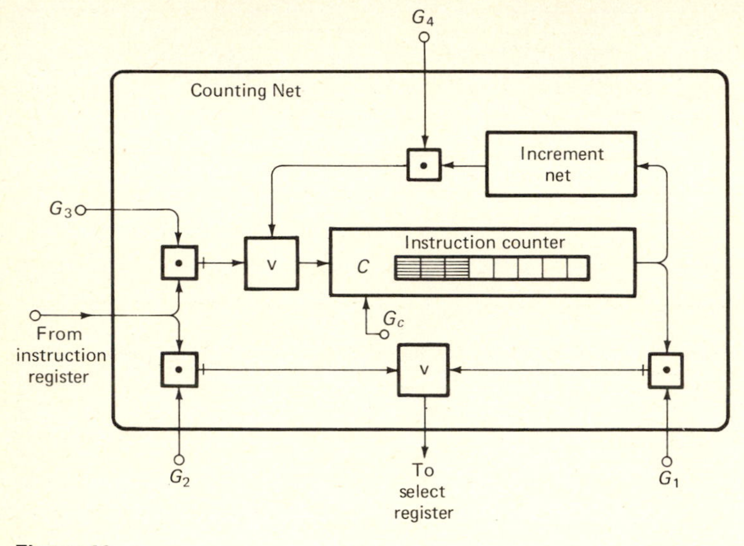

Figure 20

The corresponding control gate G_2 can be described by the expression

$$G_2 = f_0 T_4 \vee f_1 T_4 \vee f_3 T_4 \vee f_4 T_4 = (f_0 \vee f_1 \vee f_3 \vee f_4)\, T_4$$

Similarly gate G_3 controls the transfer from the instruction register to the instruction counter for either of the branch operations at time T_4. The network operating this control gate is described by

$$G_3 = (f_5 \vee \bar{N} f_6) T_4$$

The gate G_4 controls the incrementing of the instruction counter. The contents of C are simply added sequentially to a constant value of 1. The gate is described by

$$G_4 = (f_0 \vee f_1 \vee f_2 \vee f_3 \vee f_4 \vee N f_6) T_7$$

Finally, the instruction counter register C shifts itself only when gate G_1 or G_3 or G_4 is operated. This is described by

$$G_C = G_1 \vee G_3 \vee G_4$$

This process of specifying the control-gate functions is continued for all the units of the entire computer as shown in Figure 21.

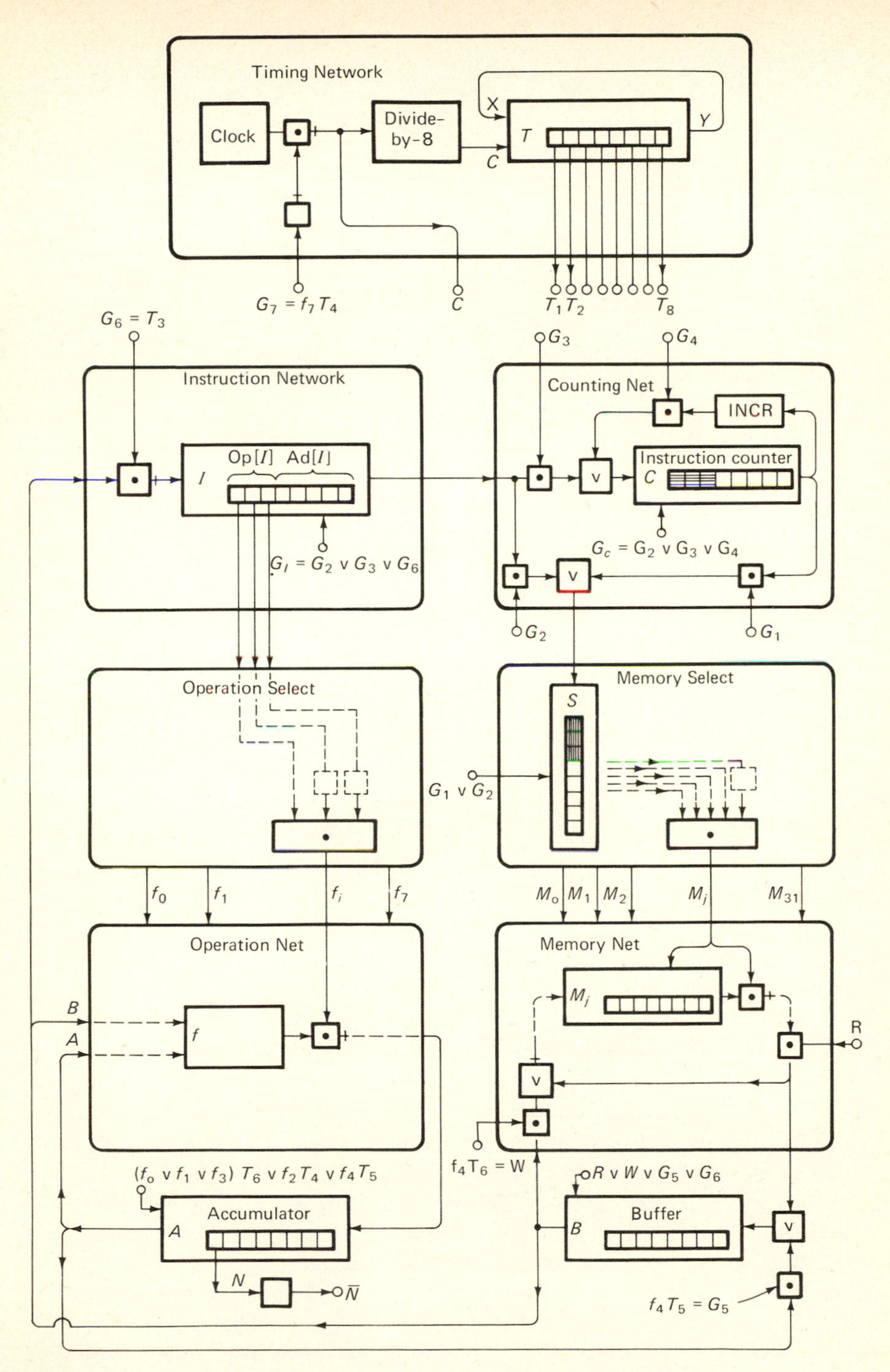

Timing Network
Clock
Divide-by-8
X
Y
T
C
$G_7 = f_7 T_4$
C
$T_1 T_2$
T_8
$G_6 = T_3$
G_3
G_4
Instruction Network
Op[I] Ad[I]
I
$G_I = G_2 \vee G_3 \vee G_6$
Counting Net
INCR
Instruction counter
C
$G_c = G_2 \vee G_3 \vee G_4$
G_2
G_1
Operation Select
Memory Select
S
$G_1 \vee G_2$
f_0
f_1
f_i
f_7
M_0
M_1
M_2
M_j
M_{31}
Operation Net
B
A
f
Memory Net
M_j
R
$f_4 T_6 = W$
$(f_0 \vee f_1 \vee f_3)\ T_6 \vee f_2 T_4 \vee f_4 T_5$
Accumulator
A
N
$\bar{N}$
$R \vee W \vee G_5 \vee G_6$
Buffer
B
$f_4 T_5 = G_5$

Figure 21

As another example, consider the memory unit. The W terminal, controlling the writing into the memory, is given as

$$W = f_4 T_6$$

and the read control R is given by

$$R = T_2 \vee (f_0 \vee f_1 \vee f_3) T_5$$

The buffer B then shifts itself for these two cases and also when the buffer B is loading the instruction register I (at T_3) or when the accumulator A is loading B ($f_4 T_5$). This can be described by the expression

$$G_B = R \vee W \vee G_5 \vee G_6 \qquad \text{where } G_5 = f_4 T_5 \text{ and } G_6 = T_3$$

You may wish to check some of the other gate specifications given in Figure 21. Notice that the gating networks are not shown, but they can be constructed from the given switching algebraic specifications.

6. EXTENDING THE COMPUTER

It is important to realize that the computer we have considered is an extremely simple one. It could be modified and extended in many ways.

The simplest modification would be to increase the size of all registers (and therefore all words). In a practical computer each word could be from 16 to 64 bits long, and there may be from 2,000 to 64,000 such words. There may be dozens or hundreds of operations, each of which (such as subtraction) could be done in many ways. The register transfers are often performed in a parallel manner as opposed to our serial or sequential transfer. Parallel transfer increases the speed of computation, but it also increases the complexity and cost of the computer. Of course we must also have input-output units such as teletypewriters, punched-card readers, and printers.

Other modifications could involve different interpretations or meanings of the contents or words. For example a four-address system could have an instruction word consisting of an operation followed by four addresses A, B, C, D. It could indicate that the operation is performed on the contents of register A and B, the result is stored in C, and the next instruction fetched from D. Notice that the computer corresponding to this type of instruction would not require an accumulator or an instruction counter.

Many other extensions and concepts were not considered. The important concepts include indexing, microprogramming, compiling, and higher-level languages.

SUMMARY

In this chapter we considered very briefly one example of a large system, its structure, and behavior. This system is known as a computer, but that is mainly because computation is all it has been used for. It can also be used for communication and control. More complex systems, such as economies, cities, ships, and the human body, can be described by probabilistic, stochastic, and other models; but these are beyond the scope of this book.

PROBLEMS

1. Construct a subtractor for the computer described in this chapter.

2. Subtraction of two positive numbers A and B of fixed maximum length n can be done by a method called *ones complement* according to the following algorithm:

(*a*) Find the complement of B by NOTting every digit in B.

(*b*) Add number A and the above complement of B.

(*c*) If the resulting number has $n + 1$ digits, eliminate the most significant digit and then add 1 to this result. The resulting number will be positive.

(*d*) If the resulting number does not have $n + 1$ digits, take its complement (and the resulting number is negative).

Check this method for the numbers $21 - 13$, and $13 - 21$ when expressed as binary numbers of length 8. Prove this result in general. Construct such a subtractor for our computer.

3. What computation is performed by the following program:

(*a*) When it is started in location 0?

(*b*) When it is started in location 1?

Location	*Word*
0000	11100001
0001	01100110
0010	00000000
0011	10000101
0100	11000000
0101	11100000
0110	01000000

4. Construct a program in machine language to select the maximum value of N integers and to store this value in location 31. What is the maximum limit for N in our simple computer?

5. How does our computer stop? How can this computer be speeded up? Construct the device which corresponds to the operation of a right shift (or divide by 2).

6. Multiplication can be performed by repeated shifting, ANDing, and adding. Illustrate this operation, showing how it can be done by registers (hardware) and also by a program (software).

7. If a computer similar to ours, with three-digit operation code and five-digit address, was constructed from three-valued components and a corresponding base 3 system:

(*a*) How many operations would be possible?

(*b*) How many locations could be addressed?

(*c*) How many integers could be represented?

(*d*) How could the integers be represented?

8. Construct the state diagram or transition table for a three-valued sequential adder.

9. Construct a program in a symbolic language for the balloon-payment problem, using a two-address system, a three-address system, and a four address system.

10. A two-address instruction format specifies the memory locations of two operands, whereas a (1 + 1)-address format gives the address of the operand and the address of the next instruction. Compare briefly the possible structures, advantages, speeds, etc., of these two types of systems.

Appendix
Electronic Digital Systems

The principles of digital systems apply to physical components of many types—electrical, mechanical, fluidic, and others. However, their greatest application has been in the area of electronics, mainly because of the great speed, reliability, and convenience of electronic components. This Appendix provides a brief survey of digital electronic systems.

1. BINARY DIGITAL LEVELS

Most digital systems operate at two distinct ranges or levels of voltage such as 0 and 5 volts, or -3 and -11 volts, or -10 and $+10$ volts. A typical voltage waveform is shown in Figure 1.

2. LOGIC CONVENTION

The voltage levels may be designated by the symbols H (high) and L (low) or by the binary digits 0 and 1. If the highest voltage level is assigned the digital value 1 (and the lowest 0), this is known as the *positive-logic* con-

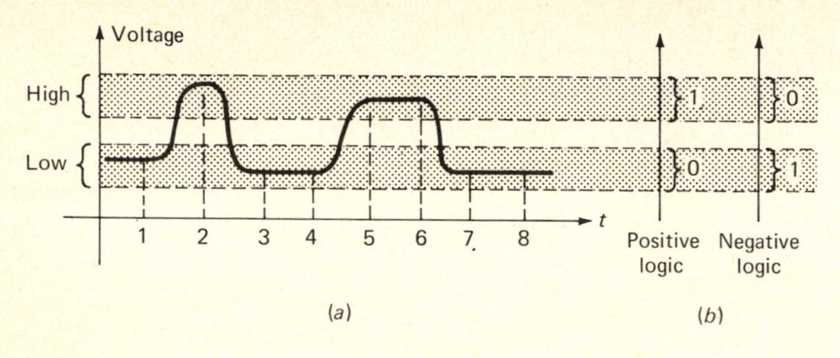

Figure 1

vention. In the *negative-logic* convention the lowest voltage level corresponds to the digital value of 1. The logic conventions are illustrated in Figure 1*b*.

Most complex digital systems such as computers are synchronized; i.e., they are operated at discrete time intervals. There is usually a clock which controls the sampling of a sequence. For example, the voltage waveform of Figure 1 sampled at each integral time interval 1, 2, 3, . . . would be read as the binary sequence

0100110 · · · in the positive logic

or

1011001 · · · in the negative logic

3. BASIC ELECTRONIC COMPONENTS

The basic components of electronic digital systems are devices such as resistors, diodes, and transistors of various sorts as shown in Figure 2. They can operate at speeds measured in nanoseconds (billionths of a second). This is quite fast when it is realized that light travels about 1 foot in a nanosecond.

These devices (when used in digital systems) behave in the following way. A resistor simply resists or impedes the flow of current through it. A diode allows current flow in one direction (from anode *A* to cathode *C*)

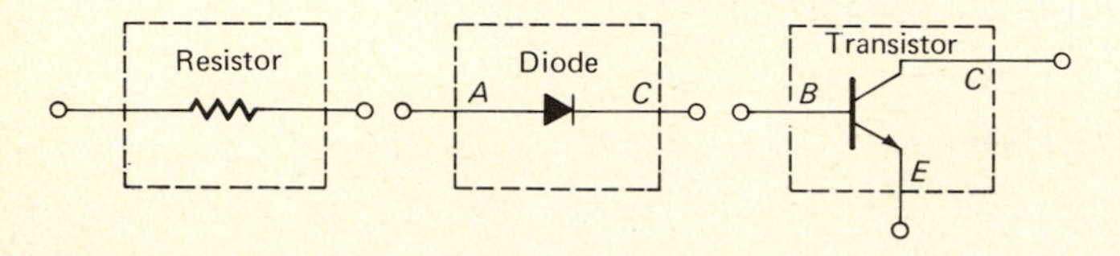

Figure 2

while preventing the flow in the opposite direction. A transistor has current flow from C (collector) to E (emitter) controlled by another current in B (base). A more detailed description of these devices can be found in any electronics text. This description is not necessary for what follows since our emphasis is not on how to interconnect them but, once they are interconnected, on how this interconnected system can be used to construct larger and larger systems.

4. BASIC DIGITAL COMPONENTS

Electronic components may be interconnected in many ways to make up digital components such as ORs and NORs and flip-flops. A digital component is usually viewed as a black box, as in Figure 3, with usually two or three inputs X_i at the left, one or two outputs Y_j at the right, some input power connections on top, and a common ground at the bottom.

For example, a common diode logic circuit is shown in Figure 4 with a typical voltage waveform. The behavior of this network can also be summarized by the table of Figure 5*a*. Notice that the output voltage is approximately equal to the maximum of the two input voltages. If the higher voltage is given the value 1 (positive-logic convention), Figure 5a can be redrawn as Figure 5*b*, which corresponds to an OR element. The output level is high if one or the other of the input levels is high.

However, if the lower voltage is given the value 1 (negative convention), Figure 5*a* can be redrawn as Figure 5*c*, which corresponds to an AND element. The output level is low if one and the other of the input levels are low.

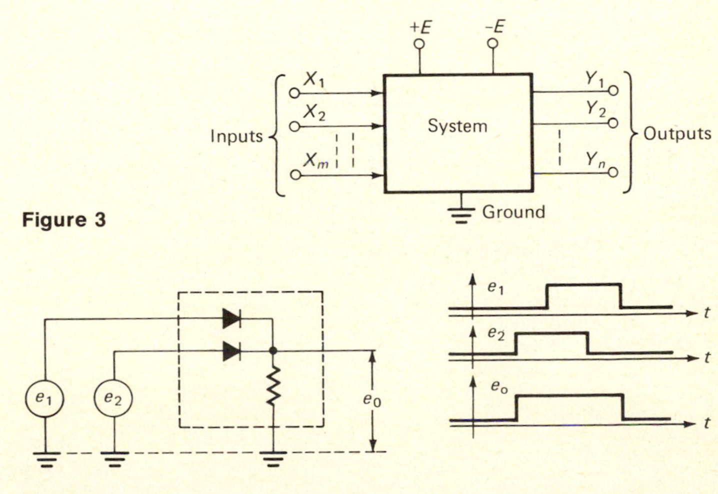

Figure 3

Figure 4

Voltage

e_1	e_2	e_0
0	0	0
0	10	8
10	0	8
10	10	9

(*a*)

Positive Logic OR

e_1	e_2	e_o
0	0	0
0	1	1
1	0	1
1	1	1

(*b*)

Negative Logic AND

e_1	e_2	e_o
1	1	1
1	0	0
0	1	0
0	0	0

(*c*)

Figure 5

This example indicates that the logical designation of the behavior (OR or AND) depends on the convention used. We will generally use the positive convention unless otherwise indicated; however, this consistency cannot always be assumed in other places. In a similar manner the British are consistent in driving on the left side of the street; the Canadians are consistent in driving on the right side. There is no confusion provided you know the convention and are consistent in applying it.

5. CONFIGURATIONS (OR FAMILIES)

Digital components are often classified according to the devices within them and how these devices are interconnected. Some typical families follow.

A *diode logic* (abbreviated DL) OR component is shown in Figure 4, and an AND component is shown in Figure 6. A problem with such diode logic systems is that the output voltage levels are always slightly lower than the input levels, as shown in Figure 5*a*. If many such components are interconnected, this attenuation of levels may cause problems.

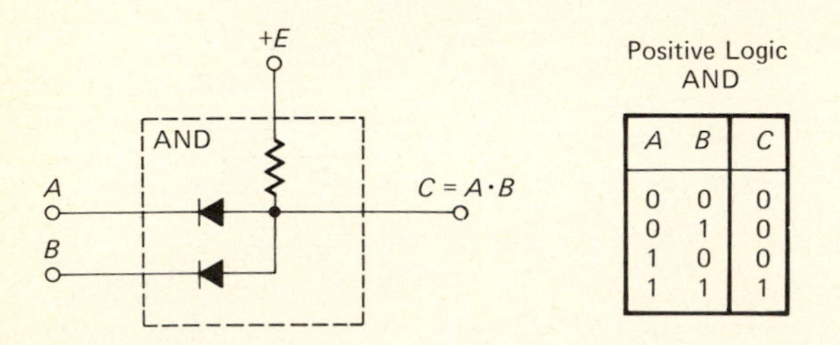

Positive Logic AND

A	B	C
0	0	0
0	1	0
1	0	0
1	1	1

Figure 6

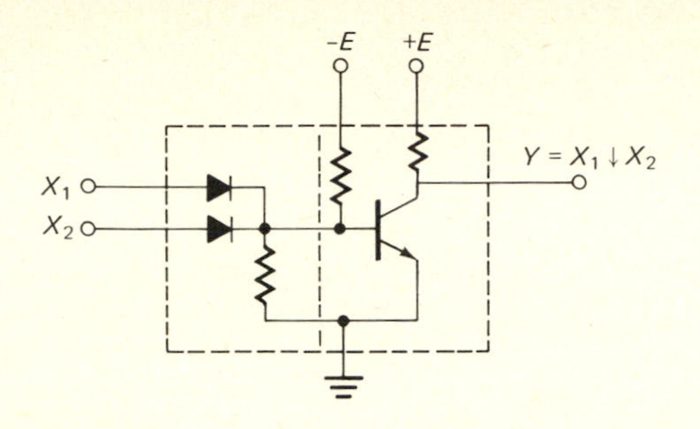

NOR

X_1	X_2	$X_1 \downarrow X_2$
0	0	1
0	1	0
1	0	0
1	1	0

Figure 7

The levels can be restored by adding a transistor network, as in the *diode transistor logic* (DTL) of Figure 7. The transistor network actually serves as an invertor or NOT network following the OR, and so this network is a NOT-OR, usually called a NOR.

Another analogous NOR component is given by the *resistor transistor logic* (RTL) component of Figure 8.

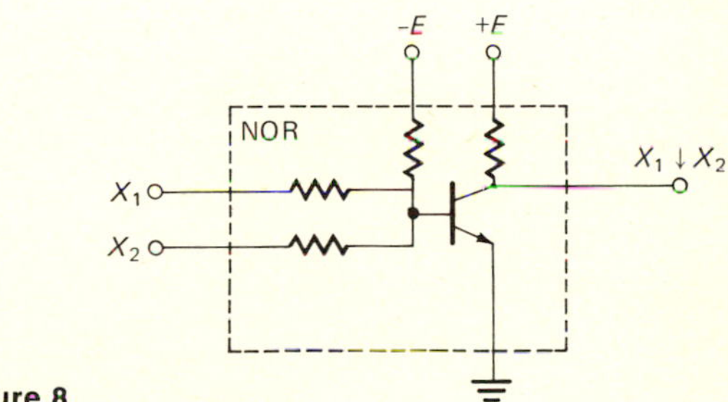

Figure 8

Still another interesting family of logic components is the *direct-coupled transistor logic* (DCTL) of Figure 9. Notice that these components require only one power supply, compared to the two previous families RTL and DTL, which require two supplies.

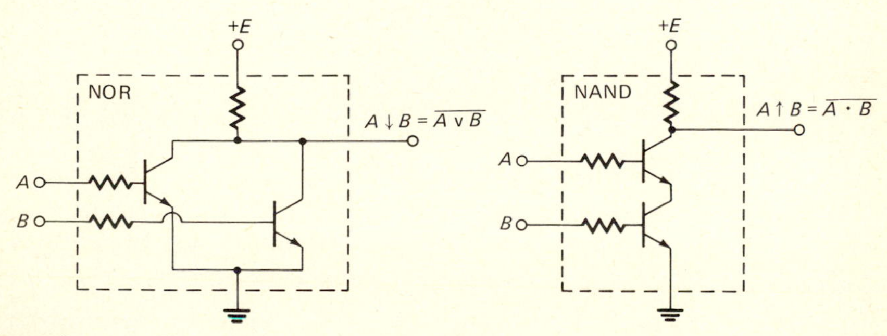

Figure 9

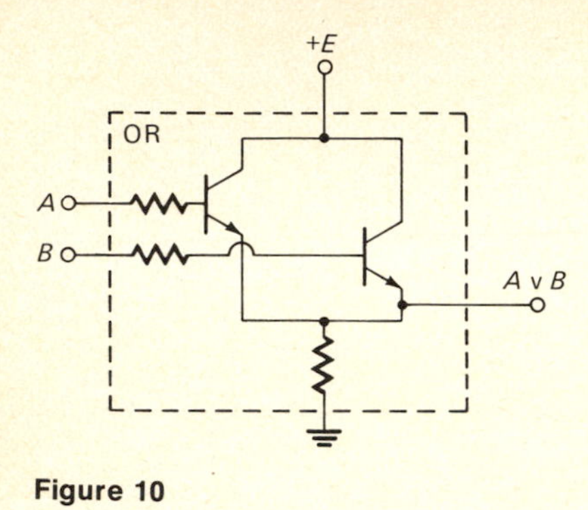

Figure 10

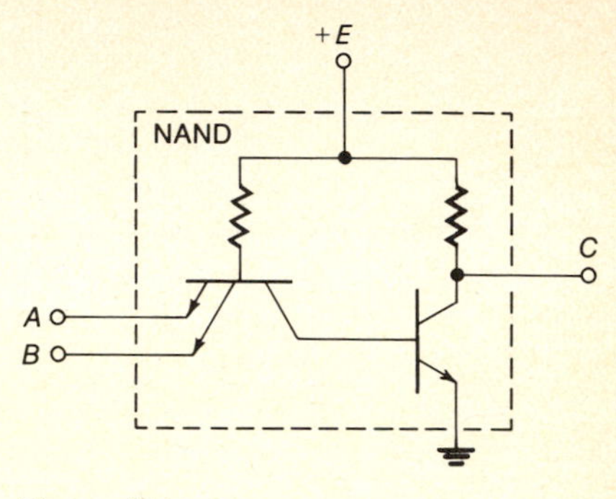

Figure 11

The *emitter-coupled transistor logic* (ECTL), also called *current mode logic* (CML), component of Figure 10 operates at very high speeds.

Another configuration, called *transistor-transistor logic* (TTL), in Figure 11, makes use of a transistor with two emitters.

We have briefly surveyed six types of electronic logic configurations (DL, DTL, RTL, DCTL, ECL, TTL). There are, and undoubtedly will be, more configurations. The choice of which configuration to use in a particular application depends on considerations of cost, speed, reliability, size, and power.

6. INTEGRATED CIRCUITS

Electronic digital components of various configurations are readily available in small units through a development known as integrated circuitry. For example, the package of Figure 12*a* is shown about full size ($\frac{1}{4}$ by $\frac{3}{4}$ inch) and contains the four NOR gates as shown in Figure 12*b*. Such components are available in most electronics stores at a cost of a few dollars.

The given component is from the RTL family. Other families may have different voltage levels and different terminals to apply the power. For example, DTL components use power terminals 7 and 14 rather than 4 and 11.

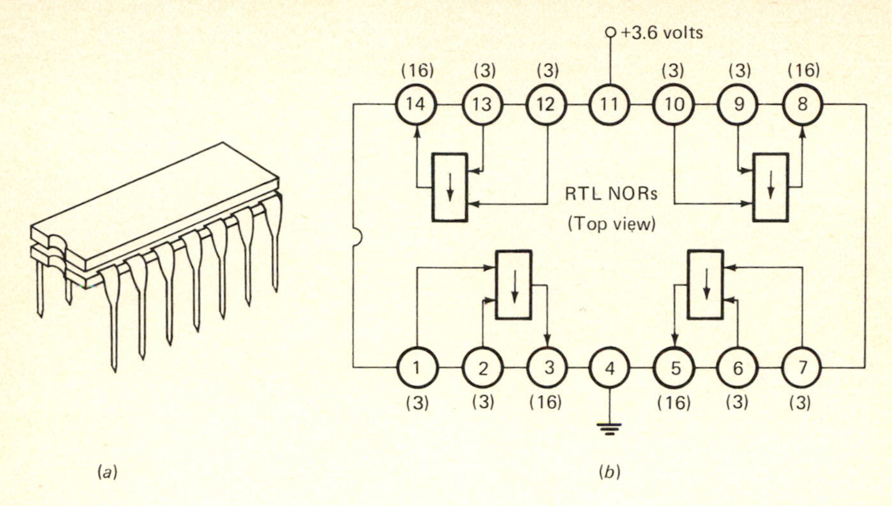

Figure 12

7. LOADING: FAN-IN AND FAN-OUT

The numbers (3 and 16) associated with the inputs and outputs of Figure 12*b* are to prevent overloading of these components. Each output is assigned a number called *fan-out* (in this case 16), which is related to the current it is capable of supplying. Similarly each input is assigned a number called *fan-in* (in this case 3), which is related to the amount of current it requires. The numbers are so chosen so that the output of one component can be connected to the inputs of any number of components provided that the sum of the fan-ins is less than or equal to the fan-out. Thus the output of the previous NOR can be connected to the inputs of at most five other NORs since

$$3 + 3 + 3 + 3 + 3 \leq 16$$

There are other ways of specifying proper loading, but as these ways become more precise, they also become more complex.

8. INTERCONNECTING DIGITAL COMPONENTS

Digital components can be interconnected in many ways to make larger systems such as shown in Figure 13. The majority vote system of Figure 13*a* provides a high output value if the majority of the inputs (two or

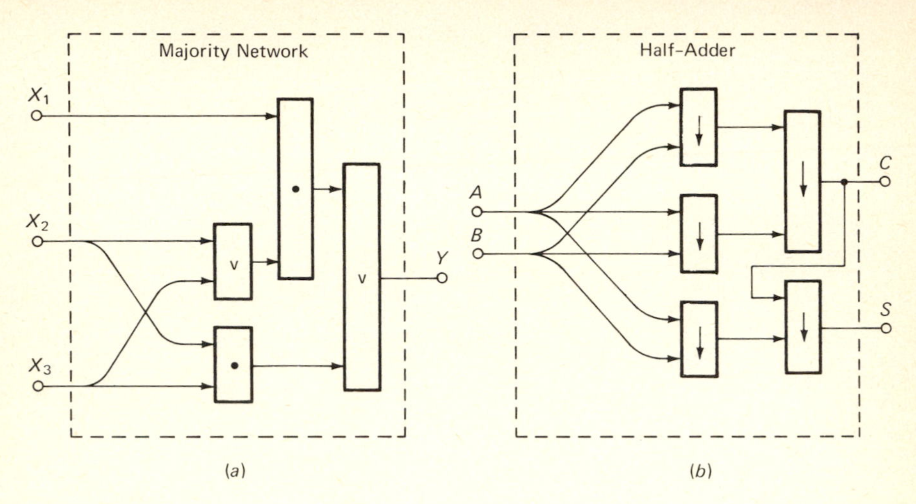

Figure 13

three) have a high value. The adder of Figure 13*b* produces a sum *S* and a carry *C* corresponding to inputs *A* (augend) and *B* (addend). The methods of designing such systems are developed in Chapter 1.

9. FLIP-FLOPS

Digital components can also be interconnected using feedback, in which the output of a component can ultimately influence its own input. For example, two NOR components could be connected as in Figure 14 to make a memory component known as a set-reset flip-flop. The behavior of this flip-flop is described by the table of Figure 15. The four rows of this table are interpreted as follows:

1. If no inputs are applied, the output does not change.
2. If the reset alone is applied, the output becomes 0.
3. If the set alone is applied, the output becomes 1.
4. If both set and reset are simultaneously applied, the output is undetermined; sometimes 0, sometimes 1.

In practice we can usually avoid this fourth case.

A typical input-output time sequence is shown in Figure 16. Notice that the output indicates which of the two inputs *S* or *R* was applied last. If the set was applied last, the output is high; it the reset was applied last, the output is low. Thus the flip-flop could serve as a storage or memory element.

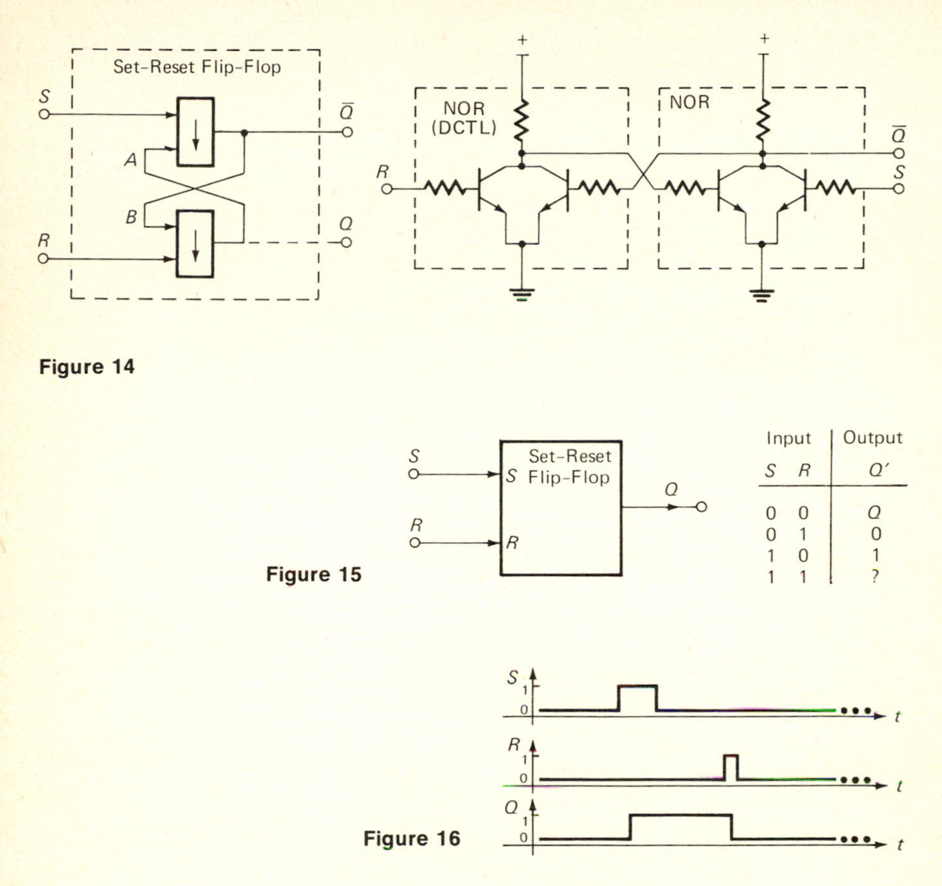

Figure 14

Figure 15

Figure 16

It is often useful to control a flip-flop by adding a gate or clock input so that the inputs can affect the outputs only when this clock pulse is applied. Such a clocked *SR* flip-flop is easily made by adding some AND components, as in Figure 17, to gate the inputs into the flip-flop only when the clock pulse is applied.

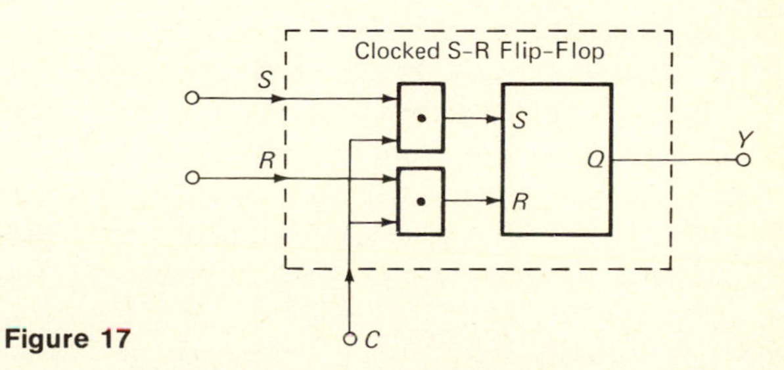

Figure 17

Another useful sequential device is the clocked *JK* flip-flop of Figure 18. The output changes only when the voltage on the clock switches from a high value to a low value. This flip-flop behaves much the same as the *SR* flip-flop except in one case. When both inputs have value 1, the output changes (or toggles) to the opposite value $\bar{Q}$.

The diagram of Figure 19 shows two clocked *JK* flip-flops as a single integrated circuit in the DTL family. Notice that the NOT of each output is available on a separate terminal (5 and 9). There are also direct clear inputs (terminals 4 and 10) for separate direct control of the flip-flops.

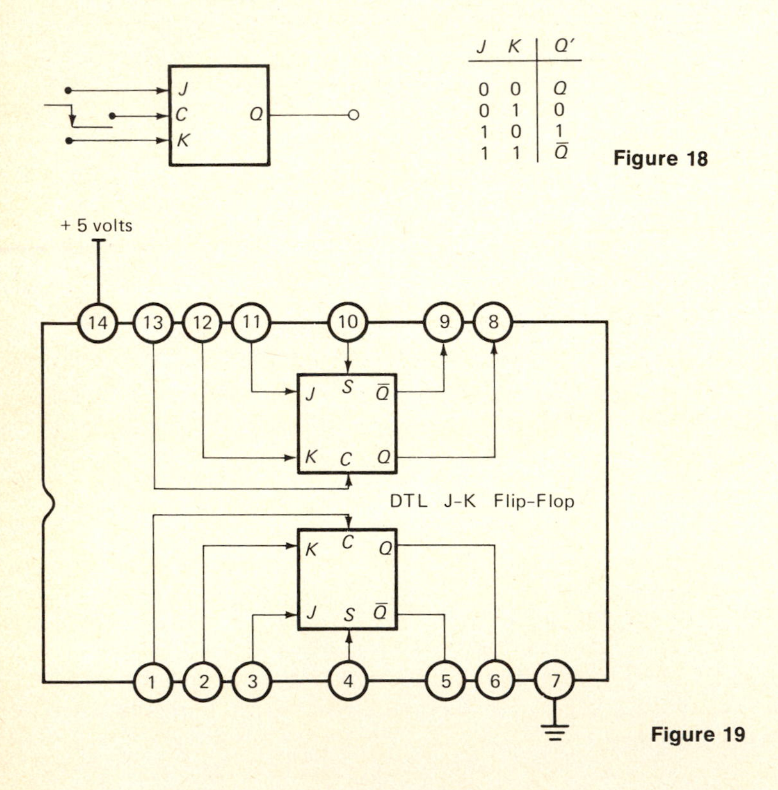

J	K	Q′
0	0	Q
0	1	0
1	0	1
1	1	$\bar{Q}$

Figure 18

Figure 19

Flip-flops can be interconnected with other switching devices to form sequential systems. For example, the sequential adder of Figure 20*a* can be constructed from two half adders and a clocked *SR* flip-flop.

The waveforms of Figure 20*b* illustrate the addition of the numbers 12 and 6 to obtain the resulting sum of 18. These numbers are represented in a binary positional notation. For example,

$$10010 = 1 * 2^4 + 0 * 2^3 + 0 * 2^2 + 1 * 2^1 + 0 * 2^0 = 18$$

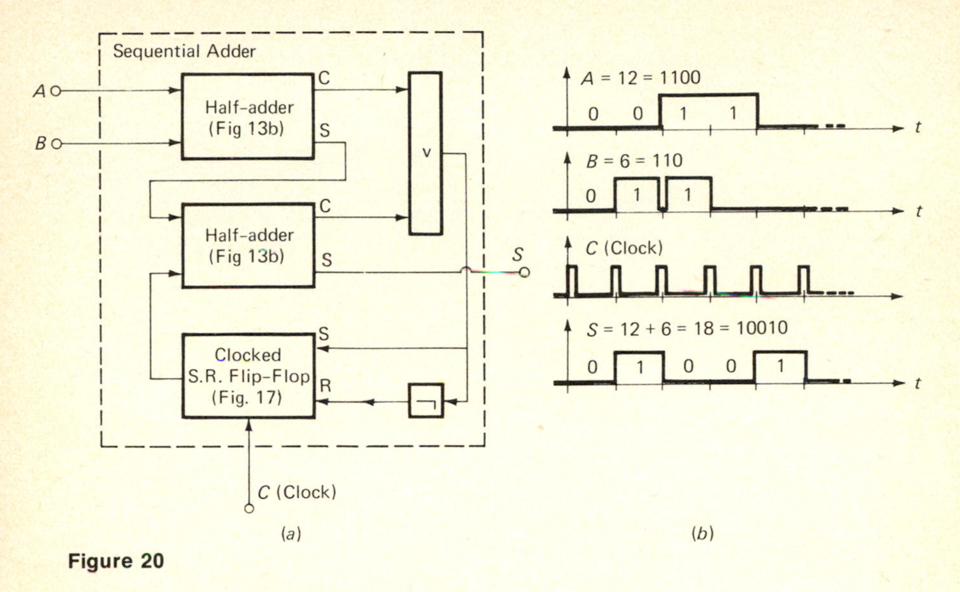

Figure 20

Notice that the numbers are added with the least significant digits considered first, and that there is no limit to the length of binary numbers which can be used. Other such sequential systems are considered in Chapter 5.

10. SHIFT REGISTERS

One of the most useful components of large systems are shift registers. They are used in Chapter 7 to construct digital computers. Shift-right registers can be made from either clocked *JK* or *SR* flip-flops, as shown in Figure 21.

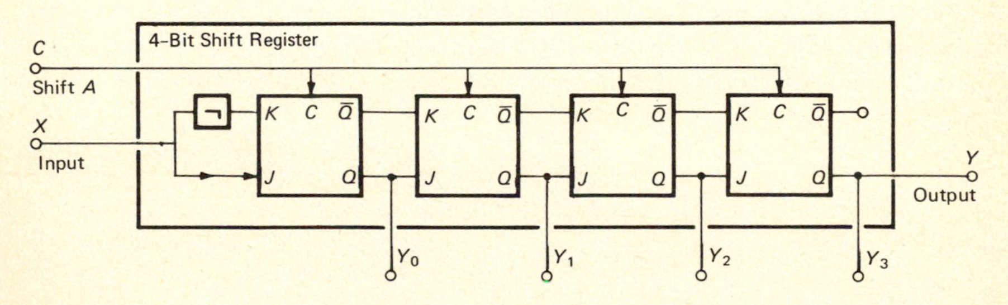

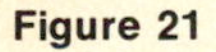
Figure 21

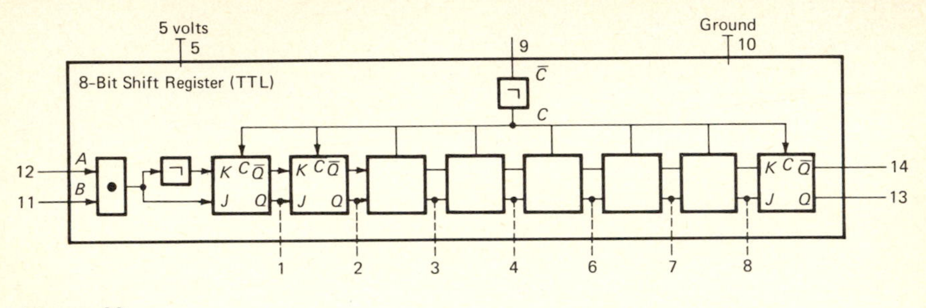

Figure 22

At each clock pulse (shift command) the value of the input X enters the first flip-flop, and the value of each flip-flop is shifted to the next flip-flop to its right. After n shifts an input value appears at the output having been shifted through all the n flip-flops in sequence. So the values at the previous four time intervals are stored in order on the four flip-flops. Using such shift registers it is possible to store and manipulate sequences of digits for computation, communication, and control.

A larger shift register in the TTL family is shown in Figure 22. Through a process known as medium-scale integration (or MSI) it is made to occupy the same size of package as shown in Figure 12*a*.

SUMMARY

This has been a brief survey of some electronic binary digital devices (diodes, transistors), components (ANDs, NORs, flip-flops), and systems (majority voter, adder, shift register). It has introduced some technical developments (integrated circuits) and conventions (positive and negative logic, fan-in, etc.).

This field develops so quickly that such a survey is quickly obsolete because of other devices (field-effect transistors, FETs), larger components (read-only memories, ROMs), more families (high threshold logic, HTL), other technical developments (large-scale integration, LSI), and possibly many-valued logics (rather than just the binary). It is hoped that some of the concepts of this Appendix are sufficiently "portable" to be applied to any new developments.

References

The subject matter of this book is usually available only at a higher, less accessible level or scattered in many other books, so there are few references for corequisite or parallel reading. However, after this book has been read, many alternative references in many fields are possible. Some of these are indicated below.

The deterministic systems of Chapters 1 and 5 are discussed in many books on switching theory, sequential machines, and automata theory; for example:

Booth, T.L.: "Digital Networks and Computer Systems," Wiley, New York, 1971.

Gill, A.: "Introduction to the Theory of Finite-state Machines," McGraw-Hill, New York, 1962.

Hennie, F.C.: "Finite-state Models of Logical Machines," Wiley, New York, 1967.

Hill, F.J., and G.R. Peterson: "Switching Theory and Logical Design," Wiley, New York, 1968.

McCluskey, E.J.: "Introduction to the Theory of Switching Circuits," McGraw-Hill, New York, 1965.

The probabilistic and stochastic aspects of systems in Chapters 4 and 6 are considered in many books on mathematics and engineering, such as:

Drake, A.W.: "Fundamentals of Applied Probability Theory," McGraw-Hill, New York, 1967.
Papoulis, A.: "Probability, Random Variables, and Stochastic Processes," McGraw-Hill, New York, 1965.
Pfeiffer, P.E.: "Concepts of Probability Theory," McGraw-Hill, New York, 1965.

Other references related more directly to computation, communication, or control are:

Abramson, N.: "Information Theory and Coding," McGraw-Hill, New York, 1967.
Birkhoff, G., and T.S. Bartee: "Modern Applied Algebra," McGraw-Hill, New York, 1970.
Feldbaum, A.A.: "Optimal Control Systems," Academic, New York, 1965.
Minsky, M.: "Computation: Finite and Infinite Machines," Prentice-Hall, Englewood Cliffs, N.J., 1967.

There are many books on other nondigital models, especially the linear continuous systems. Two of the most noteworthy are:

Dertouzos, M.L., M. Athans, R.N. Spann, and S.J. Mason: "Systems, Networks, and Computation," McGraw-Hill, New York, 1972.
Huggins, W.H., and D.R. Entwhistle: "Introductory Systems and Design," Blaisdell, Waltham, Mass., 1968.

A treatment of systems at a much higher level can be found in:

Klir, G.J.: "An Approach to General Systems Theory," Van Nostrand Reinhold, New York, 1969.
Wymore, A.W.: "A Mathematical Theory of Systems Engineering," Wiley, New York, 1967.
Zadeh, L.A., and C.A. Desoer: "Linear Systems Theory," Mc-Graw-Hill, New York, 1963.

Index

A posteriori probability (*see* Reverse probability)
A priori probability (*see* Forward probability)
Absorbing state, 323
Absorption property, 26
Accumulator, 349
Addend, 56
Adder:
  parallel: design of, 56
    full adder, 56
    half adder, 55, 188, 374
  sequential, 257
    electronic, 376–377
    from *JK* flip-flop, 290–291
    from *MN* flip-flop, 293–294
    from *T* flip-flop, 288–289
Addition:
  of matrices, 130
  of numbers, 55
Addition operation of computer, 359
Additive property, 221
Address, 346
Algebra:
  of EXCLUSIVE-OR, 59
  as representation of switching network, 15
  of sets, 83
  of switches, 23, 26
Algorithms, 133
  for analysis of probabilistic systems, 171
  for computing variance, 226
Algorithms:
  for operation of computer, 358
  for optimizing sequential systems, 278
  for simulating stochastic systems, 321
  for sorting, 136–139
Ambulance example, 343
Analysis, 10, 68
  of deterministic systems, 21–27
  of probabilistic systems, 160–176, 188–192
  of sequential systems, 259–273
  of stochastic systems, 312–330
AND operator:
  flow, 16
  level, 44
    electronic, 370
    mechanical, 48–49
Arithmetic unit (*see* Operation unit of computer)
Arrangements, 99
Array, 116
  operations, 128–132
    product, 131, 138
    sum, 130
  representation, 116
  (*See also under* Matrix)
Assembly Language, 138, 355
Assignment:
  of states, 267
  of values, 4
Associative property, 26

Asynchronous sequential system, 257
Attached diagram, 15
Attributes, 3
Augend, 56
Average value (*see* Expected value)

Balloon payment example, 133–135, 138, 353–356
Barometer example, 170, 235–237
Base, 53
Basic events, 146, 153
Basic language, 137–138
Baye's theorem, 172
BCD (*see* Binary coded decimal)
BCD sequential code, 280
BEC (*see* Binary erasure channel)
Behavior of computer, 355–356
Behavioral view, 6, 67
Binary coded counter, 298–299
Binary coded decimal (BCD), 88, 280
Binary erasure channel (BEC), 178
Binary sequential adder, 257
Binary system, 4
  channels, 176
  numbers, 53
  operators, 58, 109
  relations, 113, 121
  symmetric channels, 184
  values, 14, 19
Binits, 346
Binomial density function, 221–214
Birth and death process, 341
Black box, 4
Blood type example, 92
Boolean algebra, 23
Branch type circuits, 42
Branching operation of computers, 359
  conditional, 360
  unconditional, 359
Buffer, 350
Bursts, 307
Bytes, 305

Cancellation property, 62
Cancer and smoking, 161–162
Cardinality of sets, 92
Cartesian product, 97
Cascade interconnection, 180, 296
Causality property, 269
Channels:
  BEC type, 178
  BSC type, 184
  crummy, 104, 174
  representation as relation, 114
  symmetric, 184
  TSC type, 184
  Z type, 178
Characteristic table, 284
Chebyshev's theorem, 246
Classical probability, 144, 157
Classifier, 179–180
Clock, 256, 287
Clocked delay element, 286
Clocked *JK* flip-flop, 376
Clocked *SR* flip-flop, 375
Coding:
  BCD, 88–89, 280–282
  comma, 257–258
  crummy channels, 104–106, 176–178
  NOT, 176–178
  parity, 104–106
Coin:
  as an information instrument, 239
  simulation, 320
  stochastic, 311
  thick, 98, 202
Combinations, 102
  of information instruments, 238–239
  (*See also* Arrangements)
Combinatorics, 99
Comma coding, 257–258
Communication, 10
  deterministic, 99–102
  probabilistic, 104–106, 174–178, 183
  sequential, 257–258
  stochastic, 339, 342
Commutative property, 26
Commutator, 303

Comparator, 49, 259
Compiling, 364
Complement of a set, 79
Complementarity property, 25
Complete induction, 22
Complexity, a measure of, 31, 45
Composite events, 148, 153
Composition of relations, 119
Computation, 134
Computer, 348
Computing system, 10
Conditional branch, 360
Conditional expected value, 232
Conditional probability, 160
Configurations, 370
Contact, 13
Content of register, 346
Control, 11
  of deterministic systems, 40
  of probabilistic systems, 234
  of sequential systems, 258
  of stochastic systems, 333
Control unit of computer, 353, 356–364
Controller, 234, 336
Controller policy, 235
Convolution, 273
Coordinates of *n*-tuples, 95
Correlation, 227
Correlation coefficient, 227
Counters:
  binary coded, 298
  ternary scalor, 252
Counting, 92, 99
Counting network of computer, 360–362
Covariance, 226
Craps example, 341
Crummy system, 114, 126
Current mode logic, 372

Data word, 352
DCTL (*see* Direct-coupled transistor logic)
Decision making, 231–239
Decoder (*see* Coding)
Definiteness of sets, 75
Delay, 252, 259
  clocked, 286
  flip-flop, 286
  formal, 287
  ideal, 252
Demons, 237
De Morgan's law, 27
Dependence, degree of, 228
Design, 68
  (*See also* Optimization; Synthesis)
Detached diagram, 15
Deterministic system:
  analysis, 21–27
  identification, 50–51
  implementation, 40–49
  modeling, 5, 13–14
  optimization, 30–33, 34–40
  representation, 14–21
  synthesis, 28–30, 34–40
Diagrams:
  flow, 135
  Karnaugh, 77, 90
  sagittal, 113
  state, 250
Dice:
  ace-six flats, 151
  in craps, 341
  honest, 151
  simulation of, 321
  stochastic, 341
Diode, 368
Diode logic, 370
Diode transistor logic (DTL), 371
Direct-coupled transistor logic (DCTL), 371
Disagreement detector, 46
Disjoint (*see* Mutually exclusive property)
Dispenser example, 217, 258–259
Distinctness in sets, 75
Distinguishability of states, 277
Distinguishable arrangements, 99
Distributive property, 25–26, 45
Domain of a relation, 118
Don't care condition, 38, 89
  bound values, 293

Don't care condition:
in excitation tables, 289
free values, 293
using set diagrams, 89
DTL (*see* Diode transistor logic)
Dualization property, 26–27
Duals, 27
Dynamic programming, 334–337

ECL (*see* Emitter-coupled transistor logic)
Efficiency of information system, 230, 237
Electronic components, 368
flip-flops, 375
logic circuits, 45, 46, 369–372
shift registers, 377–378
Elevator example, 37–38
Emitter-coupled transistor logic (ECL), 371–372
Encoding:
comma, 257–258
NOT, 176–178
parity, 104
Enumeration, 102
Equilibrium condition, 315
Equilibrium probability, 315
Equivalence:
of deterministic systems, 22
of n-tuples, 96
of probabilistic systems, 180–181
relation, 120
of sequential systems, 304
of sets, 83
of states, 274
of systems in general, 9
Ergodicity, 322–323, 326–327
Error, 170, 177, 198
Error bursts, 306–307
Error detector, 104
Evaluation, 10, 68
Event space, 153
Events, 146, 152
basic, 146, 153
composite, 148–149, 153
Events:
independent, 188–190
mutually-exclusive, 150
as sets, 152–153
Everywhere-defined relation, 123
Excitation table, 288
EXCLUSIVE-OR, 46
in parity coding, 105–106
properties of, 58–59
Execution of instruction, 358
Expected value, 208–210
conditional, 232
of a function of random variables, 217
properties of, 219–222
Experiment, 146, 326
multiple, 326
simple, 326
Extended probability product principle, 204
Extended product principle, 165–167
Extension of sets, 75

f-complete, 61
Factorial, 100
Family of events, 153
Family of logic, 370
Fan-in, 373
Fan-out, 373
Feedback, 260, 264
Fetch of instruction, 358
Field-effect transistor, 378
Finite memory length, 271, 327
Flags, 100, 141
Flip-flop, 283
counting of types, 293
electronic configuration, 374–377
types: delay, 286
JK, 285
MN, 293
set-reset, 283–284
toggle, 285
Flow diagram, 135
Flow graph, 117
Flow implementation, 41

Fluid flow system, 41
Fluid storage example, 39, 71
Fluidic systems, 302–303
Formal delay, 252, 259, 287
Fortran, 137
Forward probability, 168–170
Four-address computer, 364
Freeway congestion example, 162
Full adder, 57
Function-on-a-relation, 128–130
Functional completeness, 61
Functions:
  definition, 123–124
  probability, 153
  probability density, 207–208
  transition, 254
  (*See also* specific types)

Gambler's ruin, 341
Gate-type circuits, 42
Gear replacement example, 309, 329, 333
Geometric density function, 216
Goals, 3, 11, 38, 234

Half adder, 55, 127, 188, 374
Half-life, 323
Halt operation of computer, 360
Hammock network, 191–192, 196–197
Hazard, 302
Hierarchy, 6
High-level languages, 137
High threshold logic (HTL), 378
Homogeneity property, 220
HTL (*see* High threshold logic)
Hydraulic networks, 41–42
Hypochondriac example, 241

Ideal delay, 252, 259
Idempotent property, 25
Identification, 9, 68
  of deterministic system, 50–51
  of probabilistic system, 170
  of stochastic system, 306, 327
Implementation, 10, 68
  of deterministic instantaneous systems, 40
  of sequential systems, 283–296
Impulse function, 273
Impulse response, 273
INCLUSIVE-OR, 16
Incomplete specification, 37–39, 89
Incompletely specified sequential system, 258
Independence, 188–205
Independent repeated trials, 202–204
Indexing, 364
Induction, complete, 22
Infinite memory length, 271
Information, mutual, 229
Information instruments, 234
INHIBIT-AND, 62
Input-output pair, 96, 146, 249
Input-output-state equations:
  for sequences, 256
  for symbols, 254
Input-output-unit, 348
Input sequence (tape), 255
Inputs, 3, 125
  of stochastic system, 330
Inspector (classifier) example, 179
Instantaneous description, 134
Instantaneous property, 13–14
Instruction counter, 353
Instruction register, 353
Instruction word, 352
Integrated circuits, 372
Intension of sets, 76
Intersection of sets, 80
Inverse of relation, 118
Inversion method of analysis, 317
Inverter:
  three-valued, 64
  two-valued (*see* NOT)
Isomorphism, 84
Iteration, 314-316
Iteration instruction, 136

JK flip-flop, 285, 376
Joint probability, 167

Karnaugh map, 77
  for deterministic systems, 84–91
  extending systematically, 90
  for probabilistic systems, 155–156
  for sequential systems, 267, 268, 290
  for stochastic systems, 311

Languages, 137
Large scale integration (LSI), 378
Level implementation, 42
Linearity property, 271
Linguistic example, 311
Loading of digital components, 373
Loading operation of computer, 358
Location, 346
Logic, 111
Logic convention, 367
Logical design, 28
LSI (*see* Large scale integration)

Machine language, 354–355
Macroscopic view, 322
Mahoney map (*see* Karnaugh map)
Maintenance system example, 322–323
Majority indicator:
  design of, 34–35
  level implementation, 44
  mechanical, 49
  probabilistic, 188
  in reliability, 198–199
  sagittal diagram of, 127
Many-valued systems (*see* Nonbinary systems)
Mapping (*see* Functions)
Marginal probability, 168
Markovian system (*see* Stochastic system)
Matrix, as a relation, 116
  (*See also* Array)
Matrix interpretation:
  of probabilistic interconnection, 181–183
  of stochastic behavior, 316
Matrix operations, 130–132
Matrix product, 131, 137–138
Max operator, 64
Meally model of sequential systems, 254
Mean time before failure (MTF), 217
Mean value (*see* Expected value)
Mechanical level networks, 48–49
Medium scale integration (MSI), 378
Memory:
  of computer, 350–351
  elements (*see* Flip-flops)
  length, 271
Memory address selector, 351
Memory unit, 350
Merging of states, 275–276, 278
Micro operations, 359
Micro programming, 364
Micro programs, 359
Microscopic view, 322
Min operator, 64
Minimization (*see* Optimization)
Minority operator, 63
Minterm form (*see* Standard form)
Model, 3
Modeling, 9
  deterministic systems, 13–14
  probabilistic systems, 145–154
  sequential systems, 248–254
  stochastic systems, 306–311, 337–338
Modulo system, 73, 140
Monte Carlo simulation, 319–322, 325
Moore model of sequential systems, 254
MSI (*see* Medium scale integration)
MTF (*see* Mean time before failure)
Multi-input systems, 126, 186–188
Multi-output networks, 39, 47–48
Multi-stage systems, 333
Multiple experiment, 326
Multiplication, 365
Mutual information, 229
Mutually exclusive property, 81, 150

n-ary relations, 121

n-tuple, 95–97
NAND, 6, 62, 371–372
Negative logic convention, 368
Neuron, 72, 303
Next-state function, 250
Noise, 218, 340
Nonbinary systems, 4–5, 63–65
Nonergodic systems, 322–323
NOR, 59–61
Normality, 153
NOT operator:
  flow, 16
  level, 44
Null set, 81
Numbers, 52–54
  BCD, 88, 280
  binary, 52
  decimal, 52
  octal, 58

Octal numbers, 58
Odds, 144
Ones-complement, 365
Op codes, 352
Operation:
  of unit property, 26
  of zero property, 26
Operation codes, 352
Operation decoder, 350
Operation selector, 350
Operation unit of computer, 349
Optimal controller, 235–236, 336–337
Optimization, 10
  of instantaneous deterministic systems, 30–33, 68, 84–91
  of probabilistic systems, 176–183, 193–201
  of sequential systems, 274–282
  of stochastic systems, 330–337
OR operator:
  crummy, 126
  delayed, 262
  electronic, 45, 369, 372
  EXCLUSIVE (*see* EXCLUSIVE-OR)
  flow, 16–19
OR operator:
  level, 42–43
Oracles, 237
Order relation, 140
Ordered pair, 95
Orientation, 3, 125, 145
Oriented system, 125
Outcome (*see* Events)
Output of stochastic system, 324
Output function, 250
Output (response) sequence, 255
Outputs, 4, 125

Painting example, 333
Parallel adder, 56
Parallel interconnection, 16
Parallel transfer, 364
Parity, 104
Parity encoder, 104
Partition, 81
Partitioning method of optimizing states, 277–280
Payoff matrix, 235
pdf (*see* Probability density function)
Permutation, 99
Poisson density function, 215
Policy, 235
Population movement example, 142
Positional notation, 53
Positive logic convention, 367
Positivity property, 153
Power set, 103
Precedence of operators, 17
Principle of optimality, 337
Probabilistic system:
  analysis, 160–176, 188–192
  identification, 170
  modeling, 5, 143–153
  optimization, 176–183, 193–201
  representation, 154–155, 169
  synthesis, 176–188, 193–199
Probability:
  conditional, 160
  equilibrium, 315
  forward, 168
  joint, 167
  marginal, 168

Probability:
reverse, 171–173
transition, 307
Probability density function (pdf), 207
binomial, 211–214
geometric, 216–217
Poisson, 215
uniform, 243
Probability mass function (*see* Probability density function)
Probability matrix, 167
Probability product property, 160
extended, 165, 204
Product of matrices, 131, 138
Product of sets, 97
Product principle, 100
Programming, 137, 353–356
Programming languages, 137
Properties (in general), 8
of instantaneous deterministic systems, 23–27
of probabilistic systems, 163–166
of sequential systems, 269–270
of stochastic systems, 310, 337
Properties (particular):
absorption, 26
additivity, 221
associativity, 26
cancellation, 62
causality, 269
commutativity, 26
distributivity, 25
dualization, 26
ergodicity, 322
finite memory length, 271
homogeneity, 220, 322
linearity, 271–273
response separation, 269
state separation, 270
superposition, 272
Protocol, 251
Punched cards, 28

Queues, 309

Race, 302
Random numbers, 319
Random variable, 205
Random voltage sources, 208
Range of relation, 118
Read-only memory (ROM), 378
Readout of registers, 346
Recirculating register, 346
Redundant components, 198
Reflexive relation, 120
Register transfers, 347, 359
Registers:
in computer, 345–347
electronic, 377–378
shift, 297–298
Relations, 112–122
Relative frequency, 147
Relay, 2, 5, 12
deterministic, 13–15
probabilistic, 145–150, 158, 167, 191, 193
sequential, 248–252, 338
stochastic, 307, 338
Repeated trials, 202–205
conditional, 204–203
independent, 202–203
Replacement example, 309
Replicated system, 198
Representation, 9
of deterministic systems, 14–21
of probabilistic systems, 154–155, 169
of sequential systems, 253–254
of stochastic systems, 308, 330
Resistor, 368
Resistor transistor logic (RTL), 371
Response separation property, 269
Restricted additivity, 153
Reverse probability, 171
Ring counter, 357
ROM (*see* Read-only memory)
RTL (*see* Resistor transistor logic)

Sagittal diagram, 113
Scalor (*see* Counters)

Sequencing register transfers, 357
Sequential adder, 251
Sequential comparator, 257
Sequential subtractor, 259
Sequential systems:
  analysis, 259–265, 269–274
  implementation, 283–295
  modeling, 5, 248–253
  optimization, 274–282
  representation, 253–254
  synthesis, 265–269
Series interconnection, 16
Set algebra, 83
Set diagram, 77
  (*See also* Karnaugh map)
Set product, 97
Set reset flip-flop, 283–284, 374–375
Sets, 74
Shift registers, 297–298, 345, 377–378
Shifting operation of computers, 348
Simple experiment, 326
Simulation, 319–322
Single-address organization, 350–353
Single-valued relation, 123
Sink state, 323
Size of sets, 92–93
Smog example, 312, 342
Smoking and cancer example, 161–162
Snapshot (*see* Instantaneous description)
Sneak path, 42
Sorting algorithm, 136–139
Stage, 310
Standard deviation, 225
Standard form, 35
State, 250, 310
  assignment, 267
  diagram, 250, 308
  distinguishability, 277
  equivalence, 274
  optimizing, 274–282
  of register, 345
  separation property, 270
State-determined system, 300
Stationary system, 326
Stochastic systems:
  analysis, 312–329
  modeling, 5, 306–310
  optimization, 330–337
  representation, 308
Store operation of computer, 359
Structural view, 6
Structure of computer, 360–364
Sub-algorithms, 137
Sub-routines, 137
Subset, 78
Subtractor:
  instantaneous, 58
  sequential, 259
Sums of random variables, 218–221
Super state, 277
Superposition property, 272
Survey example, 77, 93–95, 132, 156–158, 162, 189–190
Switching algebra, 23
Switching operators:
  AND, 16, 44
  EXCLUSIVE-OR, 46, 58
  INHIBIT-AND, 62
  NAND, 62
  NOR, 60
  OR, 16, 43
Symmetric channel, 184
Symmetric matrix, 129
Symmetric relation, 120
Synchronous systems, 256, 376–377
Synthesis, 10
  of deterministic systems, 28, 34
  of probabilistic systems, 176–188, 193–199
  of sequential systems, 265–269, 283–295
System, 1, 124–127

Table-of-combinations:
  for deterministic systems, 20
  for probabilistic systems, 148
  for sequential systems, 253
  for stochastic systems, 311

Taxicab example, 308, 323, 343
Ternary (*see* Nonbinary systems)
Ternary counter, 252–253, 266–268, 290–292
Ternary symmetric channels, 184
Three-valued, 64
Threshold network, 72
Time-dependent activity example, 311–312
Timing in sequential systems, 295
Timing network of computer, 356–357
Transfer of shift registers, 347
Transfer contact, 13
Transistor, 368
Transistor-transistor logic (TTL), 371
Transition equations, 313
Transition matrix, 308
Transition probabilities, 307
Transition table, 253–254
Transitive relation, 120
Transportation system example, 128–130, 308
Trees:
  in probabilistic systems, 155
  representing relations, 118
  in sequential systems, 280–281
  in stochastic systems, 205
Triplication, 198
Truth table (*see* Table-of-combinations)
TTL (*see* Transistor-transistor logic)

Unconditional branch operation, 359
Union of sets, 80
Universe, 78

Variance, 223–228
Veitch diagram (*see* Set diagram)
Venn diagram (*see* Set diagram)

Weather example, 310, 327–329
Words in computers, 352

Z channel, 178